DATE DUE			
MAR 2 1977			
MAR 1 5 1977 BR			
MAR 1 5 1977			

DEMCO 38-297

METHODS IN COMPUTATIONAL PHYSICS
Advances in Research and Applications

Volume 16

Controlled Fusion

Methods in Computational Physics

Advances in Research and Applications

1. STATISTICAL PHYSICS
2. QUANTUM MECHANICS
3. FUNDAMENTAL METHODS IN HYDRODYNAMICS
4. APPLICATIONS IN HYDRODYNAMICS
5. NUCLEAR PARTICLE KINEMATICS
6. NUCLEAR PHYSICS
7. ASTROPHYSICS
8. ENERGY BANDS OF SOLIDS
9. PLASMA PHYSICS
10. ATOMIC AND MOLECULAR SCATTERING
11.* SEISMOLOGY: SURFACE WAVES AND EARTH OSCILLATIONS
12.* SEISMOLOGY: BODY WAVES AND SOURCES
13.* GEOPHYSICS
14. RADIO ASTRONOMY
15.† VIBRATIONAL PROPERTIES OF SOLIDS
16.‡ CONTROLLED FUSION

* Volume Editor: Bruce A. Bolt.
† Volume Editor: Gideon Gilat.
‡ Volume Editor: John Killeen.

METHODS IN COMPUTATIONAL PHYSICS

Advances in Research and Applications

Series Editors

BERNI ALDER

Lawrence Livermore Laboratory
Livermore, California

SIDNEY FERNBACH

Lawrence Livermore Laboratory
Livermore, California

MANUEL ROTENBERG

University of California
La Jolla, California

Volume 16

Controlled Fusion

Volume Editor

JOHN KILLEEN

Department of Applied Science–Davis/Livermore
and CTR Computer Center
Lawrence Livermore Laboratory
University of California
Livermore, California

1976

ACADEMIC PRESS   NEW YORK   SAN FRANCISCO   LONDON

A Subsidiary of Harcourt Brace Jovanovich, Publishers

Copyright © 1976, by Academic Press, Inc.
ALL RIGHTS RESERVED.
NO PART OF THIS PUBLICATION MAY BE REPRODUCED OR
TRANSMITTED IN ANY FORM OR BY ANY MEANS, ELECTRONIC
OR MECHANICAL, INCLUDING PHOTOCOPY, RECORDING, OR ANY
INFORMATION STORAGE AND RETRIEVAL SYSTEM, WITHOUT
PERMISSION IN WRITING FROM THE PUBLISHER.

ACADEMIC PRESS, INC.
111 Fifth Avenue, New York, New York 10003

United Kingdom Edition published by
ACADEMIC PRESS, INC. (LONDON) LTD.
24/28 Oval Road, London NW1

LIBRARY OF CONGRESS CATALOG CARD NUMBER: 63-18406

ISBN 0-12-460816-7

PRINTED IN THE UNITED STATES OF AMERICA

Contents

CONTRIBUTORS ix
PREFACE xi

NUMERICAL MAGNETOHYDRODYNAMICS FOR HIGH-BETA PLASMAS

Jeremiah U. Brackbill

I. Introduction	1
II. Numerical Methods	3
III. The Computation of Convective Transport	10
IV. A Generalized Mesh Method for MHD	17
V. Applications	29
VI. Conclusions	38
Appendix	38
References	39

WATERBAG METHODS IN MAGNETOHYDRODYNAMICS

David Potter

I. The Waterbag Concept	43
II. Equilibrium Properties of One Waterbag	49
III. Equilibria of Current Distributions	60
IV. Adiabatic Constraints	69
V. Further Applications	76
References	82

SOLUTION OF CONTINUITY EQUATIONS BY THE METHOD OF FLUX-CORRECTED TRANSPORT

J. P. Boris and D. L. Book

I. Introduction	85
II. Elements of FCT Algorithms	93
III. Optimization of FCT Algorithms	105
IV. Applications of Flux-Corrected Transport	115
References	128

MULTIFLUID TOKAMAK TRANSPORT MODELS
John T. Hogan

I. General Remarks	131
II. Plasma Models	142
III. Suprathermal Plasma: Injected Ions and Alpha Particles	150
VI. Neutral Gas	153
V. Impurities	158
VI. Summary	161
References	162
Appendix: Bibliography	164

ICARUS—A ONE-DIMENSIONAL PLASMA DIFFUSION CODE
M. L. Watkins, M. H. Hughes, K. V. Roberts, P. M. Keeping, and J. Killeen

I. Introduction	166
II. The Physical Model	169
III. The Numerical Model	176
IV. Programming Techniques	179
V. Applications	190
VI. Summary	206
References	207

EQUILIBRIA OF MAGNETICALLY CONFINED PLASMAS
Brendan McNamara

I. Introduction	211
II. Toroidal Equilibrium	215
III. Anisotropic Pressure Equilibria	231
References	249

COMPUTATION OF THE MAGNETOHYDRODYNAMIC SPECTRUM IN AXISYMMETRIC TOROIDAL CONFINEMENT SYSTEMS
Ray C. Grimm, John M. Greene, and John L. Johnson

I. Introduction	253
II. Formulation of the Problem	257
III. Representation of the Normal-Mode Equations	262
IV. Application	272
V. Discussion	276
References	278

COLLECTIVE TRANSPORT IN PLASMAS

John M. Dawson, Hideo Okuda, and Bernard Rosen

I. Introduction	282
II. The Simulation Model	283
III. Elementary Theory of Convective Diffusion in a Uniform Thermal Plasma	289
IV. The Simulation of Plasma Diffusion across a Magnetic Field (Uniform Thermal Plasma)	295
V. Simulation of Diffusion in Nonuniform Plasmas	310
References	325

ELECTROMAGNETIC AND RELATIVISTIC PLASMA SIMULATION MODELS

A. Bruce Langdon and Barbara F. Lasinski

I. Introduction	327
II. Simulation of Collisionless Plasmas	328
III. Electromagnetic Codes Working Directly with E and B	330
IV. Algorithms with Special Stability Properties	361
V. SUPERLAYER	363
References	364

PARTICLE-CODE MODELS IN THE NONRADIATIVE LIMIT

Clair W. Nielson and H. Ralph Lewis

I. Introduction	367
II. The Darwin Model	368
III. Hamiltonian Formulations	371
IV. Lagrangian Formulation	374
V. Solution of the Field Equations	379
VI. One-Dimensional Comparisons	381
VII. A Two-Dimensional Example	384
VIII. Summary	386
References	387

THE SOLUTION OF THE KINETIC EQUATIONS FOR A MULTISPECIES PLASMA

John Killeen, Arthur A. Mirin, and Marvin E. Rensink

I. Introduction	389
II. Mathematical Model	392
III. Solutions Using Angular Eigenfunctions	401
IV. Finite-Difference Solution in a Two-Dimensional Velocity Space	411
References	430

AUTHOR INDEX	433
SUBJECT INDEX	441
CONTENTS OF PREVIOUS VOLUMES	446

Contributors

Numbers in parentheses indicate the pages on which the authors' contributions begin.

D. L. BOOK, *Plasma Physics Division, Naval Research Laboratory, Washington, D.C.* (85)

J. P. BORIS, *Plasma Physics Division, Naval Research Laboratory, Washington, D.C.* (85)

JEREMIAH U. BRACKBILL, *University of California, Los Alamos Scientific Laboratory, Los Alamos, New Mexico* (1)

JOHN M. DAWSON, *Physics Department, University of California at Los Angeles, Los Angeles, California* (281)

JOHN M. GREENE, *Plasma Physics Laboratory, Princeton University, Princeton, New Jersey* (253)

RAY C. GRIMM, *Plasma Physics Laboratory, Princeton University, Princeton, New Jersey* (253)

JOHN T. HOGAN, *Oak Ridge National Laboratory, Oak Ridge, Tennessee* (131)

M. H. HUGHES, *Culham Laboratory, UKAEA Research Group, Abingdon, Oxfordshire, England* (165)

JOHN L. JOHNSON,* *Plasma Physics Laboratory, Princeton University, Princeton, New Jersey* (253)

P. M. KEEPING, *Culham Laboratory, UKAEA Research Group, Abingdon, Oxfordshire, England* (165)

JOHN KILLEEN, *Lawrence Livermore Laboratory, University of California, Livermore, California* (165, 389)

A. BRUCE LANGDON, *Lawrence Livermore Laboratory, University of California, Livermore, California* (327)

* On loan from Westinghouse Research Laboratories, Pittsburgh, Pennsylvania.

BARBARA F. LASINSKI, *Lawrence Livermore Laboratory, University of California, Livermore, California* (327)

H. RALPH LEWIS, *Los Alamos Scientific Laboratory, University of California, Los Alamos, New Mexico* (367)

BRENDAN MCNAMARA, *Lawrence Livermore Laboratory, University of California, Livermore, California* (211)

ARTHUR A. MIRIN, *Lawrence Livermore Laboratory, University of California, Livermore, California* (389)

CLAIR W. NIELSON, *Los Alamos Scientific Laboratory, University of California, Los Alamos, New Mexico* (367)

HIDEO OKUDA, *Plasma Physics Laboratory, Princeton University, Princeton, New Jersey* (281)

DAVID POTTER,* *University of California at Los Angeles, Los Angeles, California* (43)

MARVIN E. RENSINK, *Lawrence Livermore Laboratory, University of California, Livermore, California* (389)

K. V. ROBERTS, *Culham Laboratory, UKAEA Research Group, Abingdon, Oxfordshire, England* (165)

BERNARD ROSEN, *Department of Physics, Stevens Institute of Technology, Hoboken, New Jersey* (281)

M. L. WATKINS, *Culham Laboratory, UKAEA Research Group, Abingdon, Oxfordshire, England* (165)

---

* On leave of absence from Imperial College, London, England.

Preface

NUMERICAL CALCULATIONS HAVE HAD an important role in controlled thermonuclear research since its beginning in the early 1950s. In the last few years the application of computers to plasma physics has advanced rapidly. This is due to the increasing sophistication of the mathematical models that made more realistic numerical simulation of plasmas possible by taking advantage of the increased speed and memory of computers. In the next few years there should be a substantial increase in the development and use of numerical models in order to meet the needs of the fusion power program.

The behavior of a plasma confined by a magnetic field is simulated by a variety of numerical models. Some models compute the macroscopic properties of the plasma dynamics. In order to simulate the time evolution of a plasma in a magnetic confinement device over most of its lifetime—from tens to hundreds of milliseconds—a set of partial differential equations of the diffusion type must be solved. Typical dependent variables are the number densities and temperatures of each particle species, current densities, and magnetic fields. On the other hand, particle codes compute on a microscopic time-scale the motion of particles under the influence of their self-consistent electric and magnetic fields, as well as any externally imposed fields. These codes give the most detailed results, such as phase–space distribution functions, fluctuation and wave spectra, and orbits of individual particles, as well as information on the growth and saturation of strong instabilities and the effects of turbulence. Prior to 1973, most calculations were with particle codes that simulated collisionless plasmas, with the objective of understanding velocity space instabilities and wave–particle interactions. Thus, Volume 9 of this series, published in 1970, is concerned with the solution of the Vlasov or collisionless Boltzmann equation by, among others, the many-particle simulation techniques.

This volume differs from Volume 9 in that the full variety of computer models needed for the simulation of realistic fusion devices is considered. In order to simulate the behavior of such devices, many new macroscopic codes such as magnetohydrodynamic, Tokamak transport, and equilibrium codes are being developed and have already given much insight into the understanding of experimental results from both low- and high-beta devices. The following types of computer models are considered in this volume:

1. Time-dependent magnetohydrodynamics.
2. Plasma transport in a magnetic field.
3. MHD and guiding-center equilibria.
4. MHD stability of confinement systems.

5. Vlasov and particle models.
6. Multispecies Fokker–Planck codes.

The first three articles consider various aspects of the numerical solution of the equations of magnetohydrodynamics (MHD). There are a great variety of such MHD codes being developed with different degrees of complexity. The so-called ideal MHD is an infinite conductivity approximation. In some models, the pressure is a scalar function; in others it is considered a tensor with different values along and perpendicular to the magnetic field. The more realistic models include the transport coefficients, e.g., thermal conductivity and electrical resistivity, and these can also be scalars or tensors. Two-dimensional codes are now fairly standard, and there are several efforts to develop three-dimensional codes. The choice of coordinate systems varies among these codes. A fixed Eulerian grid is the usual choice, but Lagrangian descriptions, particularly using magnetic flux surfaces as coordinate surfaces, as in Potter's article, are proving useful in certain problems. In the work of Brackbill, a moving grid is used which is not a Lagrangian grid. In Eulerian representations, the treatment of the convective terms is emphasized by Boris and Book.

In the past few years, a considerable effort has been devoted, at several laboratories, to the numerical solution of transport equations for toroidal plasmas. This effort provides an excellent means of comparing theory with experiment. Recent developments of these codes have concentrated on the inclusion of neutrals and impurities in the models, and the use of empirical transport coefficients. These models and their application to Tokamaks are presented in the articles by Hogan and Watkins *et al.*

It is normally necessary to develop time-independent codes to support the design and operation of each major CTR experiment. Possible equilibrium plasma configurations and their stability are investigated by these codes; the latter by calculating the eigenvalues of the linearized perturbed equations. Recent research in this area is reviewed in the articles by McNamara and Grimm *et al.*

Particle codes are usually classified as either "electrostatic" or "electromagnetic." In the first type, only the self-consistent electric field is computed via Poisson's equation, and the magnetic field is either absent or constant in time. Electrostatic codes dominated computational plasma physics in the early years and are covered extensively in Volume 9 of this series. However, some important recent results on collective transport across a given magnetic field in two- and three-dimensional systems are reviewed in the article by Dawson *et al.* Fully electromagnetic codes in which the particle equations of motion are relativistic and the electric and magnetic fields are obtained from the full Maxwell equations (wave equations) are described in the article by Langdon and Lasinski. Electromagnetic codes in the nonradiative limit

(Darwin model) where the equations are nonrelativistic and displacement currents are neglected are discussed in the article by Nielson and Lewis.

In the simulation of magnetically confined plasmas where the ions are not Maxwellian and where a knowledge of the distribution functions is important, kinetic equations must be solved. The appropriate kinetic equations are Boltzmann equations with Fokker–Planck collision terms. Numerical methods for the solution of multispecies Fokker–Planck equations are reviewed by Killeen *et al.* in the last article.

J. KILLEEN
B. ALDER
S. FERNBACH

Numerical Magnetohydrodynamics for High-Beta Plasmas

JEREMIAH U. BRACKBILL

UNIVERSITY OF CALIFORNIA
LOS ALAMOS SCIENTIFIC LABORATORY
LOS ALAMOS, NEW MEXICO

I. Introduction	1
II. Numerical Methods	3
A. A Description of Eulerian, Lagrangian, and Generalized Computation Meshes	4
B. A Survey of Numerical Methods and Applications	8
III. The Computation of Convective Transport	10
A. Properties of Approximations to the Convective Derivative	10
B. Nonlinearly Stable Approximations to Convective Transport	13
IV. A Generalized Mesh Method for MHD	17
A. Difference Equations for the Lagrangian Phase of a Generalized Mesh Calculation	18
B. The Rezone Phase of a Generalized Mesh Calculation	26
V. Applications	29
A. A Sharp Boundary Calculation of the Theta Pinch	29
B. The Rotating Theta Pinch	31
C. The Internal Kink Mode Instability	34
VI. Conclusions	38
Appendix	38
References	39

I. Introduction

MAGNETOHYDRODYNAMICS (MHD) is the least sophisticated model of a magnetically confined plasma which describes the interaction between a magnetic field and a plasma self-consistently. The MHD model treats the plasma as though it were a charge-neutral fluid in local thermodynamic equilibrium, and thus neglects all but a small part of the physics of plasmas. That small part, however, describes the transfer of momentum and energy between the plasma and the magnetic field, and that is sufficient to describe the effect of the geometry of the field on the gross equilibrium and stability of a high-beta plasma.

Even in their simplest form, the MHD equations comprise a coupled system of nonlinear, partial differential equations which are difficult, if not impossible, to solve analytically. Sometimes the MHD equations can be reduced, because of symmetries or because of ordering, to a single ordinary differential equation whose solution describes the linear stability of a plasma in equilibrium with a magnetic field. More often, the equations cannot be reduced, and the MHD equations in two, and often three, dimensions must be solved. In such cases, there is no other way to learn what the MHD model predicts about the stability of a particular experiment than by numerically solving the equations.

One has only to examine the list of problems of current interest compiled by Bodin (1972) to see that many of them will only yield to numerical calculation. Two of the problems are concerned with axisymmetric and non-axisymmetric toroidal confinement. Several two-dimensional numerical computations of the axisymmetric case have already been performed (Lui and Chu, 1974; Hofmann, 1974), and it is clear that fully three-dimensional computations will also be necessary before instabilities in these confinement geometries can be simulated. Three-dimensional simulations are an ambitious undertaking which the current interest in controlled thermonuclear research can support if the results warrant it.

To evaluate whether numerical computations can play a useful role in the solution of problems in CTR, one should examine not only the opportunities but also the constraints on numerical work. One constraint is that now and for the next several years, the computational power at the disposal of a researcher using numerical methods is barely sufficient to store and process enough information to describe the currently interesting magnetic confinement geometries. When one considers that a comparable amount of information has been used for numerical computations in two dimensions, one can see that more accuracy is required of numerical approximations in three dimensions than in two if useful results for time-dependent problems are to be obtained.

A review of the accuracy of many numerical solution algorithms in current use reveals that the computation of convective transport is significantly less accurate than the computation of the other terms in the equations describing magnetohydrodynamic flow. Furthermore it is clear from the literature of numerical methods for hydrodynamic flows that convective transport has long been a recognized problem. Why the approximation of the convective derivative should cause problems is the subject of a discussion in Section III which calls the reader's attention to the relevant literature, and reviews the current work on improving the accuracy and the stability of numerical approximations to this term. Sections IV and V present a numerical method which avoids, as much as possible, having to approximate the convective derivative.

The approximation of convective transport is a major theme in this article, both because the work described in Sections III–V has significantly improved the accuracy of numerical solutions of the MHD equations, and because it is a fundamental yet not generally recognized problem.

The article begins with a brief introduction to numerical methods for magnetohydrodynamics, including a survey of published methods and applications.

II. Numerical Methods

There has evolved a standard approach to the numerical solution of the partial differential equations describing time-dependent magnetohydrodynamic flow. Finite-difference approximations to the differential equations are solved on a computation mesh at a sequence of time-steps. Other methods, such as particle descriptions of the fluid, have been used to simulate the plasma focus experiment (Butler *et al.*, 1969; Roberts and Potter, 1968) and the theta pinch end loss problem (Tuck, 1968). However, because of the current interest in the gross stability of various confinement geometries (Bodin, 1972) and in the transport processes in plasmas, finite-difference approximations are now used almost exclusively.

There is also a standard approach to another problem. In pinch discharges, for example, the bulk of the plasma is swept inward by an incoming magnetic field leaving behind a low density plasma immersed in a strong magnetic field. In the low density plasma, the characteristic signal speed, the Alfven speed, is very large and consequently the Courant condition on the time-step (Richtmyer and Morton, 1967) is very restrictive. Several solutions have been offered for this problem. Boris (1970) offered the suggestion that a relativistic mass correction be applied in which an artificially low maximum signal speed is substituted for the speed of light. By choosing the appropriate value for the signal speed, the desired minimum time-step can be set. Lui and Chu (1974) address the same problem in a pinch discharge calculation by setting the minimum allowable density in the mesh to 15% of the initial average density. There is general agreement now that an implicit formulation of the equations of motions offers the best solution. With an implicit formulation, the time-step is no longer determined by the maximum signal speed in the mesh because the equations are unconditionally stable. Implicit equations require the solution of systems of simultaneous equations for which various iterative methods are used, including time-step splitting, alternating direction implicit, and even successive overrelaxation methods (Richtmyer and Morton, 1967). All of these methods are evidently economical.

Finally, there is the problem of the accurate computation of convective transport. As will be discussed in Section III, there is some evidence that the accuracy of many approximations to the convective derivative must be deliberately reduced to avoid nonlinear instabilities. Stable schemes are often obtained by adding considerable diffusion to the difference equations making meaningful three-dimensional calculations more difficult to do with present-day computers. More accurate approximations which are nonlinearly stable are described in Section III. In Sections IV and V, a second approach is discussed in which the convective transport, and therefore the need to add stabilizing diffusion, is much reduced by using the generalized mesh described in the following section. (In addition, the equations for Eulerian, Lagrangian, and generalized meshes are given, as well as a survey of MHD computations using each kind of mesh, in order that the discussion in the remaining sections may proceed in a reasonably informed manner.)

A. A Description of Eulerian, Lagrangian, and Generalized Computation Meshes

The first step in formulating a numerical algorithm for solving the MHD equations is to choose in a way to represent the plasma. Typically, the plasma is represented by a finite number of numbers which give the position, velocity, density, temperature, and magnetic field intensity over the physical domain. These numbers are stored at an array of points, called mesh points, comprising a computation mesh. Each mesh point is associated with a cell which is used as a control volume in constructing conservative difference equations. A computation mesh may be Eulerian, Lagrangian, or generalized. One kind is distinguished from another by the relative motion of the mesh points and the fluid, and by the complexity of the difference equations approximating spatial derivatives.

An Eulerian mesh is stationary with respect to the laboratory frame, and a Lagrangian mesh is stationary with respect to the plasma. A generalized mesh may be Eulerian, Lagrangian, or neither: the motion of the grid points with respect to both the laboratory frame and the plasma is arbitrary. The grid motion determines how much convective transport must be computed.

In the following sections, brief descriptions are given of the Eulerian, Lagrangian, and generalized meshes with the appropriate equations for each mesh. After the descriptions, a brief survey is presented of the calculations employing each kind of mesh.

1. *The Eulerian Computation Mesh*

Eulerian meshes are simple and, in many cases, economical to use. Since the mesh is stationary, the zoning may be chosen to simplify the numerical approximation of spatial derivatives. When the mesh is orthogonal, a rapid solution algorithm such as time-step splitting (Richtmyer and Morton, 1967) can be used. Most important, the conservation form of the ideal MHD equations is differenced, assuring the rigorous conservation of mass, momentum, magnetic flux, and energy. In conservation form, the continuity, induction, momentum, and energy equations are written

$$\partial \rho/\partial t + \nabla \cdot (\rho \mathbf{u}) = 0, \tag{1}$$

$$\partial \mathbf{B}/\partial t + \nabla \times (\mathbf{u} \times \mathbf{B}) = 0, \tag{2}$$

$$\partial \rho \mathbf{u}/\partial t + \nabla \cdot (\rho \mathbf{u}\mathbf{u} - \mathbf{Q}) = 0, \tag{3}$$

$$\partial \rho e/\partial t + \nabla \cdot (\rho e \mathbf{u} - \mathbf{Q} \cdot \mathbf{u}) = 0, \tag{4}$$

where ρ is the mass density of the plasma, $\mathbf{u}$ the plasma velocity, $\mathbf{B}$ the magnetic field intensity, $\mathbf{Q}$ the stress tensor, and e the specific total plasma energy. The stress tensor, $\mathbf{Q}$, and plasma energy, e, are given in mks units by

$$\mathbf{Q} = (1/\mu)\mathbf{B}\mathbf{B} - \tfrac{1}{2}[(1/\mu)\mathbf{B} \cdot \mathbf{B} + P]\mathbf{I}\mathbf{I} \tag{5}$$

and

$$e = \tfrac{1}{2}(\mathbf{u} \cdot \mathbf{u}) + \tfrac{1}{2}(\mathbf{B} \cdot \mathbf{B})/\mu\rho + i, \tag{6}$$

where μ is the permeability, i is the specific internal energy of the plasma, and P is the plasma pressure equal to a function of ρ and i. When there are large gradients in velocity, viscosity must be added to the stress tensor (Zel'dovich and Raizer, 1967). A viscosity formulation is given by Landau and Lifschitz (1958). Other transport terms, such as mass diffusion, thermal conductivity, and resistive diffusion must sometimes be added. Extensive discussions of the numerical problems encountered in adding these terms are contained in review articles by Roberts and Potter (1968) and Killeen (1972).

2. Lagrangian Computation Meshes

A Lagrangian mesh may be built up from cells of arbitrary shape and size. The cells in two dimensions, however, are usually arbitrary quadrilaterals because such cells have simple logical relationships with one another. Triangular cells have some advantages; they are always convex, and interpolations in their interior are linear. However, with triangular zones, a table of nearest neighbors must be kept because there is no longer a simple relationship between the physical and logical position of a cell with respect to its neighbors.

Deriving difference equations for a Lagrangian mesh is not as automatic as it is for a regular Eulerian mesh. The method of derivation requires a coordinate transformation from the nonrectilinear physical space to a rectilinear natural or logical space before analogs to ordinary different equations can be constructed. A derivation of generalized difference equations for three-dimensional MHD flow is given in Section IV.

The Lagrangian equations for ideal MHD, corresponding to the Eulerian equations, Eqs. (1)–(4), are written

$$d\rho/dt + \rho(\nabla \cdot \mathbf{u}) = 0, \tag{7}$$

$$d\mathbf{B}/dt + \mathbf{B}(\nabla \cdot \mathbf{u}) - (\mathbf{B} \cdot \nabla)\mathbf{u} = 0, \tag{8}$$

$$p\, d\mathbf{u}/dt - \nabla \cdot \mathbf{Q} = 0, \tag{9}$$

and

$$\rho\, di/dt + P(\nabla \cdot \mathbf{u}) = 0. \tag{10}$$

The most obvious difference between these equations and the Eulerian equations is the absence of convective transport terms. This is characteristic of the Lagrangian equations, in that they describe the evolution of a particular element of the fluid, rather than the time variation of fluid variables at a particular point in space.

3. *The Generalized Computation Mesh*

In a generalized mesh, the relative motion of the plasma and the computation mesh is completely arbitrary. That is, it is possible to specify the velocity of the computation mesh separately from the velocity of the plasma. The equations for a generalized mesh are derived either by adding to a Lagrangian calculation a rezone phase in which the convective transport due to relative motion between the plasma and the mesh is computed, or by rewriting the Eulerian equations, Eqs. (1)–(4), in a moving frame.

There are many advantages to be gained from a generalized mesh. Convective transport may be reduced, in many problems of interest, by transforming away the bulk motion of the plasma as suggested by Roberts and Potter (1968). Contact surfaces, as between a plasma and a vacuum, may be resolved easily by a Lagrangian interface within a generalized mesh. For other applications, the mesh points may be constrained to lie on flux surfaces to allow the accurate calculation of anisotropic thermal conduction (Hertweck and Schneider, 1970). Finally, the mesh points can be concentrated in a region of the mesh where resolution is required, even as that region moves through space (Hirt et al., 1974). In summary, a generalized mesh offers enormous flexibility in designing the calculation to fit the application.

As mentioned above, the equations for a generalized mesh can be obtained by transforming the Eulerian equations to a moving coordinate system. In a frame moving with velocity $\mathbf{u}'$, the equations corresponding to Eqs. (1)–(4) are written

$$\partial \rho/\partial t + \nabla \cdot [\rho(\mathbf{u} - \mathbf{u}')] + \rho(\nabla \cdot \mathbf{u}') = 0, \tag{11}$$

$$\partial \mathbf{B}/\partial t + \nabla \cdot (\mathbf{u} - \mathbf{u}')\mathbf{B} - \nabla \cdot \mathbf{B}\mathbf{u} + \mathbf{B}(\nabla \cdot \mathbf{u}') = 0, \tag{12}$$

$$\partial \rho \mathbf{u}/\partial t + \nabla \cdot [\rho \mathbf{u}(\mathbf{u} - \mathbf{u}') - \mathbf{Q}] + \rho \mathbf{u}(\nabla \cdot \mathbf{u}') = 0, \tag{13}$$

and

$$\partial \rho e/\partial t + \nabla \cdot [\rho e(\mathbf{u} - \mathbf{u}') - \mathbf{Q} \cdot \mathbf{u}] + \rho e(\nabla \cdot \mathbf{u}') = 0, \tag{14}$$

where, in the induction equation, terms of the form $\nabla \cdot (\mathbf{AB})$ mean ∇ is scalar multiplied with $\mathbf{A}$ but differentiates both $\mathbf{A}$ and $\mathbf{B}$. When $\mathbf{u}'$ is zero, these equations reduce to the conservation equations, Eqs. (1)–(4), and when $\mathbf{u}'$ and $\mathbf{u}$ are equal, they reduce to the Lagrangian equations, Eqs. (7)–(10).

Alternatively, the equations can be separated into Lagrangian and convective transport phases as proposed by Hirt et al. (1974). In this formulation, the time variation in the moving frame is computed in two steps. In a Lagrangian phase, Eqs. (7)–(10) are solved, and in a convection phase, the convective transport is computed from the integral equations

$$\frac{\partial}{\partial t} \int_V \rho dV = \int_{s(V)} [\hat{\mathbf{n}} \cdot (\mathbf{u} - \mathbf{u}')\rho]\, ds, \tag{15}$$

$$\frac{\partial}{\partial t} \int_V \rho \mathbf{u} dV = \int_V \rho \frac{d\mathbf{u}}{dt} dV - \int_{s(V)} [\hat{\mathbf{n}} \cdot (\mathbf{u} - \mathbf{u}')\rho \mathbf{u}]\, ds, \tag{16}$$

$$\frac{\partial}{\partial t} \int_V \mathbf{B} dV = \frac{d}{dt} \int_V \mathbf{B} dV - \int_{s(V)} [\hat{\mathbf{n}} \cdot (\mathbf{u} - \mathbf{u}')\mathbf{B}]\, ds, \tag{17}$$

and

$$\frac{\partial}{\partial t} \int_V \rho e dV = \frac{d}{dt} \int_V \rho e dV - \int_{s(V)} [\hat{\mathbf{n}} \cdot (\mathbf{u} - '\mathbf{u})\rho e]\, ds, \tag{18}$$

where V is a control volume, and $s(V)$ is its boundary with outward directed unit normal $\hat{\mathbf{n}}$. Both the conservation form and the two-phase form permit arbitrary relative motion between the plasma and the mesh, and both reduce to the Lagrangian form when the velocity of the mesh and the velocity of the fluid are equal. The two-phase formulation does permit implicit equations of motion and explicit convection terms. Since the more stringent stability condition must be met by the equations of motion, gains in running speed are made by making them implicit even when the convection terms are explicit.

B. A Survey of Numerical Methods and Applications

1. *Eulerian Computations*

Since the review by Roberts and Potter (1968), there have been many applications of Eulerian computations to MHD problems. Freeman and Lane (1969) reported an MHD method for two-dimensional axisymmetric flows using an explicit, Lax–Wendroff time advancement algorithm (Richtmyer, 1963), with a Lapidus diffusion term (Lapidus, 1967) added to each of the difference equations corresponding to Eqs. (1)–(4). This method was later applied to the simulation of the interaction of a plasmoid with an axisymmetric magnetic field (Freeman, 1971). Roberts and Boris (1969) reported a method for three-dimensional flow in which automatic finite difference equation generators for spatial derivatives were incorporated into an explicit time advancement algorithm. Lindemuth and Killeen (1973) have published an algorithm for two-dimensional, axisymmetric flow in which all terms appearing in the equation, including transport terms, are differenced implicitly. The solution of the implicitly formulated equations is performed with an alternating direction implicit method (Peaceman and Rachford, 1955). As in the earlier code of Freeman and Lane, diffusion terms are added to each of the equations to smooth the solution. However, only the mass diffusion coefficient is purely numerical (I. Lindemuth, private communication, 1975).

Duchs (1968) has performed two-dimensional studies of the rotation induced by transverse multipole fields in a plasma confined by a theta pinch field. The equations of motion for a two-fluid plasma in a $z =$ const. plane of an axisymmetric theta pinch are solved, including the Hall term, but neglecting anisotropic pressure effects and electron inertia terms. Lui and Chu (1974) have applied a two-dimensional numerical method to a time-dependent, boundary value problem in an axisymmetric torus. The motion is computed in a poloidal plane with equations which include thermal and resistive diffusion, but no explicit viscosity. The value of the magnetic flux on the boundary is prescribed as a function of time. As the plasma is pinched inward by an increasing, applied magnetic field, a low density plasma region is left behind. To avoid problems with large Alfven speeds and the consequent restrictions on the maximum time-step consistent with accurate computation, the density in the "vacuum" region is not permitted to fall below 15% of the initial, uniform density. The diffusion equations are advanced by means of a two-step scheme where values of the diffusion coefficients are advanced in time on the first step, and implicit equations are used to advance the dependent variables on the second step. The equations of motion are also written implicitly. Both the diffusion equations and the equations of motion are solved by means of an alternating direction implicit algorithm.

A two-dimensional, MHD, multifluid code which includes transport due to microturbulence is reported by Wagner and Manheimer (1973). Anisotropic heat conduction and thermoelectric magnetic field generation are also included in the model. The code has been applied to the simulation of the return current from a relativistic electron beam flowing through a plasma.

Three-dimensional, nonlinear, ideal MHD calculations have been reported by Wooten *et al.* (1974). Calculations of the nonlinear evolution of fixed boundary kink modes were performed with an explicit, leapfrog advancement algorithm. The method is an extension of a computational method for the solution of the linearized MHD equations reported by Bateman *et al.* (1974).

2. *Lagrangian Computations*

One-dimensional Lagrangian computations of pinch discharges have been performed for many years (Hain *et al.*, 1960), but there is only one recently reported Lagrangian calculation in two dimensions. The calculation of the nonlinear evolution of the kink mode in a Tokamak by White *et al.* (1974) follows the motion of the boundary of an incompressible plasma due to the perturbation of an unstable equilibrium. A cylindrical approximation to the toroidal geometry of the plasma is used, and helical symmetry is assumed. In this geometry and because of the incompressibility of the plasma, the field and velocity may be computed from potential functions. A closed sequence of straight line segments lying on the plasma boundary form the Lagrangian computation mesh.

3. *Computations with a Generalized Mesh*

Hertweck and Schneider (1970) reported a method for calculating end losses in a theta pinch using flux coordinates. The mesh is stationary in the axial direction, but moves radially so that the relative velocity between the flux surfaces and the mesh is zero when the conductivity is infinite. The differential equations are transformed into the nonorthogonal, time-dependent coordinates and then differenced. The differential equations appear quite complex in these coordinates, but are simply Eqs. (11)–(14) with reference to the moving frame. As Hertweck and Schneider (1970) point out, the accurate calculation of field aligned thermal conduction is only possible using flux coordinates. The results of a calculation of end losses including thermal conduction were later reported by Schneider (1972).

A method developed by Anderson (1975) also uses flux coordinates. However, the computation mesh is orthogonal. Zone lines perpendicular to the flux surfaces are reconstructed each computation time-step making possible the use of a time-step splitting algorithm to advance the solution in time (Richtmyer and Morton, 1967). With time-step splitting a multidimensional calculation is performed by a sequence of one-dimensional operations so that it is convenient to apply the flux corrected transport algorithm of Boris and Book (1973) to reduce computational diffusion.

An implicitly formulated method in which the computation is split into Lagrangian and convective transport phases has been applied to MHD computations by Brackbill and Pracht (1973). Two-dimensional calculations of a z-pinch in r–θ coordinates and axisymmetric calculations of a hot diamagnetic pellet expanding into a magnetic field are performed using an almost-Lagrangian mesh. In an almost-Lagrangian mesh, the relative motion between the mesh and the plasma is kept to a minimum to reduce computational diffusion. An axisymmetric calculation of a sharp boundary theta pinch (Brackbill, 1973) was reported in which a vacuum field solution was coupled to plasma flow across a Lagrangian interface.

A similar method for resolving the plasma–vacuum interface was reported by Lui (1973). Lui used an Eulerian mesh with arbitrarily shaped cells to resolve curved boundaries, and Lagrangian mesh points to resolve a moving vacuum–plasma interface. The method was applied to the study of the reflection and focusing of shocks by curved walls.

A method for computing three-dimensional MHD flow with a generalized mesh is presented in Section IV, and the results of its application to the rotating theta pinch and the internal kink mode are displayed.

III. The Computation of Convective Transport

A. Properties of Approximations to the Convective Derivative

The fundamental difficulty with a calculation using an Eulerian mesh is the accurate computation of convective transport. Whether the equations for an Eulerian mesh are written in conservation form, Eqs. (1)–(4), or in non-conservation form, Eqs. (7)–(10) or Eqs. (15)–(18), it is the approximation of the convective derivative, $\mathbf{u} \cdot \nabla$, which determines the overall accuracy and stability of the solution. One might reasonably ask: Simple difference schemes approximating the convective derivative can be made to have the same formal accuracy as any other difference equation, so why does the computation of the convective derivative introduce errors? The answer is that if the terms in the difference equations corresponding to the convective derivative are written in a simple way, the equations are often unstable. Other, less accurate difference equations which are stable must be used instead (Richtmyer, 1963).

1. *Nonlinear Stability*

Linear stability analysis is not always informative about the stability of difference approximations to the convective derivative. It is necessary that the difference approximations be linearly stable, but linearly stable difference equations are often nonlinearly unstable. For example, Richtmyer and Morton (1967) discuss the application of the leapfrog difference equation to a nonlinear convective transport equation,

$$\partial u/\partial t + (\partial/\partial x)(\tfrac{1}{2}u^2) = 0, \qquad 0 \leqslant x \leqslant 1, \qquad t \geqslant 0. \tag{19}$$

The "leapfrog" approximation to the equation is written

$$^{n+1}u_j = {}^{n-1}u_j + \tfrac{1}{2}(\delta t/\delta x)[({}^n u_{j+1})^2 - ({}^n u_{j-1})^2], \tag{20}$$

and is linearly stable (Richtmyer, 1963). Nevertheless, it is explosively unstable when numerically solved. That is, the solution grows exponentially in some cases with a growth rate which goes as $(\delta t)^{-1}$. An analysis of the problem indicates the leapfrog approximation is stable when δt is sufficiently small so that "the truncation error in the difference equations generates sufficiently small perturbations." The difference between linear and nonlinear problems which causes the difference in stability is that "the behavior depends markedly on the relative magnitude of the perturbations" (Richtmyer and Morton, 1967). Similar conclusions are drawn by Fornberg (1973), in an analysis of the leapfrog and Crank–Nicolson approximations to Eq. (19).

2. *The Stability of Numerical Approximations to Linear Partial Differential Equations with Variable Coefficients: Kreiss's Theorem*

The analysis of the stability of numerical approximations to linear partial differential equations with variable coefficients, such as the equation

$$\partial u/\partial t = a(x)\, \partial u/\partial x = 0, \tag{21}$$

is the next step beyond linear stability analysis in determining the stability of numerical approximations to nonlinear equations. A summary is given of an important theorem by Kreiss (1964) whose relevance to the present discussion is shown in the following section. Kreiss's theorem can be summarized, for our purposes, in the following way. Consider a linearly stable difference equation approximation to Eq. (21). If the true solution to Eq. (21) were substituted into the approximation, the right and left sides would be unequal by terms of order δx^m and δt^n, where δx is the mesh spacing and δt is the time-step. These terms are the truncation errors in the approximation. Then if a diffusionlike term exists or is added to the difference equation with positive diffusion coefficients of at least one lower order than the truncation errors, that is, of order δx^{m-1} and δt^{n-1} or less, the equation will be stable. An equation with such a diffusion term is called positively stable by Richtmyer (1963). Evidently, the role of the diffusion is to add sufficient smoothing to the solution so that perturbations due to truncation errors in the difference equation are kept at a sufficiently low level.

3. A Heuristic Stability Theory

Kreiss's theorem shows that the presence of enough diffusion to damp truncation-error-produced perturbations is a necessary and sufficient condition for the stability of difference approximations to linear differential equations with variable coefficients. That this condition is, at least, a necessary condition for the stability of difference approximations to nonlinear equations is suggested by the work of Hirt (1968) in developing a heuristic stability theory. In his study, finite difference approximations are reduced to differential equations by expanding each of their terms in a Taylor series. The lowest order terms in the expansions are the differential equations being approximated. Higher order terms in δx and δt are truncation errors. It is Hirt's observation that some truncation errors are associated with nonlinear instabilities of the difference equations. Specifically, he exhibits cases where the coefficient of the second spatial derivative of the density in the differential equation derived from the difference equation approximating the continuity equation appears to determine the stability of a system of coupled equations describing fluid flow. That is, when the coefficient is negative, there is antidiffusion in the continuity equation which destabilizes the entire system of equations. The converse of Kreiss's theorem for linear equations seems to be true for non-linear equations also. Without the addition of positive diffusion of the same order as the truncation errors, the difference equations are unstable. For stability, the equations must be dissipative or positively stable in the sense described by Kreiss. Evidently, Hirt's heuristic stability theory is pointing to the existence of a second necessary condition for the stability of the difference equations. That is, not only must the difference equations be linearly stable, but they must also satisfy Kreiss's theorem.

A truncation error analysis of various difference approximations to Eq. (15), the convective derivative of the density, is given by Hirt (1968). Among the equations for which his results are applicable are three which are linearly stable (Richtmyer, 1963); upstream or donor cell differencing, leapfrog differencing, and implicit spatially-centered differencing, These difference approximations are listed in Table I, along with the order of accuracy. In Table II, the corresponding coefficients of the second derivative of the density are listed, including nonlinear terms. In each case, the nonlinear terms are proportional to velocity gradients, and would appear no matter which form of the continuity equation were approximated. (It is also true that in each case, the nonlinear terms are proportional to gradients of $\mathbf{u}''$, the relative velocity between the plasma and the mesh. Thus, it is possible to transform to a coordinate frame where nonlinear truncation errors are negligible, a fact which can be exploited with a generalized mesh method.)

As we can see from the entries in Table II, all of the mass diffusion coefficients become negative with sufficiently large velocity gradients. This result is in agreement with Richtmyer's (1963) observations. It also helps one to understand how the addition of viscosity to the momentum equation enhances the stability of difference equations, for the effect of viscosity is to decrease velocity gradients.

Similar errors are found in the difference approximations to Eqs. (16)–(18), the convective derivative of the momentum, magnetic flux, and energy.

Evidently, the difficulty with most approximations to convective transport is that they do not satisfy the second necessary condition for stability given by Kreiss's theorem. As Hirt's analysis shows, the violation of the assumptions of Kreiss's theorem is often associated with nonlinear numerical instabilities.

B. Nonlinearly Stable Approximations to Convective Transport

1. *Truncation Error Corrections*

Amsden and Hirt (1973) and Rivard *et al.* (1973) have applied Hirt's heuristic stability theory to the application of truncation error corrections of the difference equations for fluid flow. For example, to stabilize the leapfrog differencing of the continuity equation (entry b, Table I), a mass diffusion term is added with diffusion coefficient, κ, given by

$$\kappa = \alpha \tfrac{1}{2} \delta x^2 \, \partial u''/\partial x \tag{22}$$

where α is typically equal to 2. Without decreasing the order of accuracy of the original difference approximation, diffusion is added to the difference equation to stabilize it. The added diffusion should be sufficient because it

TABLE I

Difference Approximations to the Continuity Equation

Name	Difference equation	Truncation error
a. Upstream differencing	$^n p_J = {}^{n-1} p_J - \dfrac{\delta t}{\delta x}(\langle{}^n p_{J+1/2}\rangle {}^n u_{J+1/2} - \langle{}^n p_{J-1/2}\rangle {}^n u_{J-1/2})$ $\langle{}^n p_{J+1/2}\rangle = \begin{cases} {}^n p_{J+1}, & u_{J+1/2} \leq 0 \\ {}^n p_J, & u_{J+1/2} > 0 \end{cases}$	$O(\delta x^2, \delta t^2)$
b. Leapfrog differencing	$^{n+1} p_J = {}^{n-1} p_J - \dfrac{\delta t}{\delta x}({}^n(\rho u)_{J+1} - {}^n(\rho u)_{J-1})$	$O(\delta x^3, \delta t^3)$
c. Implicit, spatially-centered differencing	$^{n+1} p_J = {}^n p_J - \dfrac{\delta t}{2\delta x}({}^{n+1}(\rho u)_{J+1} - {}^{n+1}(\rho u)_{J-1})$	$O(\delta x^3, \delta t^2)$

TABLE II

MASS DIFFUSION COEFFICIENTS DUE TO TRUNCATION ERRORS

Name	Mass diffusion coefficient[a]
a. Upstream differencing	$\frac{\delta x}{2}|u''| - \frac{\delta t}{2} u''^2 - \frac{\delta x^2}{2}\frac{\partial u''}{\partial x} - \frac{\delta x^3}{8}\frac{\partial^2 u''}{\partial x^2} - \frac{\delta x^4}{24}\frac{\partial^3 u''}{\partial x^3}$
b. Leapfrog differencing	$-\frac{\delta x^2}{2}\frac{\partial u''}{\partial x} - \frac{\delta x^4}{12}\frac{\partial^3 u''}{\partial x^3}$
c. Implicit, spatially centered differencing	$\frac{\delta t}{2} u''^2 - \frac{\delta x^2}{2}\frac{\partial u''}{\partial x} - \frac{\delta x^4}{12}\frac{\partial^3 u''}{\partial x^3}$

[a] u'' is the relative velocity between the mesh and the plasma ($u'' = u - u'$).

cancels the lowest order truncation error, and adds some positive diffusion to cancel higher order errors as well. Some confidence in this approach is derived from its reducing to a prescription which satisfies Kreiss's theorem for linear equations with variable coefficients.

Rivard *et al.* (1973) apply similar truncation error corrections to all of the fluid equations. A somewhat similar approach is taken by Lapidus (1967), in that diffusion is added to every equation. However, in his approach and in its applications to MHD calculations (Freeman and Lane, 1968; Lindemuth and Killeen, 1973) the added diffusion is not computed from a truncation error analysis.

2. *Flux-Corrected Transport*

A method approximating convective transport, which is discussed in detail in another article in this volume, has been developed by Boris and Book (1973). Their method requires that the solution of the difference equation for the convective transport of mass and energy be positive everywhere. The method is composed of two steps. In the first step, a large velocity-independent diffusion is added to the difference approximation of the convective derivative. The addition of the diffusion assures that the solution obtained in the first step is positive everywhere. In the second step the diffusion added in the first step is subtracted everywhere, subject to the constraint that the solution obtained be positive everywhere.

In the first step, the convective derivative in the continuity equation is approximated in one dimension by the equation

$$\tilde{\rho}_j = {}^n\rho_j - \tfrac{1}{2}(u''\,\delta t/\delta x)({}^n\rho_{j+1} - {}^n\rho_{j-1}) \\ + [\tfrac{1}{8} + \tfrac{1}{2}(u''\,\delta t/\delta x)^2]({}^n\rho_{j+1} - 2{}^n\rho_j + {}^n\rho_{j-1}), \qquad (23)$$

where $\tilde{\rho}$ denotes the solution obtained on the first step. The difference equation is the Lax–Wendroff equation (Richtmyer, 1963) plus an additional velocity-independent diffusion. To first order in δx and δt, the difference equation approximates the differential equation

$$\partial\rho/\partial t = -u''\,\partial\rho/\partial x + \tfrac{1}{8}(\delta x^2/\delta t)(\partial^2\rho/\partial x^2). \qquad (24)$$

The diffusion coefficient for the velocity-independent diffusion is thus $\tfrac{1}{8}\delta x^2/\delta t$. This diffusion is subtracted in the second step of the algorithm, where negative diffusion is computed. The negative or antidiffusion is computed subject to the constraint that the second step create no new maxima and minima in the solution, and that it not accentuate already existing extrema. In the antidiffusion step, mass and energy fluxes between pairs of cells are computed, assuring that the total mass and total energy are conserved. The antidiffusive mass fluxes, f, are computed from the equation

$$f_{j+1/2} = \text{SGN } \Delta_{j+1/2} \max\{0, \min(\Delta_{j-1/2} \text{ SGN } \Delta_{j+1/2}, \text{``}\tfrac{1}{8}\text{''}|\Delta_{j+1/2}|,$$
$$\Delta_{j+3/2} \text{ SGN } \Delta_{j+1/2})\}, \tag{25}$$

where

$$\Delta_{j+1/2} = \tilde{\rho}_{j+1} - \tilde{\rho}_j.$$

The symbol "$\tfrac{1}{8}$" signifies that higher order corrections are made so that the antidiffusion in the second step cancels the diffusion added in the first step to fourth order in δx and δt wherever possible. The new density is then computed from the equation

$$^{n+1}\tilde{\rho}_j = \tilde{\rho}_j - f_{j+1/2} + f_{j-1/2}. \tag{26}$$

Similar equations are written for the energy.

The flux equation, Eq. (25), allows no adjustments in the second step to local maxima and minima. Where the variation of $\tilde{\rho}_j$ is monotonic, the antidiffusion will continue until either $\Delta_{j-1/2}$ or $\Delta_{j+3/2}$ is zero, that is, until $\tilde{\rho}_j$ equals either $\tilde{\rho}_{j+1}$ or $\tilde{\rho}_{j-1}$. One can see from this how the positivity of the solution, $^{n+1}\rho$, is preserved, for the minimum value of $\tilde{\rho}_j$ over the entire domain cannot be decreased toward zero, and no other value can be made smaller than the minimum.

Flux-corrected approximations to the convective derivative have been used in computations of two- and three-dimensional flow employing a time-step splitting algorithm to advance the solution in time. In such computations, one-dimensional difference equations in each orthogonal coordinate direction are solved in a cyclic fashion. Thus, a one-dimensional formulation of flux-corrected transport can be used.

IV. A Generalized Mesh Method for MHD

In this section, a method is presented for computing three-dimensional MHD flow on a generalized mesh. The equations of motion are split into two phases as proposed by Hirt *et al.* (1974) and discussed in Section II. In the first phase, the Lagrangian equations of motion, Eqs. (7)–(10), are solved exactly as one would solve them for a pure Lagrangian computation. In the second phase, the transport of mass, momentum, flux, and energy due to relative motion between the plasma and the mesh is computed from Eqs.

(15)–(18). (The second phase is characterized as a rezone, where the rezone is with respect to the Lagrangian frame, rather than the fixed, Eulerian frame.)

The prescription for the grid velocity, $\mathbf{u}'$, is completely arbitrary. By choosing $\mathbf{u}'$ properly, the generalized mesh may be made Eulerian, Lagrangian, flux coordinate (Hertweck and Schneider, 1970), or almost-Lagrangian (Brackbill and Pracht, 1973). The almost-Lagrangian prescription for $\mathbf{u}'$ will be outlined in a brief paragraph in this section.

The bulk of this section will be devoted to developing generalized difference equations for the Lagrangian phase of the calculation. These equations describe the dynamical evolution of the system, and must be solved no matter which rezoning prescription is employed in the rezone phase. It will be shown how the equations are derived, how they are solved, how boundary conditions are applied, and what their properties are.

A. Difference Equations for the Lagrangian Phase of a Generalized Mesh Calculation

Difference equations for nonrectilinear meshes may be derived by the finite-element method. As the method is outlined by Zienkiewicz (1971), the flow field is subdivided into elements, where each element is associated with a number of mesh points at which the values of the flow variables are stored. At every point in each element, the flow variables are determined by interpolation so that they are continuous between elements. To derive equations of motion, a functional is defined and minimized over the flow field with the value of variables at mesh points as parameters. In MHD flow, the appropriate functional is the Lagrangian.

In contrast with this approach the derivation presented here proceeds directly from the equations of motion. Variables are defined over the mesh as in the finite element method, but the kinematical equations are approximated by a coordinate transformation like that used by Schulz (1964). A derivation for the dynamical equation after Goad (1960) using d'Alembert's principle and the principle of virtual work completes the formulation of Lagrangian difference equations.

1. *Kinematical Equations*

The kinematical evolution of the thermodynamic variables is contained in the continuity, induction, and internal energy equations, Eqs. (7), (8), and (10). These equations describe the evolution of the thermodynamic variables ρ, $\mathbf{B}$, and i as the plasma dilates and shears. When the momentum equation, Eq. (9), is solved simultaneously and consistently with the kinematic equations, the full dynamics of the system described by these equations are obtained.

It is convenient, when deriving the complete system of difference equations, to consider the kinematical and dynamical equations separately. That is, difference equations approximations to Eqs. (7), (8), and (10) are derived separately from the difference equation for Eq. (9).

The difference equations are solved on a computation mesh composed of the six-surfaced conforming cells described by Pracht (1975). These cells are holomorphic to the unit cube; the coordinate transformation from a point in the unit cube (ξ, η, γ) to a point in physical space $\mathbf{x}$ is trilinear, and is given by

$$\mathbf{x} = [\mathbf{x}^1 \xi(1-\eta) + \mathbf{x}^2 \xi \eta + \mathbf{x}^3 (1-\xi)\eta + \mathbf{x}^4 (1-\xi)(1-\eta)](1-v)$$
$$+ [\mathbf{x}^5 \xi(1-\eta) + \mathbf{x}^6 \xi \eta + \mathbf{x}^7 (1-\xi)\eta + \mathbf{x}^8 (1-\xi)(1-\eta)]v, \qquad (27)$$

where $\mathbf{x}^l$ is the position of the lth vertex of the computation cell within which $\mathbf{x}$ lies. When both $\mathbf{x}^l$ and $\dot{\mathbf{x}}^l = \mathbf{u}^l$ are stored, the elements of the rate of strain tensor, $S_{\alpha\beta}$, written

$$S_{\alpha\beta} = \partial u_\alpha / \partial x_\beta, \qquad \alpha, \beta = 1, 2, 3, \qquad (28)$$

are determined in the interior of a cell. A consistent kinematical description of the flow follows when the thermodynamic variables are constant within each cell and discontinuous at cell boundaries, and $S_{\alpha\beta}$ is replaced by its average value over the cell in Eqs. (7), (8), and (10). The average value of $S_{\alpha\beta}$ over a cell is computed from the equation

$$\langle S_{\alpha\beta} \rangle = \frac{1}{V} \int_V dV S_{\alpha\beta}. \qquad (29)$$

When $\langle S_{\alpha\beta} \rangle$ is substituted for $S_{\alpha\beta}$ the kinematical equations are written

$$d\rho/dt + \rho \langle S_{ii} \rangle = 0, \qquad (30)$$

$$\rho \, di/dt + P \langle S_{ii} \rangle = 0, \qquad (31)$$

and

$$dB_\alpha/dt + B_\alpha \langle S_{ii} \rangle - B_i \langle S_{i\alpha} \rangle = 0, \qquad (32)$$

with the convention that the repeated subscripts i, j, k are to be summed over. The time derivatives and the time levels of other terms appearing in these equations are not yet specified or approximated.

The integral in Eq. (29) is evaluated by rewriting the integrand so that it depends explicitly on ξ, η, and ν,

$$dV(\partial u_\alpha/\partial x_\beta) = d\xi\, d\eta\, d\nu\, \partial(u_\alpha, x_j, x_k)/\partial(\xi, \eta, \nu)\, J \tag{33}$$

where βjk is a cyclic permutation of 1, 2, 3. The integration volume in ξ, η, ν space is the unit cube. Integrating Eq. (29) over this volume yields a linear equation for each element of the rate of strain tensor written

$$^n\langle S_{\alpha\beta}\rangle = \frac{1}{V}\sum_{l=1}^{8} c_\beta^l u_\alpha^l, \tag{34}$$

where l labels the vertices of the cell, and c_β^l are coefficients which depend entirely on the geometry of the cell.

The geometric coefficients, c_β^l, of which there are twenty-four for each cell, summarize in storable form the geometry of the mesh. They appear in all equations which refer to this geometry, as in the calculation of the cell volume.

$$V = \sum_{l=1}^{8} c_\alpha^l x_\alpha^l, \quad \alpha = 1, 2, \text{ or } 3, \tag{35}$$

and the calculation of gradients of thermodynamic variables at cell vertices. They are recomputed and stored only when the computation mesh moves.

The geometric coefficients are written in a compact vector form in the Appendix.

2. The Dynamical Equations

The system of difference equations in space is completed by deriving an approximation to Eq. (9), the momentum equation, by generalizing Goad's (1960) derivation of difference equations for two-dimensional hydrodynamics. This derivation employs the computation of virtual work and d'Alembert's principle to form a relation between changes in the potential energy of the computation cells and the corresponding acceleration of their vertices.

Consider a small elaboration of the description of the flow developed in the previous section. To each vertex, there will now be assigned a mass, m^l, which is computed by some yet to be specified rule from the distributed masses of the cells neighboring vertex l. In addition, the potential energy of a cell is defined by the equation

$$W \equiv \int_V dV \rho w, \tag{36}$$

where w is the specific potential energy

$$w = i + \mathbf{B} \cdot \mathbf{B}/2\mu\rho. \tag{37}$$

An evolution equation for w, corresponding to the equation for the internal energy, Eq. (10), is given by

$$\rho \, dw/dt - (\mathbf{Q} \cdot \nabla) \cdot \mathbf{u} = 0. \tag{38}$$

This equation is now used to compute changes in the cell potential energy corresponding to displacements of a vertex labeled l.

The change in W over a time interval, δt, is computed by integrating Eq. (38) over the volume of a cell, and over δt,

$$\delta W = W(t+\delta t) - W(t) = \int_v dV' \int_t^{t+\delta t} dt' Q_{ij} S_{ji}. \tag{39}$$

The stress tensor is constant within a cell and may be factored from the volume integral. Combining Eqs. (34) and (39) yields an equation for the remaining volume integral for δW written

$$\delta W = \int_t^{t+\delta t} dt' \left(Q_{ij} \sum_{l=1}^{8} c_j^l u_i^l \right). \tag{40}$$

When some average value over the time interval δt is substituted for Q_{ij} and c_j^l in Eq. (40), an equation between changes in W and displacements of the vertices results, and is written

$$\delta W = \left(\bar{Q}_{ij} \sum_{l=1}^{8} c_j^l \right) \delta s_i^l, \tag{41}$$

where δs_i^l is given by

$$\delta s_i^l = \int_t^{t+\delta t} dt' u_i^l. \tag{42}$$

In words, Eq. (41) says that the change in the potential energy of a cell is equal to the negative of the product of the force exerted by the cell on each vertex and the displacement of that vertex,

$$\delta W = -\sum_{l=1}^{8} \mathbf{F}^l \cdot \delta \mathbf{s}^l, \tag{43}$$

where F^l is the force exerted on the vertex l by the cell. The force on each vertex is related to the acceleration experienced by that vertex by d'Alembert's equation,

$$(\mathbf{F}^l - m^l \ddot{\mathbf{u}}^l) \cdot \delta \mathbf{s}^l = 0. \tag{44}$$

Therefore, the components of the acceleration of vertex l due to stresses acting interior to a cell are given by the ratio of δW to the product of m^l and the components of $\delta \mathbf{s}^l$,

$$\dot{u}_\alpha^l = -(1/m^l)\bar{Q}_{\alpha i} c_i^l. \tag{45}$$

The resultant acceleration is computed by summing the contributions from all eight neighboring cells, and is given by

$$\dot{u}_\alpha^l = -\frac{1}{m^l} \sum_k (\bar{Q}_{\alpha i}^k c_i^{lk}), \tag{46}$$

where c_i^{lk} is the ith component of the geometric coefficient assigned to the lth vertex of cell k.

Some of the properties of the equation of motion are easily shown. For example, it can be shown that when the geometric coefficients are summed over the vertices of a cell, l, the result is zero. Thus, the contribution to the change in the momentum of plasma from stresses acting interior to each cell is zero, and momentum is conserved. Similarly, it is easily shown that when the geometric coefficients at a vertex are summed over the surrounding cells, k, the result is zero. Thus, when the stress is constant, a vertex experiences no acceleration.

3. *Differencing in Time and the Conservation of Energy*

Unconditionally stable difference equations in time are obtained when the equations are made implicit. When they are centered in time, the magnetic flux and the total energy are conserved exactly. Since the evolution of internal energy is calculated separately, there is no spurious exchange of energy between the magnetic field and the plasma, and local energy conservation in the sense described by Fromm (1961) is obtained.

However, time-centered equations are not used. In the one-dimensional studies which preceded the formulation of the two- and three-dimensional algorithms, it was found that implicit, first order equations were less dispersive than second order equations, and not appreciably more diffusive. To obtain exact energy conservation with the first order equations, Eq. (38)

for the potential energy is solved rather than Eq. (31) for the plasma internal energy. The forward biased time and space difference approximations for the continuity, potential energy, and magnetic induction equations are written

$$^{n+1}\rho - {}^n\rho + {}^{n+1}\rho \, {}^{n+1}\langle S_{ii}\rangle \, \delta t = 0, \tag{47}$$

$$^{n+1}\rho({}^{n+1}w - {}^nw) - \bar{Q}_{ij}{}^{n+1/2}\langle S_{ji}\rangle \, \delta t = 0, \tag{48}$$

and

$$^{n+1}B_\alpha - {}^nB_\alpha + ({}^{n+1}B_\alpha{}^{n+1}\langle S_{ii}\rangle - {}^{n+1}B_i{}^{n+1}\langle S_{i\alpha}\rangle) \, \delta t = 0, \tag{49}$$

where the components of the stress tensor are given by

$$\mu\bar{Q}_{\alpha\beta} = {}^{n+1/2}B_\alpha{}^{n+1/2}B_\beta - ({}^{n+1}P\mu + {}^{n+1}B_i{}^nB_i) \, \delta\alpha\beta. \tag{50}$$

The time level of each term in these equations is denoted by the leading superscript, with n denoting time level t and $n + 1$ time level $t + \delta t$. The time levels of the rate of strain tensor correspond to the time levels assigned to the velocities appearing in them. Time-centered terms are denoted by $n + \frac{1}{2}$, and are evaluated by linear averaging between time level n and $n + 1$. The momentum equation, Eq. (46), is approximated by

$$^{n+1}u_\alpha{}^l - {}^nu_\alpha{}^l = -\sum_k (\bar{Q}_{\alpha i}^k c_i^{lk}) \, \delta t / m^l, \tag{51}$$

where the vertex mass, m^l, is equal to the linear average of the masses of the eight neighboring cells.

Total energy is exactly conserved because the sum over cells of the change in potential energy, δW, is equal to the sum over vertices of the change in kinetic energy, δKE. Consider the change in the kinetic energy of vertex l, given by

$$\delta \text{KE}^l = {}^{n+1/2}u_j{}^l \sum_k (\bar{Q}_{ji}^k c_i^{lk}) \, \delta t. \tag{52}$$

The sum of the associated changes in the potential energy in the eight neighboring cells is given by

$$\delta W = \sum_k (\bar{Q}_{ij}^k V^{k\,n+1/2} \langle S_{ji}\rangle^k) \, \delta t. \tag{53}$$

The sum of δW and δKE is given by

$$\delta W + \delta \text{KE} = \sum_k \bar{Q}_{ij}^k(-{}^{n+1/2}u_j{}^l c_i^{lk} + V^{k\,n+1/2}\langle S_{ij}\rangle^k) \, \delta t \equiv 0, \tag{54}$$

where Eq. (34) has been used. By construction, total energy is exactly conserved over the computation mesh.

The internal energy is computed by subtracting the magnetic field energy from the potential energy,

$$^{n+1}i = {}^{n+1}w - \tfrac{1}{2}{}^{n+1}[(B \cdot B)/\mu\rho]. \tag{55}$$

Because the position of a vertex is advanced with the following first order equation

$$^{n+1}x_\alpha^l = {}^nx_\alpha^l + {}^{n+1}u_\alpha^l \, \delta t, \tag{56}$$

rather than a time-centered equation, errors of order δt^2 are being absorbed by the internal energy. Were this to prove troublesome, Eqs. (47)–(49) and (55) could be centered in time to remove this error. The time levels in the stress tensor, given by Eq. (50), are already appropriately chosen.

4. The Solution of the Implicit Equations of Motion

In incompressible flow, the fluid velocity is divergence-free. When $\nabla \cdot \mathbf{u}$ is initially zero, a pressure field which assures that $\nabla \cdot \mathbf{u}$ remains zero is a solution of Poisson's equation,

$$\nabla^2 P' = R, \tag{57}$$

where P' is equal to P/ρ and R contains body forces and perhaps viscous stresses. In a method developed by Harlow and Welch (1965), the equation is solved for P' for each cycle before solving an explicit difference approximation to the momentum equation. The resulting velocity field is divergence-free.

It was later noted (Harlow and Amsden, 1971) that a relatively minor modification of the incompressible flow equations results in equations which can be used for flow at all speeds. The method using these equations treats the density in the equation of state and the density and velocity in the continuity equations implicitly. For incompressible flow, the resulting equations reduce to those used for incompressible flow. For compressible flow, the equations are simply implicit formulations of the usual Eulerian equations.

In a variation of the original Eulerian method for flow at all speeds (Hirt et al., 1974), the solution of Poisson's equation for the pressure is obtained by iteratively solving the continuity and momentum equations simultaneously. This method is conveniently applied to Lagrangian difference equations. It is also easily extended to MHD flow.

In the extension of the method to MHD, the magnetic field is treated implicitly in the induction and momentum equations. A consistent solution of the continuity, induction, and momentum equations is obtained iteratively by an adaptation of the successive overrelaxation scheme developed by Hirt et al. (1974). In this scheme, the iteration variables are the density and magnetic field intensity. Within a cell, each of these variables is adjusted to reduce the residual error in the solution of the continuity and induction equations, Eqs. (7) and (8). Subsequently, changes in the velocity at the vertices of the cell due to changes in the stress tensor are computed from the equations of motion. The changes in the velocity at the vertices of a cell affect the residual error in neighboring cells through the rate of strain tensor so that the residual error propagates from cell to cell during an iteration pass as in a Gauss–Seidel iteration.

The adjustment within each cell is performed by means of a Newton–Raphson iteration. The derivatives of the residual errors with respect to the iteration variables are computed from analytical expressions obtained by formally differentiating the equations of motion, and the complete system of four simultaneous equations is solved. The derivatives include not only the direct variation of the residuals with the iteration variables, but also the implicit variations through the rate of strain tensor. As is discussed by Hirt et al. (1974), the inclusion of the implicit variation makes the iteration stable for all signal speeds because it causes the adjustment of the iteration variable to be bounded. When a number of equations are being solved simultaneously, it is necessary for the same reason to compute all terms in the Jacobian matrix.

Even though the implicit equations are stable for a larger time-step, in practice, the time-step is limited to one which satisfies the inequality,

$$|u|\,\delta t < f\,\delta x, \qquad 0 < f < 1/4. \tag{58}$$

When this inequality is satisfied, the relative changes in the thermodynamic quantities over a time-step, δt, are of the order of f. When f is 0.25 or less, about ten iterations are all that are necessary for the relative error in the solution of the continuity and induction equations to be reduced to one part in 10^5 for low speed, compressible flows.

5. Boundary Conditions

When the boundary is a closed conducting surface and the normal component of the magnetic field at the boundary is zero, the boundary condition is given by

$$\mathbf{u} \cdot \hat{\mathbf{n}} < 0, \tag{59}$$

where $\hat{n}$ is the outward directed normal to the surface and $\mathbf{u}$ is the velocity of a boundary vertex. In general, however, it is not sufficient to simply readjust the normal velocity to satisfy Eq. (59), for two reasons. First, when stresses on a boundary are computed from one side, fictitious tangential stresses due to truncation errors are exerted on boundary vertices. Second, the equation of motion is only first order accurate in δt so that a boundary vertex drifts off the boundary by an amount of order δt^2 at each time-step. A general boundary treatment is derived which is an extension of free-surface boundary conditions (Hirt et al., 1970) to a free-slip boundary. The boundary of the plasma is treated as a free surface upon which the conducting boundary exerts a force of constraint.

The appropriate boundary condition for the stress at a free surface is given by

$$\mathbf{Q} \cdot \hat{n} = 0. \tag{60}$$

In order to satisfy Eq. (60), the mesh is extended as described by Hirt et al. (1970), and the stress, $\mathbf{Q}$, is extrapolated beyond the boundary. The extrapolation of the field components is such that $\mathbf{B} \cdot \hat{n}$ on the boundary is zero, and that $\mathbf{B} \times \hat{n}$ exterior to the mesh includes geometric corrections for the curvature of the boundary.

Next, the condition on the normal velocity, Eq. (59), is satisfied by computing a normal force of constraint. Where the equation for the boundary is given by

$$f(\mathbf{x}) = 0, \tag{61}$$

the displacement of a boundary vertex, $n\lambda$, due to the action of the force of constraint, is given by the solution of the equation

$$f(\mathbf{x} + \tilde{\mathbf{u}}\delta t + \lambda \hat{n}) = 0, \tag{62}$$

where $\tilde{\mathbf{u}}$ includes contributions from all the stresses acting interior to the boundary. For example, when the boundary is a right circular cylinder, λ is given by the equation

$$\lambda^2 + 2(\hat{n} \cdot \tilde{\mathbf{u}} - R/\delta t)\lambda + [\tilde{\mathbf{u}} \cdot \tilde{\mathbf{u}} 2R(\hat{n} \cdot \tilde{\mathbf{u}})]/\delta t = 0, \tag{63}$$

where $\lambda \leqslant 0$ consistent with Eq. (59). The smaller magnitude root is always chosen. The larger root corresponds to moving the vertex to a diametrically opposed point on the surface.

B. The Rezone Phase of a Generalized Mesh Calculation

The transport of mass, momentum, magnetic flux, and energy from cell to cell is computed from Eqs. (15)–(18). The difference equations for the convection phase in two dimensions are given by Hirt et al. (1974) and Pracht

(1975). The extension to include the transport of magnetic flux is described in Brackbill and Pracht (1973). In addition, a description of a two-dimensional computer code is given in Amsden and Hirt (1973). With this number of detailed references for what is essentially a standard and well-known method for the computation of convective transport, there is no need for another description here.

When the grid velocity, $\mathbf{u}'$, is zero so that the mesh is Eulerian, the generalized mesh method is prey to all the ills which beset any other Eulerian calculation. That is, the numerical diffusion introduced by the donor cell equation seriously reduces the overall accuracy of the calculation. An application of the truncation error correction procedure developed by Rivard et al. (1973) to the calculation of the convective transport should significantly improve the accuracy of the rezone phase. Applying this procedure to two- and three-dimensional MHD calculations is relatively straightforward.

Our usual way of dealing with the accuracy problem in rezoning is to do as little of it as possible. By reducing the relative velocity between the plasma and the mesh, the truncation errors in all the approximations to the convective derivative listed in Section III are reduced. Thus, an almost-Lagrangian prescription for the grid velocity, that is, a prescription in which $\mathbf{u} - \mathbf{u}'$ is small everywhere, reduces transport to the point where the errors introduced by its approximation are insignificant. An almost-Lagrangian prescription using the smoothing algorithm is outlined in the following section.

1. *The Almost-Lagrangian Mesh*

An almost-Lagrangian calculation exploits the ability of a Lagrangian mesh to be distorted without affecting the accuracy of the generalized difference equations. Some consideration of the coordinate transformation given by Eq. (27) will convince the reader that the formal accuracy of the difference equations is preserved as long as every cell in the mesh in convex. The formal accuracy is lost when a cell becomes concave, for a concave cell has no image in natural coordinates. Thus, the duty of the rezone in an almost-Lagrangian calculation is to prevent concave cells. For many flows, this requires little relative motion between the plasma and the mesh, provided an appropriate mesh prescription is used. One such prescription is the smoothing algorithm, which is outlined in the next paragraph.

2. *The Smoothing Algorithm*

Our most successful prescription for an almost-Lagrangian calculation is the smoothing algorithim, which was first suggested to us by P. Browne (unpublished notes, 1972) and is described in Brackbill and Pracht (1973).

This algorithm is derived by minimizing an integral I, given by

$$I = \int_V dV' [(\nabla \xi)^2 + (\nabla \eta)^2 + (\nabla v)^2], \tag{64}$$

over the computation mesh. The Euler equation corresponding to this minimization principle can be solved directly by interchanging the dependent and independent variables (Thompson et al., 1974).

Our approach is to apply the finite-element method. The integral I is evaluated by constructing elements in which the natural coordinates, ξ, η, and v, are piecewise linear functions of the physical coordinates, x, y, and z. The integral is then minimized by treating the position of each mesh point as a parameter. The result of a single sweep over the mesh is a trial value for the displacement of each vertex, $\delta \mathbf{x}$, from the position at which I would be minimized. The essence of the almost-Lagrangian mesh is that this displacement is reduced by a fractional amount, ω, each cycle, given by

$$\omega = \delta t / \tau, \tag{65}$$

where τ is a mesh relaxation time, which is typically ten time-steps or more. Thus, at each mesh point, the grid velocity to be substituted into Eqs. (15)–(18) to compute transport is given by

$$\mathbf{u}' = \delta \mathbf{x} / \tau. \tag{66}$$

This velocity is small compared with the velocity determining the time-step when τ is large compared with δt (cf. Section IV, A, 4).

The effect of the smoothing algorithm on a distorted mesh is best shown by an example. In Fig. 1, the application of the algorithm to a distorted,

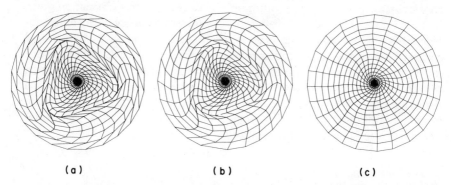

(a) (b) (c)

FIG. 1. The effect of the smoothing algorithm on a distorted Lagrangian mesh can be seen in these plots of a computation mesh. The Lagrangian mesh is shown in 1(a), and the mesh after 10 and 50 applications of the smoothing algorithm is shown in 1(b) and 1(c). All of the cells are already convex in the mesh shown in 1(b), demonstrating that a Lagrangian mesh is transformed into an acceptable mesh by only a small amount of rezoning.

Lagrangian mesh is shown. In Fig. 1a, the mesh is clearly pathological. Some zones even overlap. After 10 iterations of the algorithm with $\omega = 0.25$, every cell in the mesh shown in Fig. 1b is convex. Finally, after 50 iterations, the mesh appears in Fig. 1c to have converged to a symmetric, polar mesh.

V. Applications

The results of calculations using the generalized mesh method are presented. These calculations demonstrate how the flexibility of the method can be exploited. For example, in the calculation of the implosion of an axisymmetric theta pinch, the Lagrangian character of the mesh is used to resolve the boundary between the plasma and the vacuum, and thus to define the boundaries of the region over which the vacuum field is computed. In addition, the results show the accuracy which is achieved by presenting a direct comparison between linear theory and a nonlinear calculation of the rotating theta pinch.

A. A Sharp Boundary Calculation of the Theta Pinch

The compression of a theta pinch is driven by an applied azimuthal current on a conducting cylindrical wall surrounding the plasma. The current produces a magnetic field at the wall which drives the plasma toward the axis of symmetry. Some magnetic field diffuses into the plasma, and some plasma is not picked up by the incoming field. That is, the boundary between the plasma region and the field region is not sharp. Nevertheless, the inertial properties of the plasma left behind are not important, nor is any current carried by this low density plasma. Thus, the sharp boundary model mentioned by Roberts and Potter (1968) can be used. In this model, there is an interface separating the plasma and the vacuum. In the vacuum, the equation for the azimuthal component of the vector potential (Lewis, 1966), is solved. In the plasma, the ordinary MHD equations, including resistive diffusion but not mass diffusion, are solved. At the interface, the continuity of stress condition across the free surface of the plasma is maintained to first order in δt, the time-step.

The equation for the vector potential is solved by an application of the finite-element method. The integral to be minimized is written (Morse and Feshback, 1953)

$$I = \int_V [(\nabla A_\theta)^2 + (A_\theta/r)^2] r \, dr \, dz, \tag{67}$$

where r and z are the radial and axial coordinates. The integral is minimized with the value of A_θ at each mesh point as a parameter. The resulting system of linear equations is solved by successive overrelaxation (Young, 1962), but an alternating direction implicit method can also be used.

The boundary conditions for the vacuum region are the value of A_θ on the vacuum interface, and either A_θ or $\partial A_\theta/\partial n$, where n is the normal coordinate to the wall, at each point on the wall. The vector potential at the interface is determined by integrating the change in A_θ away from the axis, where $d(rA_\theta)$ is given by

$$d(rA_\theta) = r(B_z\,dr - B_r\,dz). \qquad (68)$$

The field components, B_r and B_z, are known in the interior of the plasma. Because the value of the normal component of the magnetic field on either side of the interface is determined by the variation of A_θ along the interface, $\mathbf{B}\cdot\hat{\mathbf{n}}$ is continuous across the interface.

Some results of a sharp boundary theta pinch calculation are shown in Fig. 2. In these calculations the current at the wall is given by,

$$\partial A_\theta/\partial n = (\alpha + \beta \cos^2 kz)\,t + \gamma, \qquad (69)$$

where α, β, and γ are constants of order unity. One-half the wavelength corresponding to k is computed, and periodic boundary conditions are applied at $z = 0$ and $z = \pi/k$.

Initially, the computation mesh appears as shown in Fig. 2a, with large zones resolving the plasma and small zones the vacuum. The axis of the pinch is on the left and the conducting wall is on the right. In Fig. 2b, the plasma is

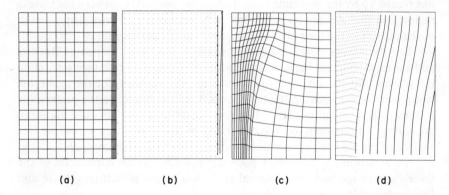

(a) (b) (c) (d)

FIG. 2. The results of a sharp boundary calculation of a theta pinch are shown. The computation mesh and magnetic field lines at the initial time in 2(a) and 2(b), and just before maximum compression in 2(c) and 2(d), show the plasma, depicted by dots in 2(b) and 2(d), being compressed by the magnetic field entering on the right in each frame onto the axis of symmetry on the left. The leftmost magnetic field line in 2(d) nearly coincides with the zone line in 2(c) which delineates the vacuum–plasma interface.

denoted by marker particles, and flux surfaces by contours in the vacuum region. At a later time, the computation mesh appears as shown in Fig. 2c. The interface, which coincides with the left boundary of the fifth zone from the right, has moved to the left. The mesh has in fact followed the plasma, shown at the corresponding time in Fig. 2d, as it is compressed by the magnetic field. (The value of magnetic flux associated with each contour is constant.)

The advantages of a generalized Lagrangian mesh are evident in this calculation. First, the interface is resolved by the mesh simply by causing points on the interface to move with the local fluid velocity. Second, the plasma is resolved at the final time even as it occupies a smaller and smaller fraction of the domain, because it carries its zones with it as it moves. It would not be impossible to use fluid markers (Amsden and Harlow, 1970) to resolve the interface, but it would be difficult to do it as easily or as well.

B. The Rotating Theta Pinch

The rotating theta pinch is initially in uniform rotation within a cylindrical, conducting shell. A one-dimensional equilibrium is given by the equation

$$(d/dr)(p + \mathbf{B}^2/2\mu) = \Omega^2/r, \tag{70}$$

where Ω is the angular velocity of the plasma, and $\mathbf{B}$ is the axial field. For $0 \leq |k| < k_c$ and $m \neq 0$, where k and m are the axial and azimuthal wavenumbers, respectively, the equilibrium is unstable to perturbations in radial position of the form

$$\xi(r) = \sum_n \sum_k \xi_{k,n} \cos(kz + m\theta). \tag{71}$$

When $\xi(r)$ is the radial eigenfunction given by linearizing and solving the MHD equations, the variation in time of the eigenfunction is given by

$$\xi(t) = \xi(0) \exp(\gamma t), \tag{72}$$

where γ is the growth rate of the instability.

The domain is periodic in θ and z, and is bounded at r equal to R by a conducting, cylindrical shell.

When k equals zero, gradients in the axial direction are zero and the flow is confined to the r–θ plane. A three-dimensional computation of the two-dimensional flow is performed by reducing the number of zones in the axial direction to one.

A detailed comparison between linear theory and the results of the computation is made possible by the use of trace particles (Pracht, 1975), which move with the local fluid velocity so that their trajectory is also the trajectory of the fluid element within which they lie. In stability calculations, trace particles are used to form a shadow grid of Lagrangian coordinates which is initially identical with the computation mesh but which is no longer identical when relative motion between the fluid and the mesh occurs because of rezoning. If R_0 is the initial radial coordinate of a trace particle in cylindrical coordinates, its position at any time t can be written

$$R(R_0, \theta, z) = R_0 + \sum_{k,m} \xi_{k,m} \exp[i(kz + m\theta + \Phi_{k,m})], \qquad (73)$$

where $\xi_{k,m}$ and $\Phi_{k,m}$ are the amplitude and phase of the displacement of R from R_0. When R is known, the amplitude and phase are computed from the equations

$$\xi_{k,m} = (II^*)^{1/2} \qquad (74)$$

and

$$\Phi = \tan^{-1}[\text{Im}(I)/\text{Re}(I)], \qquad (75)$$

where I is given by

$$I = \frac{1}{2\pi L} \int_0^{2\pi} \int_0^L d\theta \, dz \, [\mathbf{R}(R_0, \theta, z) - \mathbf{R}_0] \cdot \hat{\mathbf{r}} \exp[-i(kz + m\theta)]. \qquad (76)$$

The $\xi_{k,m}$ computed from Eq. (74) can be compared with the $\xi_{k,m}$ of Eq. (72), which are predicted by linear theory. In Fig. 3, the growth of $\xi_{0,3}$ in time is shown for three different computations.

Curve 1, the lowest curve, is the result of an Eulerian calculation with donor cell differencing of the convection terms. Donor cell differencing is accurate only to first order and extremely diffusive, and is clearly inadequate for stability calculations where flow is occurring, for it radically reduces the growth rate from that obtained with more accurate calculations. For example, the results displayed in curve number two, the middle curve, were obtained by using interpolated donor cell differencing of the convection terms (Amsden and Hirt, 1973), a second order accurate scheme. Beyond one-half rotation period, the growth rate is nearly constant and comparable to the growth rate measured from the upper curve, which is the result of an almost-Lagrangian calculation. In the almost-Lagrangian calculation, the mesh relaxation time, τ, is approximately $20\delta t$ and a donor cell approximation to the convection terms is employed. A linear growth phase can be seen between 0.25 and 1.2 rotation periods, where the growth is exponential with a very slowly declining growth

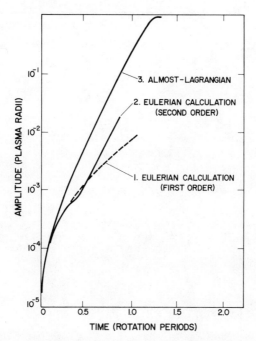

FIG. 3. The growth of a perturbation of an unstable equilibrium in a rotating theta pinch is plotted as computed with Eulerian and almost-Lagrangian computation meshes.

rate. Saturation can be seen after 1.2 rotation periods when the growth rate rapidly declines to zero.

The radial eigenfunction at one rotation period is shown in Fig. 4. The curve from an almost-Lagrangian calculation is somewhat broader but similar in shape to the initial perturbation. On the other hand, the radial eigenfunction for the Eulerian calculation has developed a double-humped structure which suggests that dispersion errors have spatially separated modes contained in the initial eigenfunction. The Eulerian calculation is terminated by an instability of the type described in Section III, at about one rotation period.

An essential difficulty with three-dimensional calculations is that the amount of information it is possible to store is only enough to finely resolve portions of the flow. In the case of the $m = 3$ perturbation, the radial eigenfunction is most efficiently resolved by clustering the points radially about $r = 1$. When the almost-Lagrangian mesh is used, the eigenfunction is resolved at all times because the mesh points move with the flow. By contrast, the Eulerian mesh provides fine resolution of a portion of the flow only until it moves out of that region of the mesh which can provide it. In general, optimum overall resolution with an Eulerian mesh is given by a uniform distribution

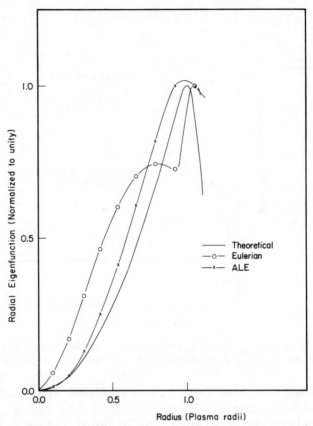

Fig. 4. The computed and linear theoretical perturbation amplitudes for a rotating theta pinch are plotted at a time corresponding to one rotation period of the pinch.

of points in space even though such a distribution is not optimum at any given time during the calculation.

C. The Internal Kink Mode Instability

The internal kink mode instability in a diffuse screw pinch is discussed by Goedbloed and Hagebeuk (1972). A recent paper by Rosenbluth et al. (1973) includes both linear and nonlinear results.

In these results, the growth of a perturbation of an unstable equilibrium is computed. The perturbation is given by

$$\xi_r = \xi_0 \exp[i(kz + m\theta)]. \tag{77}$$

The plasma equilibrium is defined by the radial force balance equation

$$\partial P/\partial r = J_z B_\theta, \tag{78}$$

where J_z is the axial current density in the plasma, and B_θ is the azimuthal field. The plasma is infinitely conducting, and fills the interior of a conducting cylinder. The radial variation of the current density is written

$$J_z = \begin{cases} J_0(r/a), & r \leqslant a \\ 0, & r > a \end{cases} \tag{79}$$

where a, the radius of the current carrying plasma, is one-half the radius of the conducting cylinder. The wavenumber, k, is chosen so that the minimum value of the safety factor, q, is 0.65. The safety factor is equal to 1 at $r = 0.65a$, and $r = (0.65)^{1/2}$. The total beta, the ratio of the plasma pressure at the axis to the total magnetic field pressure at the surface of the plasma, is 0.4.

In Figs. 5–9, some of the results are shown of a computation in three dimensions of the evolution of an $m = 1$ perturbation. The initial shape of the plasma column is represented by an interior surface of the mesh in Fig. 5. In Fig. 6a, a cross section of the mesh is shown. In Figs. 7a and 8a, contours of constant density and velocity vectors in a cross section are shown. The density contour values form a geometric progression from the minimum density,

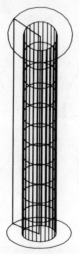

FIG. 5. The surface of a cylindrical plasma in equilibrium is represented by an interior surface of the computation mesh. The lines exterior to the plasma surface outline the boundary of the mesh.

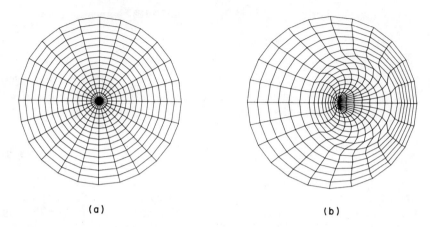

FIG. 6. A cross section of a cylindrical computation mesh is shown for an internal kink mode calculation. In 6(a), the initial mesh is shown. In 6(b), the mesh is shown at a time when an $m = 1$ perturbation has carried the plasma, which occupies the inner nine radial zones, almost to the wall.

$\rho = 10^{-2}$, to the maximum density, $\rho = 1$. The velocity, represented by vectors pointing from a mesh point in the direction of flow, are nonzero initially. As Bateman *et al.* (1974) note, it is easier to generate consistent initial conditions when a velocity perturbation is used. The pattern of the velocity vectors is typical of an $m = 1$ perturbation for incompressible flow, with the plasma

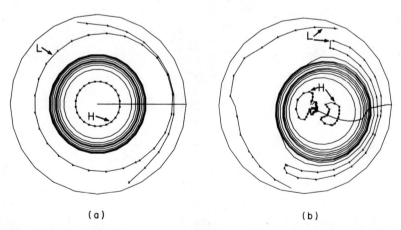

FIG. 7. The isodensity contours corresponding to the mesh shown in Fig. 6 are plotted. The contours labeled H and L represent densities of 1.0 and 0.01 (in arbitrary units), respectively. The values of the density associated with intermediate contours form a geometric progression.

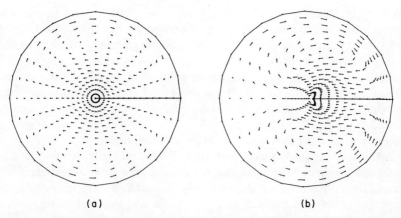

FIG. 8. The velocity vectors corresponding to the mesh in Fig. 6 are shown. Each vector being at a vertex and points in the direction of flow. The maximum velocity represented by a vector in 8(b) is roughly twice the maximum in 8(a).

interior to the singular surface (where $q = 1$) moving bodily toward the wall, and the plasma exterior to the singular surface being displaced.

In Figs. 6b and 8b, the fully developed instability is shown at a time corresponding to one Alfven transit time. The surface of the plasma in Fig. 9, corresponding to Fig. 5, is circular in cross section, but has suffered a helical

FIG. 9. The plasma surface is represented by an interior surface of the mesh, as in Fig. 5, at a time corresponding to that drawn in Figs. 6b, 7b, and 8b. The helical distortion of the plasma column characteristic of an $m = 1$ perturbation is clearly shown.

displacement from the geometric axis. The computation mesh in Fig. 6b, corresponding to Fig. 6a, shows the distortion of the almost-Lagrangian mesh due to the plasma flow. The current carrying plasma, which occupied the inner eleven rows of zones in the mesh initially, has carried zones with it as it moves toward the wall. The density gradient at the edge of the current carrying plasma is shown in Fig. 7b. Its size, as indicated by the spacing of the contours, is not diminished substantially from its initial value, which can be deduced from Fig. 7a. Finally, the velocity vectors in Fig. 8b show similar flow patterns to those shown in Fig. 8a at the initial time.

There is not enough space here for a detailed analysis of these results, but we can conclude from them that the instability at large amplitude is not qualitatively different from the instability at small amplitude. Eventually, flux trapping between the wall and the plasma will slow the motion of the plasma toward the wall, but, other than flux trapping, there seems to be no evident saturation mechanism operating. A more careful analysis of the internal kink mode instability is planned with the particle diagnostics described in the last section.

VI. Conclusions

The emphasis in this article has been on a fundamental problem in the solution of the equations for ideal magnetohydrodynamic flow. This problem, namely, the accurate computation of convective transport, evidently arises because of nonlinear instabilities in numerical approximations to the convective derivative. There is evidence that these instabilities are due to certain truncation errors which produce negative diffusion. These errors can be computed, and their effect can be suppressed either by adding compensating positive diffusion or by reducing the relative motion between the computation mesh and the plasma with a generalized mesh. A comparison among the results of linear stability theory, a generalized mesh calculation, and an uncorrected Eulerian calculation supports the conclusion that more accurate computation of convective transport leads to significant increases in overall accuracy. Similar improvements can be expected with corrected Eulerian calculations.

Appendix

A compact vector form for the geometric coefficients, c_α^l, discussed in Section IV, A, 1, has been developed by D. C. Barnes (unpublished notes, 1975). Where $\mathbf{r}^l$ is the position vector for vertex l, the geometric coefficient corresponding to vertex 1 [in a cell whose vertices are labeled as in Eq. (27)] is written

$$c^1 = \tfrac{1}{12}(r^2 \times r^5 + r^5 \times r^4 + r^4 \times r^2 + r^2 \times r^6 + r^6 \times r^5$$
$$+ r^5 \times r^8 + r^8 \times r^4 + r^4 \times r^3 + r^3 \times r^2).$$

The coefficients for vertices 2–4 are obtained by a cyclic permutation of the indices within the two groups 1–4 and 5–8. Similarly, the geometric coefficient for vertex 5 is written

$$c^5 = \tfrac{1}{12}(r^1 \times r^6 + r^6 \times r^8 + r^8 \times r^1 + r^1 \times r^2 + r^2 \times r^6$$
$$+ r^6 \times r^7 + r^7 \times r^8 + r^8 \times r^4 + r^4 \times r^1).$$

From c^5, the coefficients for vertices 6–8 are obtained by a cyclic permuattion of the indices within the two groups 1–4 and 5–8.

Acknowledgments

The author has received substantial help with this article and with the methods described in it from many people. Among them are W. E. Pracht, whose generously provided three-dimensional hydrodynamics program simplified enormously the task of writing a code for magnetohydrodynamics; D. C. Barnes, who checked and made many improvements and extensions to the equations described in Section IV; J. P. Freidberg, who suggested many of the problems to which the generalized mesh method has been applied; and C. W. Nielson, whose critical reading of the manuscript is responsible for many improvements. The author also thanks W. B. Goad and C. W. Hirt for many helpful conversations and suggestions.

This work was supported by USERDA, Contract No. W-7405-ENG. 36.

References

Amsden, A. A., and Harlow, F. H. (1970). "The SMAC Method: A Numerical Technique for Calculating Incompressible Fluid Flows," Rep. No. LA-4370, Los Alamos Sci. Lab., Los Alamos, New Mexico.

Amsden A. A., and Hirt, C. W. (1973). "YAQUI: An Arbitrary Lagrangian-Eulerian Computer Program for Fluid Flow at All Speeds," Rep. No. LA-5100, Los Alamos Sci. Lab., Los Alamos, New Mexico.

Anderson, D. V. (1975). *J. Comput. Phys.* **17**, 246.

Bateman, G., Schneider, W., and Grossmann, W. (1974). *Nucl. Fusion* **14**, 669.

Bodin, N. A. B. (1972). *Nucl. Fusion* **12**, 721.

Boris, J. P. (1970). "A Physically Motivated Solution of the Alfven Problem," Memo. Rep. No. 2167, Naval Res. Lab., Washington, D.C.

Boris, J. P., and Book, D. L. (1973). *J. Comput. Phys.* **11**, 38.

Brackbill, J. U. (1973). *Proc. Conf. Numer. Simul. Plasmas, 6th, 1973* Lawrence Livermore Lab. Conf. Rep. 730804.
Brackbill, J. U., and Pracht, W. E. (1973). *J. Comput. Phys.* **13**, 455.
Butler, T. D., Henins, I., Jahoda, F. C., Marshall, J., and Morse, R. L. (1969). *Phys. Fluids* **12**, 1904.
Duchs, D. (1968). *Phys. Fluids* **11**, 2010.
Fornberg, B. (1973). *Math Comput.* **27**, 45.
Freeman, J. R. (1971). *Nucl. Fusion* **11**, 425.
Freeman, J. R., and Lane, F. O. (1968). *Proc. APS Top. Conf. Numer. Simul. Plasmas, 1968* Los Alamos Sci. Lab. Rep. No. LA–3990.
Fromm, J. E. (1961). "Lagrangian Difference Approximations for Fluid Dynamics," Rep. No. LA–2535, Los Alamos Sci. Lab., Los Alamos, New Mexico.
Goad, W. B. (1960). WAT: A Numerical Method for Two-Dimensional Unsteady Fluid Flow," Rep. LAMS–2365, Los Alamos Sci. Lab., Los Alamos, New Mexico.
Goedbloed, J. P., and Hagebeuk, H. J. L. (1972). *Phys. Fluids* **15**, 1090.
Hain, K., Hain, G., Roberts, K. V., Roberts, S. J., and Koppendorfer, W. (1960). *Z. Naturforsch. A* **15**, 1039.
Harlow, F. H., and Amsden, A. A. (1971). *J. Comput. Phys.* **8**, 197.
Harlow, F. H., and Welch, J. E. (1965). *Phys. Fluids* **8**, 842.
Hertweck, F., and Schneider, W. (1970). "A Two-Dimensional Computer Programme for Solving the MHD Equations for the Theta Pinch in a Time Dependent Coordinate System, Rep. No. IPP 1/110. Inst. Plasma Phys., Garching.
Hirt, C. W. (1968). *J. Comput. Phys.* **2**, 339.
Hirt, C. W., Cook, J. L., and Butler, T. D. (1970). *J. Comput. Phys.* **5**, 103.
Hirt, C. W., Amsden, A. A., and Cook, J. L. (1974). *J. Comput. Phys.* **14**, 227.
Hofmann, J. (1974). *Nucl. Fusion* **14**, 438.
Killeen, J. (1972). In "Information Processing 71" (C. V. Freiman, J. E. Griffith, and J. L. Rosenfeld, eds.), p. 1191. North-Holland Publ., Amsterdam.
Kreiss, H. O. (1964). *Comm. Pure Appl. Math.* **17**, 335.
Landau, L. D., and Lifschitz, E. M. (1959). "Fluid Mechanics." Addison-Wesley (Pergamon), Reading, Massachusetts.
Lapidus, A. (1967). *J. Comput. Phys.* **2**, 154.
Lewis, H. R. (1966). *J. Appl. Phys.* **37**, 2541.
Lindemuth, I., and Killeen, J. (1973). *J. Comput. Phys.* **13**, 181.
Lui, H. C. (1973). "FLIC Codes of Shock Focusing in a Coaxial Electromagnetic Shock Tube," Lab. Rep. No. 60, Columbia University, New York.
Lui, H. C., and Chu, C. K. (1974). *Int. Conf. Numer. Methods Fluid Dyn. 4th, 1974* p. 263.
Morse, P. M., and Feshback, H. (1953). "Methods of Theoretical Physics," Chapter 3. McGraw-Hill, New York.
Peaceman, D. W., and Rachford, H. H. (1955). *J. Soc. Ind. Appl. Math.* **3**, 28.
Pracht, W. E. (1975). *J. Comput. Phys.* **17**, 132.
Richtmyer, R. D. (1963). "A Survey of Difference Methods for Non-steady Fluid Dynamics," NCAR Tech. Note 63–2. Nat. Cent. Atmos. Res., Boulder, Colorado.
Richtmyer, R. D., and Morton, K. W. (1967). "Difference Methods for Initial Value Problems." Wiley (Interscience), New York.
Rivard, W. C., Farmer, O. A., Butler, J. D., and O'Rourke, P. J. (1973). "A Method for Increased Accuracy in Eulerian Fluid Dynamics Calculations," Rep. No. LA–5426–MS, Los Alamos Sci. Lab., Los Alamos, New Mexico.
Roberts, K. V., and Boris, J. P. (1969). *3rd Annu. Numer. Plasma Simul. Conf., 1969* Paper no. 32.

Roberts. K. V., and Potter, D. E. (1968). *Methods Compt. Phys.* **9**, 339.
Rosenbluth, M. N., Dagazian, R. Y., and Rutherford, P. H. (1973). *Phys. Fluids* **16**, 1894.
Schneider, W. (1972). *Z. Phys.* **252**, 147.
Schulz, W. D. (1964). *Methods Comput. Phys.* **3**, 1.
Thompson, J. F., Thames, F. C., and Mastin, C. W. (1974). *J. Comput. Phys.* **15**, 299.
Tuck, J. (1968). *Eur. Conf. Controlled Fusion Plasma Phys. Res., 2nd, 1968* Vol. 2, p. 595.
Wagner, C. E., and Manheimer, W. M. (1973). *Proc. Conf. Numer. Simul. Plasmas, 6th 1973* Lawrence Livermore Lab. Conf. Rep. 730804.
White, R., Monticello, D., Rosenbluth, M. N., and Strauss, N. (1974). *Plasma Phys. Fusion Res., 1974*, Vol.1, p. 495.
Wooten, J., Hicks, H. R., Bateman, G., and Dory, R. A. (1974). "Preliminary Results of the 3D Nonlinear Ideal MHD Code," ORNL TM 4784. Hollifield Nat. Lab., Oak Ridge, Tennessee.
Young, D. M. (1962). *In* "A Survey of Numerical Analysis" (J. Todd, ed.), Chapter 11. McGraw-Hill, New York.
Zel'dovich, Ya. B., and Raizer, Yu. P. (1967). "Physics of Shock Waves and High Temperature Hydrodynamic Phenomena," 2nd ed., Vol. 2. Academic Press, New York.
Zienkiewicz, O. C. (1971). "The Finite-Element Method in Engineering Science." McGraw-Hill, New York.

Waterbag Methods in Magnetohydrodynamics

David Potter*
UNIVERSITY OF CALIFORNIA AT LOS ANGELES
LOS ANGELES, CALIFORNIA

I. The Waterbag Concept 43
   A. Eulerian and Lagrangian Difference Methods 44
   B. Lagrangian Contours and Surfaces 45
   C. Summary 49
II. Equilibrium Properties of One Waterbag 49
   A. The Equilibrium Equations 49
   B. The Equilibrium Model of One Contour 51
   C. The Variational Procedure 54
   D. Equilibrium of One Waterbag in Three Dimensions 56
III. Equilibria of Current Distributions 60
   A. Formulation of Axisymmetric Equilibria in Flux Space 60
   B. Variational Methods to Obtain Equilibria 63
   C. Some Illustrative Solutions 67
IV. Adiabatic Constraints 69
   A. Adiabatic Changes of Equilibrium 69
   B. Adiabatic Equations Applied to Axisymmetric Toroids . . . 71
   C. Relaxation of the Adiabatic Constraints by Diffusion . . . 74
V. Further Applications 76
   A. Remarks on "Equilibrium" and Inertial Models 76
   B. Quasi-Incompressibility in the Low-Beta Approximation . . . 77
   C. Instabilities Described in the Helical Plane 79
   D. Concluding Remarks 81
   References 82

I. The Waterbag Concept

THE STUDY OF MAGNETOHYDRODYNAMIC (MHD) phenomena has expanded dramatically in the last fifteen years. This interest has been fueled by developments in astrophysics and particularly in controlled thermonuclear fusion. The magnetohydrodynamic equations have a very wide application including, for example, such disparate phenomena as solar flares, the production of the

* On leave of absence from Imperial College, London, England,

magnetic field in the earth's core, the magnetosphere, or the magnetic containment of a plasma in the laboratory. We may regard the magnetohydrodynamic equations as a global set of laws in which only the choice of particular initial conditions, boundary conditions, or certain terms isolate very different regimes. It is to be expected therefore that for both the analytic and computational resolution of MHD phenomena an equally diverse set of techniques and methods should be developed.

Until recently in MHD simulation this has not been the case. The codes which have been available have relied heavily on conventional Eulerian conservative methods. This paper describes a new technique for MHD simulation in which the magnetic field lines or surfaces (if they exist) are isolated and their evolution followed directly. The magnetic field may be described by contours or surfaces of the magnetic flux functions. In a nonturbulent system of high conductivity, such surfaces are well defined and frequently act as incompressible but deformable "waterbags." Such an approach is found to be particularly powerful in describing quasi-static plasmas such as those of interest in magnetic containment for thermonuclear fusion.

A. Eulerian and Lagrangian Difference Methods

The conventional difference approach to MHD simulation has proved useful for studying the dynamics of near sonic or supersonic phenomena. For some time, one-dimensional calculations using point Eulerian and Lagrangian difference methods have become standard (for example, see Hain et al., 1960; Chu and Taussig, 1967, Duchs et al., 1972). More recently, two-dimensional calculations using conservative Eulerian difference methods have been successful in interpreting the properties of real experiments, particularly for sonic and high-beta fluids (Duchs, 1968; Potter, 1971, 1973; Killeen and Lindemuth, 1973). A review and comprehensive bibliography is given by Roberts and Potter (1970).

In many regimes of interest, however, severe difficulties occur with the conventional difference method. In the presence of a strong magnetic field, of prime interest in magnetic containment, the vector properties of the magnetic field play an important role. In low-beta, the equations take an elliptic form where the magnetic field is predominantly defined by the boundaries or external conductors. The propagation of Alfven waves is anisotropic, while at low densities torsional waves propagate along field lines which, of varying direction, are poorly represented on an Eulerian mesh. Furthermore, the diffusion processes become grossly anisotropic in following the field lines.

More generally, Eulerian meshes give rise to numerical diffusion (Potter, 1973) which can totally corrupt solutions for subsonic and "submagnetosonic" plasmas. In one space dimension these difficulties are overcome by the use of point Lagrangian methods (Oliphant, 1963). In view of the simplicity, resolution, and accuracy afforded by the Lagrangian method, many authors have attempted to extend this approach to two and three dimensions. However, in multidimensional problems, the sheared motion of the fluid quickly induces a distorted nonorthogonal mesh, which leads to gross inaccuracy since adjacently labeled points become increasingly separated.

B. Lagrangian Contours and Surfaces

The concept of a point Lagrangian method, useful in one dimension, does not extend to two or more dimensions. Rather than defining a set of Lagrangian points in two dimensions, it is appropriate to define a set of Lagrangian contours which move with the fluid. In three dimensions, Lagrangian surfaces may be defined. More specifically, given a state variable of interest, say f, in the plane, we may define f by a set of contours, or waterbags, where f does not alter along each contour (Fig. 1). Thus if f moves with the fluid velocity, equations of motion may be expressed for the points which make up the contours, and the function f evolves in time according to the equations of motion. However, along the contours, f does not change and thus the motion of the points around the contours does not alter the description of f. The degree of freedom therefore remains of choosing the particular distribution of points around each contour. Such an approach is Lagrangian

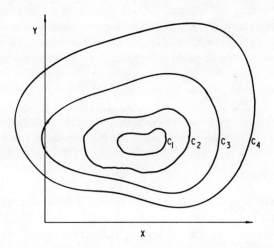

FIG. 1. The discrete representation of a function $f(x, y)$ defined in the plane by a finite set of contours $c_1, c_2, \cdots, c_J$.

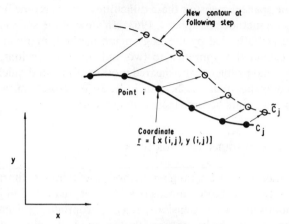

FIG. 2. Each contour, j, is defined by a finite number of points, i, with coordinates $r(i,j)$ in the plane. The function evolves according to the dynamics of the points.

but avoids a distorted or inaccurate mesh. In three dimensions, surfaces are defined and the same arguments prevail.

As in the difference method, the memory of the computer requires a finite model in the waterbag method. A finite number of contours (labeled j, $1 \leqslant j \leqslant J$) must be chosen according to the capacity of the computer at hand. Similarly, each contour must be represented by a finite set of points, labeled i, $1 \leqslant i \leqslant I$ (Fig. 2). The coordinates of each of these points, $x(i,j)$, $y(i,j)$, are stored in the computer. At each step of the calculation, the points are moved, according to the equations of motion of the contour, to $x(\tilde{i},j)$, $y(\tilde{i},j)$. While maintaining the position of the contours, it is now possible to move the points of a given contour along the contour itself. In mathematical terms, while the one coordinate j is defined, the alternate coordinates i can be freely chosen.

The question arises as to the choice of the alternate coordinate. Many possibilities arise, and the optimum choice depends on the problem in hand. In one application White *et al.* (1974) have chosen the distribution i by equal spacing along each contour. This avoids the occurrence of small space-steps.

1. *Orthogonalization*

A particularly important case is to choose the orthogonal coordinate system with metric:

$$ds^2 = (h^i\,di)^2 + (h^j\,dj)^2, \tag{1}$$

where h^i and h^j are the length elements between adjacent i and j curves,

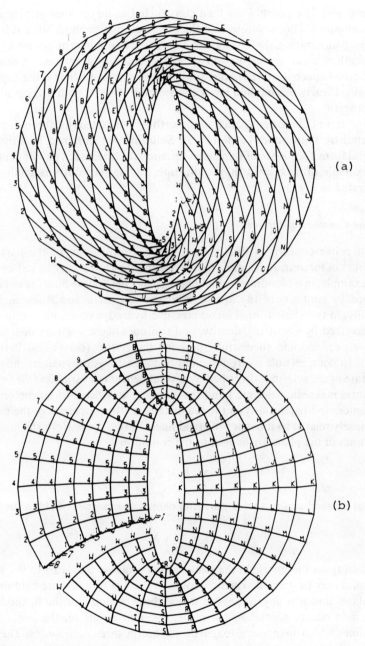

FIG. 3. The orthogonalization (b) of a sequence of sheared concentric ellipses (a). The points *i* are labeled 1 to W.

respectively. The points i are then constructed to lie on lines orthogonal to the contours j. This is unique and the most accurate method, since it has the effect of minimizing the distance in (x, y) space of adjacent points i. While the mathematical representation of the equations of interest in such an orthogonal space is quite simple, some complexity can occur if the topology changes. Clearly this method has the advantage of being both orthogonal and Lagrangian.

The technique for constructing the orthogonal system has been discussed in detail by Potter and Tuttle (1973). Solutions are obtained by solving in integral form a sequence of Laplacian equations between each pair of contours. An example of such an orthogonalization leaving the contours unaltered is illustrated in Fig. 3.

2. *The Contours of Interest*

Of prime concern in this method is the identification of the particular contours of interest. In a simple fluid problem, the choice may be self-evident. For example, in a two-dimensional phase space, the "phase fluid" is obviously defined by contours of the distribution function (Berk and Roberts, 1970). Equally, in two-dimensional incompressible hydrodynamics, the motion may be described by a vorticity density, and contours of the vorticity distribution are well defined and successfully describe the system (Navet and Bertrand, 1971). In compressible hydrodynamics or magnetohydrodynamics, however, the state of the system is described by several dependent variables (the density, pressure, magnetic field, etc.) and the choice is less clear. In magnetohydrodynamics, and particularly in the low-beta case, the properties of the plasma are closely related to the magnetic field lines. In two dimensions the magnetic field lines in the plane are subject to the condition

$$\nabla \cdot \mathbf{B} = 0, \qquad (2)$$

so that a pseudoscalar vector potential defines the magnetic field in the plane

$$\mathbf{B} = \nabla \wedge \mathbf{A}. \qquad (3)$$

Depending on the metric in the plane, a flux function ψ related to the vector potential may be defined. Then contours of the flux function are field lines in the plane, and it is appropriate to define these contours in the method. If a large field occurs normal to the plane, different points on the contour are field lines which map out a magnetic surface in three dimensions. The projection of the surface on the plane is a contour of ψ.

Alfven waves and whistler waves propagate preferentially around such contours, while in a strong field diffusion occurs rapidly along the contours but slowly across them. In the absence of resistivity, the magnetic flux within a closed contour is conserved and thus, in the presence of a strong perpendicular field, a constraint applies to the area of each contour which acts like a "waterbag."

The choice of the contours of the magnetic flux may isolate other important properties. For example, where closed contours or magnetic surfaces are formed, the pressure, density, and currents remain constant in the plane of such a surface, but vary rapidly across the surface. This is the case for equilibrium and near equilibrium problems.

C. Summary

The waterbag method has a wide application to many problems in multi-dimensional magnetohydrodynamics. Emphasis in the application of the method in this paper is given to quasi-equilibria and low-beta problems. Many of the properties of an equilibrium plasma may be illustrated by considering one free surface. In Section II, the equilibrium properties of one free waterbag are developed and an interesting application to the three-dimensional problem is shown. The model is extended to many surfaces describing current profiles in the axisymmetric equilibrium problem in Section III. In conjunction with the momentum equilibrium between field and plasma, slow changes may occur due to diffusion, to externally applied voltages, or slow instabilities, which deform and alter the magnetic surfaces and which occur subject to the magnetohydrodynamic conservation laws. These laws, discussed in Section IV, take an integral form and in the absence of diffusion appear as a series of adiabatic constraints defining the density, pressure, and magnetic fluxes on each surface. The application of these equations is illustrated for adiabatic compression and the growth of magnetic islands.

The further application of the waterbag method to the convective MHD equations is briefly discussed in Section V. Here the inertia of the plasma may no longer be ignored, and the approach has application to the understanding of higher frequency MHD instabilities in the Tokamak.

II. Equilibrium Properties of One Waterbag

A. The Equilibrium Equations

In this and the following sections, a sequence of progressively more sophisticated models of the MHD equations (Jeffrey, 1966) is described.

A simple and most important subset of the MHD equations is that which describes a static fluid or plasma in equilibrium with a magnetic field:

$$\nabla p = \mathbf{J} \wedge \mathbf{B}, \tag{4}$$

$$\nabla \cdot \mathbf{B} = 0, \tag{5}$$

$$\mathbf{J} = \nabla \wedge \mathbf{B}. \tag{6}$$

Equation (4) arises from the conservation of momentum where inertial forces are ignored. The balance between plasma pressure and magnetic forces remains valid for a system evolving in time, provided any changes occur sufficiently slowly (Section IV). Again, the use of a scalar pressure is valid for sufficiently slow changes of the system.

Some properties of Eqs. (4)–(6) are immediately apparent:

$$\nabla \cdot \mathbf{J} = 0, \tag{7}$$

$$\mathbf{B} \cdot \nabla p = 0, \tag{8}$$

$$\mathbf{J} \cdot \nabla p = 0. \tag{9}$$

If the magnetostatic equations are to be satisfied, it follows from Eq. (8) that the plasma pressure must be constant along a field line. Containment of a mass of plasma implies closed surfaces of constant pressure which accordingly always contain a given field line. If such a field line is ergodic, it maps out a "magnetic surface." It can be shown by topological arguments that the only solutions to this system are a set of nested toroids (Kruskal and Kulsrud, 1958) of which the Tokamak (Artsimovich, 1972) and Stellarator are particular examples.

The equilibrium equations (4)–(6) have conventionally been solved by reformulating the equations as a boundary value problem. The resulting elliptic equations are expressed on an Eulerian difference mesh and solved by iteration (Callan and Dory, 1972).

The alternative procedure employed in the waterbag method is to describe the equilibrium directly by the coordinates of a finite set of magnetic surfaces, each denoted by, say, the pressure

$$p = p_1, p_2, \ldots, p_J. \tag{10}$$

We may think of the solution as being obtained by the construction of a set of surfaces or waterbags, which are moved until they rest in equilibrium.

Apart from the simplicity of the formulation and the direct description of the problem, the advantage of this procedure is that it leads to an accurate and rapid solution of the problem. More generally, when the solutions of the magnetic surfaces are directly obtained, slow changes due to adiabatic variations or diffusion can be calculated (Sections IV).

To illustrate the concepts, the equilibrium of one free waterbag is considered in this section. In the axisymmetric case [the simplest class of toroids, independent of the toroidal (ϕ) direction], a magnetic surface intersects the poloidal plane (the R–Z plane in cylindrical notation) to define a contour. We first consider the properties and method of solution for such a contour. Betancourt and Garabedian (1974) have obtained solutions for a single surface in three dimensions and their method is discussed in Section II, D.

B. The Equilibrium Model of One Contour

The model for the equilibrium of one contour in the poloidal plane is illustrated in Fig. 4. A finite domain is assumed bounded by a conducting wall W (the normal component of the magnetic field is assumed zero). It is assumed that the domain contains a toroidal plasma column which intersects the plane in an area P, bounded by a contour C. The equilibrium equations are satisfied within the region P by assuming that the plasma has a uniform pressure p and that no current flows [cf. Eq. (4)]. Between C and W a vacuum region is assumed in which no currents flow. A sheet current flows in the plane of the contour C which divides the plasma from the vacuum region. If the fields in the plasma and vacuum region are denoted by $\mathbf{B}_p$ and $\mathbf{B}_v$, respectively, it may easily be seen that the condition for equilibrium between the plasma

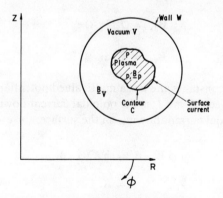

FIG. 4. A single waterbag model for the equilibrium of a plasma and field. A uniform plasma region P is bounded by a sheet current flowing through the contour C. A vacuum region V exists between the wall W and plasma boundary C.

and vacuum region is

$$\tfrac{1}{2}B_v^2 = \tfrac{1}{2}B_p^2 + p, \qquad (11)$$

where the fields are to be evaluated and the equation (11) is applied at every point on the contour C. This equilibrium equation defines the position of the surface.

To apply the equilibrium equation at each point on the contour, the magnetic fields in both the plasma and vacuum regions must be determined. According to this model the field in both the plasma and the vacuum region must satisfy Eqs. (5) and (6):

$$\mathbf{V} \wedge \mathbf{B} = 0, \qquad (12)$$

$$\mathbf{V} \cdot \mathbf{B} = 0, \qquad (13)$$

with the boundary conditions that the normal components of the magnetic field must vanish at W and C. These equations are immediately solved in the "simply-connected" plasma region,

$$\mathbf{B}_p = (I_p/2\pi R)\mathbf{e}_\phi, \qquad (14)$$

where I_p is a constant and $\mathbf{e}_\phi$ is a unit vector in the toroidal direction. There is no poloidal field since no current flows in the plasma region and the boundary conditions exclude any poloidal flux.

On the other hand, in the vacuum region both a toroidal field $B_\phi \mathbf{e}_\phi$ and a poloidal field, denoted by $\mathbf{B}_\chi$, may exist due to the sheet currents flowing on C. Satisfying Ampere's Law, Eq. (12), the vacuum fields may be expressed as

$$\mathbf{B}_{\phi v} = (I_v/2\pi R)\mathbf{e}_\phi, \qquad (15)$$

$$\mathbf{B}_{\chi v} = K\nabla\alpha, \qquad (16)$$

where I_v and K are constants and α is a multivalued potential with unit period. The difference between I_v and I_p is the poloidal current flowing on the surface while K is the toroidal current flowing on the surface, since

$$\begin{aligned}\oint \mathbf{B}_\chi \cdot \mathbf{d}l &= \oint K\nabla\alpha \cdot \mathbf{d}l \\ &= K,\end{aligned} \qquad (17)$$

where the path of the integral is taken in the vacuum region and contains the plasma column. It may be further noted that the flux ψ from the poloidal

field is given by the volume integral

$$\psi = \iiint_V \mathbf{B}_\chi \cdot \nabla \alpha \, d\tau. \tag{18}$$

To determine α and thereby the poloidal field, the divergence-free condition on the magnetic field, Eq. (13), is used to obtain Laplace's Equation

$$\nabla^2 \alpha = 0. \tag{19}$$

This equation is to be solved in the vacuum region V with Neumann boundary conditions (the normal components of the field are zero on C and W). Because the vacuum region is not simply connected a branch cut b is used between C and W. The boundary conditions across the branch cut are that the derivatives of α are continuous and that α changes by unity (Fig. 5).

The Laplacian equation (19) for the potential α in the vacuum region may be expressed in integral form using Green's Theorem. In finite form this equation becomes a matrix equation for α_i at all points i on the surfaces C and W. The poloidal field is then evaluated at each point i on the contour C:

$$B_\chi(C, i) = (K/h^i)(\partial \alpha / \partial i). \tag{20}$$

It may now be noted that this problem [Eqs. (19), (20)] is identical to that arising from the orthogonalization procedure of Potter and Tuttle (1973). In their notation the function $\tilde{p}(i)$ plays precisely the role of α, so that the

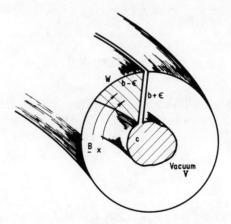

FIG. 5. A branch cut is introduced between the plasma and wall to define the vacuum region as "simply-connected."

orthogonalization procedure immediately determines the poloidal field on the contour C.

Thus, given the constants p the pressure, K the toroidal current, I_v, and I_p, the equilibrium of the simple waterbag in the plane is determined. The plasma field is given explicitly by Eq. (14) while the vacuum fields are given by Eqs. (15) and (20). The position of the waterbag is defined when Eq. (11) is satisfied.

C. The Variational Procedure

The solution for the equilibrium position of the contour C may be determined by an iterative procedure. To illustrate the procedure, a thought experiment may be imagined in which the domain, bounded by W, is regarded as an electromagnetically isolated system. At any stage in the variation an imbalance of forces occurs across the contour C. By moving the contour C these forces are allowed to perform work, thereby reducing their potential energy. We may imagine that the system is perfectly viscous so that the potential energy released into kinetic energy is immediately dissipated and released from the system. An equilibrium can therefore be found when the potential energy is minimized.

Let the poloidal flux on the wall W be ψ and the toroidal flux on W be Π. The requirement that the system be electromagnetically isolated may be satisfied if no Poynting vector exists at the wall. Thus it is required that during the variation $\delta\psi$ and $\delta\Pi$ be zero. The toroidal and poloidal fluxes between C and the wall are

$$\Pi = I_v \iint_V dS/2\pi R, \tag{21}$$

$$\psi = K \iint_V (\nabla\alpha)^2 \, 2\pi R \, dS, \tag{22}$$

where the integrals are taken over the surface V [Eq. (18)]. The integrals, Eqs. (21) and (22), are the inductances L, L' of the system

$$\Pi_v = LI_v, \tag{23}$$

$$\psi = L'K, \tag{24}$$

so that by determining the inductances L and L' during the variation, the currents I_v and K, which keep the fluxes Π_v and ψ constant, are determined. The plasma pressure p and I_p may be assumed constant.

At any step of the variation, the force **F** along the outward normal **n** of the contour is given by [Eq. (11)]

$$\mathbf{F} = (p + \tfrac{1}{2}B_p^2 - \tfrac{1}{2}B_v^2)\mathbf{n}. \tag{25}$$

If each point i of the contour is displaced a distance δr_i in the plane, the loss of potential energy is

$$\delta E = -\oint_C \mathbf{F}_i \cdot \delta \mathbf{r}_i \, 2\pi R h^i \, di. \tag{26}$$

By choosing an appropriate displacement δr_i of the contour a negative definite change of the potential energy may always be ensured. One such appropriate choice (Betancourt and Garabedian, 1974) is

$$\delta \mathbf{r}_i = \omega \mathbf{F}_i, \tag{27}$$

where ω is a relaxation parameter. Thus,

$$\delta E = -\omega \int_C F_i^2 \, 2\pi R h^i \, di, \tag{28}$$

which is negative definite. Since there is no flux of potential energy across the boundary W, the potential energy is minimized by this procedure and an equilibrium may thereby be obtained.

Using the description of the force [Eq. (25)] and the definition of the normal **n** in terms of the coordinates (R, Z) of the contour, the new coordinates are defined [Eq. (27)] at each step of the iteration:

$$\delta R_i = -\omega \left[p + \frac{I_p^2}{(2\pi R_i)^2} - \frac{I_v^2}{(2\pi R_i)^2} - \frac{K^2}{h^{i2}}\left(\frac{\partial \alpha}{\partial i}\right)^2 \right] \frac{\partial Z}{h^i \, \partial i}, \tag{29}$$

$$\delta Z_i = \omega \left[p + \frac{I_p^2}{(2\pi R_i)^2} - \frac{I_v^2}{(2\pi R_i)^2} - \frac{K^2}{h^{i2}}\left(\frac{\partial \alpha}{\partial i}\right)^2 \right] \frac{\partial R}{h^i \, \partial i}. \tag{30}$$

During each step of the variation the force F_i may be assumed constant, and the pair of equations in R and Z solved, for example, by the Lax or Lax–Wendroff method. The parameter ω must be chosen small to satisfy the assumption of small variations. Certainly, if Eqs. (29) and (30) are resolved explicitly, ω must satisfy a condition of the Courant–Friedrichs–Lewy type:

$$\omega \lesssim \mathrm{Min}\,(h^i/F_i). \tag{31}$$

D. Equilibrium of One Waterbag in Three Dimensions

The concepts discussed for a waterbag in the plane extend naturally to a single waterbag in three dimensions. Solutions by this method have recently been obtained by Betancourt and Garabedian (1974).

1. The Fields in Three Dimensions

The model is the same as in the two-dimensional problem illustrated in Figs. 4 and 5. Because of the lack of toroidal symmetry, however, we may no longer solve analytically for the toroidal field, and the Laplacian equations for the field must now be solved. Again in both the plasma and the vacuum regions, no currents flow so that the fields satisfy Eqs. (12) and (13). Hence the magnetic fields may be described in terms of multivalued potentials:

$$\mathbf{B} = \nabla\phi \tag{32}$$

in both the plasma and vacuum region. In the plasma region the volume is not simply connected so that the potential is not harmonic in the toroidal direction. It is useful to introduce the period of the potential explicitly

$$\mathbf{B}_p = I_p \nabla\beta, \tag{33}$$

where β is a potential of unit period and I_p is a constant identified physically as the equivalent current producing the toroidal field.

In the vacuum region, the space is not simply connected both in the toroidal and the poloidal directions and we therefore introduce two constants I_v and K:

$$\mathbf{B}_v = I_v \nabla\beta + K\nabla\alpha, \tag{34}$$

where the potential β is harmonic over the poloidal plane and has unit period in the toroidal direction. α is harmonic over the toroidal plane and has unit period around the poloidal plane. It follows from the integral form of Ampere's Law that I_v is the poloidal current producing the toroidal field in the vacuum region, while K is the toroidal current on the plasma–vacuum interface C producing the poloidal field. As before, the toroidal flux Π_p in the plasma region is given by the volume integral:

$$\Pi_p = \iiint_P \mathbf{B}_p \cdot \nabla\beta \, d\tau, \tag{35}$$

while the toroidal flux Π_v and poloidal flux ψ_v are

$$\Pi_v = \iiint_V \mathbf{B}_v \cdot \nabla \beta \, d\tau, \tag{36}$$

$$\psi_v = \iiint_V \mathbf{B}_v \cdot \nabla \alpha \, d\tau. \tag{37}$$

The magnetic field on the free waterbag may be determined from the potentials, which in turn are obtained by applying the divergence-free condition to the magnetic fields [Eq. (13)]:

in P:
$$\nabla^2 \beta = 0; \tag{38}$$
in V:
$$\nabla^2 \beta = 0, \tag{39}$$

$$\nabla^2 \alpha = 0. \tag{40}$$

In the plasma region, Neumann boundary conditions are to be applied at C, and in the vacuum region [Eqs. (39), (40)] Neumann boundary conditions are to be applied at both C and W (Fig. 5).

It is possible to solve these Laplacian equations in integral form using Green's theorem in three dimensions in an entirely analogous manner to that described for the plane in Section II, B. The advantage of this procedure is that no intermediate mesh is introduced and the potentials are only determined where they are required at the interface C. An alternative procedure is a mapping method described by Betancourt (1974) and we illustrate it here.

2. Mapping of the Vacuum and Plasma Regions

It is useful to describe the surface of the waterbag and the wall in terms of toroidal coordinates (Fig. 6). The axis of the toroidal coordinates, $r = 0$, is

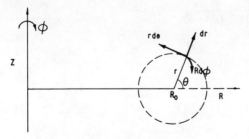

FIG. 6. Toroidal coordinates (r, θ, ϕ).

assumed to lie within the plasma region. The equations for the waterbag surface may be expressed in terms of the minor radius

$$r = c(\theta, \phi), \tag{41}$$

and equivalently the equation for the wall may be expressed by the function w

$$r = w(\theta, \phi). \tag{42}$$

Betancourt maps the vacuum region V into the rectangle ($0 \leq s \leq 1$, $0 \leq \theta \leq 2\pi$, $0 \leq \phi \leq 2\pi$):

$$\begin{aligned} s &= [r - c(\theta, \phi)]/[w(\theta, \phi) - c(\theta, \phi)], \\ \theta &= \theta, \\ \phi &= \phi. \end{aligned} \tag{43}$$

In the new nonorthogonal coordinates, $x_\mu' = (s, \theta, \phi)$, Laplace's equation for the potentials α, β [Eqs. (38)–(40)] transforms to (e.g., see McConnell, 1957)

$$(\partial/\partial x_\nu')[(1/g^{1/2})g^{\nu\lambda}(\partial\alpha/\partial x_\lambda')] = 0, \tag{44}$$

where $g^{\nu\lambda}$ is the contravariant metrical tensor

$$g^{\nu\lambda} = (\partial x_\mu'/\partial x_\lambda)(\partial x_\mu'/\partial x_\nu),$$

and g is the determinant of the elements $g^{\nu\lambda}$. Defining the function

$$f(\theta, \phi) = c(\theta, \phi)(s - 1) - sw(\theta, \phi),$$

we may determine the contravariant metrical tensor from the transformation [Eq. (43)]

$$g^{\nu\lambda} = \begin{bmatrix} \dfrac{1}{(w-c)^2} + \dfrac{f_\theta{}^2}{(w-c)^2 r^2} + \dfrac{f_\phi{}^2}{(w-c)^2 R^2} & \dfrac{f_\theta}{(w-c)r^2} & \dfrac{f_\phi}{(w-c)R^2} \\[1em] \dfrac{f_\theta}{(w-c)r^2} & \dfrac{1}{r^2} & 0 \\[1em] \dfrac{f_\phi}{(w-c)R^2} & 0 & \dfrac{1}{R^2} \end{bmatrix},$$

whence

$$1/g^{1/2} = (w - c)rR. \tag{45}$$

The boundary conditions on C and W transform to

$$g^{s\lambda} \, \partial\alpha/\partial x_\lambda' = 0.$$

The advantage of this mapping is that the rather awkwardly shaped vacuum region is mapped into a three-dimensional rectangle which is easy to index logically on the computer and on which the boundary conditions are simply specified. This rectangle is then divided by a regular rectangular finite lattice of points and "Laplace's" equation, (44), differenced on the resulting mesh. One of the difficulties is that the transformed equation describing the potentials has mixed derivatives, and a fifteen-point (rather than seven-point) difference scheme must be employed. The resulting difference equation is solved by the successive-overrelaxation method (Potter, 1973) for both potentials α and β in the vacuum region. An entirely equivalent mapping is performed in the plasma region, and the potential β is then solved.

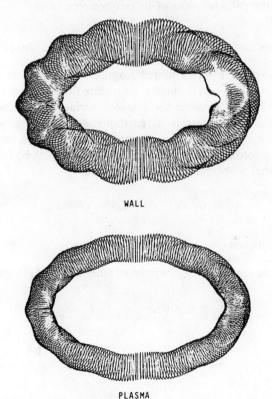

FIG. 7. A solution obtained by Betancourt and Garabedian (1974) for the equilibrium of a Stellerator plasma represented by a single waterbag.

Having solved for the potentials, the fields may now be determined on the waterbag surface C and as before the variational procedure of Section II, C is used. An example of such a solution for a Stellerator is illustrated in Fig. 7.

III. Equilibria of Current Distributions

The single waterbag models in two and three dimensions described in Section II are simple to formulate and easy to resolve on present-day computers. In a sense, though, they are one- or two-dimensional models, respectively, since they only describe a line or a surface. They contain no information on a pressure or current profile. The concept extends naturally however to a sequence of nested surfaces, across which the pressure and currents vary. We discuss here the interaction of a set of nested surfaces which intersect the poloidal plane of an axisymmetric torus.

A. Formulation of Axisymmetric Equilibria in Flux Space

We shall construct a set of nested magnetic surfaces. For the simplest topology there is a degenerate surface represented by a single toroidal field line (the magnetic axis). Between the magnetic surface and the magnetic axis the poloidal field (B_χ) intersects a radial plane to define a poloidal flux $\psi = RA_\phi$ where A_ϕ is the toroidal vector potential. Thus each surface may be denoted by the coordinate ψ. Then the "flux space" (ψ, χ, ϕ) forms a right-handed orthogonal coordinate system with metric

$$ds^2 = (h^\psi \, d\psi)^2 + (h^\chi \, d\chi)^2 + (R \, d\phi)^2. \tag{46}$$

1. *Continuous Equations*

As before (Section II), it may be inferred from Eqs. (5), (8), and (9) that

$$p = p(\psi), \tag{47}$$

$$J_\psi = 0, \tag{48}$$

$$B = B_\phi \mathbf{e}_\phi - (1/Rh^\psi) \mathbf{e}_\chi, \tag{49}$$

namely, that the pressure is constant on a surface, that there is no current normal to the surfaces, and that the field is divergence free. The currents in the poloidal plane are defined from Ampere's law [Eq. (6)] thus:

$$J_\psi = (1/Rh^\chi)(\partial/\partial\chi)(RB_\phi) = 0,$$

$$J_\chi = -(1/Rh^\psi)(\partial/\partial\psi)(RB_\phi).$$

It follows that we may define the poloidal current as a surface function $I'(\psi) = RB_\phi$, whence

$$J_\chi = -(1/Rh^\psi)(dI'/d\psi). \tag{50}$$

By considering the component of the pressure balance equation [Eq. (4)] normal to the surface, the toroidal current J_ϕ may be defined:

$$\frac{1}{h^\psi}\frac{dp}{d\psi} = J_\phi \frac{1}{Rh^\psi} - \frac{1}{Rh^\psi}\frac{dI'}{d\psi}B_\phi, \tag{51}$$

where Eqs. (47), (49), and (50) have been used. An elliptic equation is obtained for the magnetic surfaces by relating the toroidal current density J_ϕ to the toroidal vector potential A_ϕ or poloidal flux ψ.

$$\nabla^2\left(\frac{\psi}{R}\right) = -\left(R\frac{dp}{d\psi} + \frac{I'(\psi)}{R}\frac{dI'}{d\psi}\right). \tag{52}$$

The ∇^2 operator arises from the toroidal component of the curl curl operator.

Equation (52) has the form of a nonlinear Poisson's equation. If the "source" functions $p(\psi)$ and $I'(\psi)$ are specified and if ψ is specified on the boundary of the domain of interest, a solution of ψ may be obtained. The conventional approach is to iterate the solution on an Eulerian difference mesh (Callan and Dory, 1972).

In the waterbag method the elliptic equation (52) is transformed to the flux space (ψ, χ) itself. The Laplacian operator on ψ becomes

$$\nabla^2(\psi/R) = (1/h^\chi h^\psi)(\partial/\partial\psi)(h^\chi/Rh^\psi). \tag{53}$$

Noting that $h^\chi h^\psi$, the differential area in the flux space, is the Jacobian of the Eulerian cartesian coordinates, the equilibrium equation (52) transforms to

$$\frac{\partial}{\partial\psi}\left(\frac{h^\chi}{Rh^\psi}\right) = -\frac{\partial(R,Z)}{\partial(\psi,\chi)}\left(R\frac{dp}{d\psi} + \frac{I'(\psi)}{R}\frac{dI'}{d\psi}\right). \tag{54}$$

We may regard this system as a sequence of equations for h^χ/h^ψ which with the orthogonal equations (Potter and Tuttle, 1973) defines the equilibrium flux space. However, computationally, the continuous system must be represented by a finite set of waterbags and since the problem is nonlinear a variational procedure will be used to obtain the solution.

2. *The Waterbag Model for a Distribution of Pressure*

In the discrete waterbag model, the distribution is represented by a number (J) of nested surfaces with predefined fluxes,

$$\psi = (c_1, c_2, ..., c_J).$$

In the space between the surfaces, the pressure is assumed constant and the current is restricted to sheet currents flowing through the waterbag (Fig. 8). The waterbags are moved according to a variational procedure until they are in equilibrium (Section III, B).

With such a model, an exact equilibrium for the finite set of surfaces can be defined irrespective of truncation errors. Between the contours, the equilibrium equation (54) is satisfied since no pressure gradient exists and no currents flow. Across each waterbag, Eq. (54) may be integrated to yield

$$(1/h^x h^\psi)\,\delta(h^x/Rh^\psi) = -R\delta p - (1/R)\,\delta(I'^2/2),$$

where δf is the discontinuous change of the function f across the waterbag. It is to be noted that h^ψ is discontinuous while h^x and R are continuous

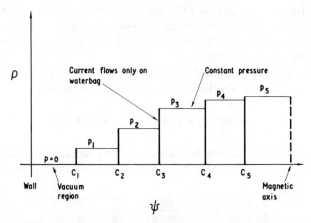

FIG. 8. To describe a current and pressure profile, a finite number of nested waterbags is used. Each waterbag carries a sheet current. Between the waterbags the plasma pressure is constant.

so that we may obtain

$$\delta(1/2R^2 h^{\psi 2}) = -\delta p - \delta I'^2/2R^2. \tag{55}$$

The term on the left-hand side may be identified as the magnetic pressure of the poloidal field, while the terms on the right are the changes of material and toroidal magnetic pressure.

B. Variational Methods to Obtain Equilibria

A similar approach is adopted to that described for one surface (Section II). Keeping the pressure p and poloidal current I' constant during the iterations, and around each contour, the waterbags are moved until they all rest in equilibrium. The question arises as to the choice of a particular algorithm by which the waterbags may be moved.

As before, the contours should be moved so as to reduce the potential energy at each iteration and, if there exists a minimum, an equilibrium will be found. We shall therefore consider the potential energy of such a system. During the motion of the contours the volume and area between a pair of contours change. However, $p(\psi)$ and $I(\psi)$ are kept constant. Regarding the contained fluid as an ideal gas, $pV^\gamma = \text{const.}$, and, since the pressure does not vary as the volume changes, we observe that for such a gas $\gamma = 0$. Now the internal energy of an ideal gas of pressure p and volume V is $pV/(\gamma - 1)$. Hence it follows that the internal energy of such a gas in the torus is

$$E_p = \iint_D -p \, 2\pi R \, dS,$$

where the integral is taken over the surface area of the domain D contained within the wall W. Similarly, since $I(\psi)$ is kept constant, and as can be shown from first principles, the toroidal magnetic energy is negative. Thus the total potential energy of the system is

$$E = 2\pi \iint_D -p(\psi) - \frac{I'^2(\psi)}{2R^2} + \frac{B_\chi^2}{2} R \, dS. \tag{56}$$

By varying the position of the surfaces and thereby the magnetic flux, the energy is varied.

$$\delta E = -2\pi \iint_D \frac{dp}{d\psi} \delta\psi + \frac{I'}{R^2} \frac{dI'}{d\psi} \delta\psi - \mathbf{B}_\chi \cdot \delta\mathbf{B}_\chi R \, dS, \tag{57}$$

Using Green's theorem, the third term in the poloidal magnetic energy is expressed in terms of the flux ψ and toroidal current J_ϕ:

$$\delta E_\chi = 2\pi \iint_D \delta\psi J_\phi \, dS - 2\pi \oint_W \delta\psi \mathbf{B}_\chi \cdot \mathbf{dl},$$

where the line integral is taken around the wall W bounding the plane D. Since the flux on the wall is held constant during the variation, the contribution on the boundary vanishes and we obtain for the variation of the total potential energy

$$\delta E = -2\pi \iint_D \left(\frac{dp}{d\psi} R + \frac{I'}{R} \frac{dI'}{d\psi} - J_\phi \right) \delta\psi \, dS. \tag{58}$$

It is evident that the energy will be minimized when the integrand of Eq. (58) is zero, which is the equilibrium equation previously obtained [Eq. (52)]. In a nonequilibrium state, however, the toroidal current J_ϕ, defined by the flux surfaces, does not satisfy a zero integrand and a choice for $\delta\psi$ must be made which ensures a negative definite change of the energy E.

A variety of choices for $\delta\psi$, defining different variational methods, suggests themselves. We may employ the same method as was used for one surface (Section II).

1. *Explicit Motion of the Surfaces*

The integral, Eq. (58), may be transformed to the flux space:

$$\delta E = -2\pi \int_0^{\psi_m} dc \oint_{c=\psi} \left(\frac{dp}{d\psi} R + \frac{I'}{R} \frac{dI'}{d\psi} - J_\phi \right) \delta\psi h^\psi h^\chi \, d\chi, \tag{59}$$

where the integrals are around each flux surface (variable χ) and over the flux surfaces $(0, \psi_m)$. $\delta\psi h^\psi$ may be recognized as the displacement of each point on the surface along the unit normal $\mathbf{n}$ to the surface:

$$\delta\mathbf{r} = \delta\psi h^\psi \mathbf{n}.$$

Writing $dl = h^\chi \, d\chi$ as the length element around the surface, we obtain

$$\delta E = -2\pi \int_0^{\psi_m} dc \oint_c \left(\frac{dp}{d\psi} + \frac{I'}{R^2} \frac{dI'}{d\psi} - \frac{J_\phi}{R} \right) \delta\mathbf{r} \cdot \mathbf{n} R \, dl. \tag{60}$$

By choosing the normal variation $\delta \mathbf{r}$ of each point on each surface [cf. Eqs. (25), (27)],

$$\delta \mathbf{r} = \omega \left(\frac{dp}{d\psi} + \frac{I'}{R^2} \frac{dI'}{d\psi} - \frac{J_\phi}{R} \right) \mathbf{n}, \tag{61}$$

the variation of the energy is negative definite,

$$\delta E = -2\pi \int_0^{\psi_m} dc \oint_{c=\psi} \frac{1}{\omega} (\delta \mathbf{r})^2 R \, dl, \tag{62}$$

thus ensuring convergence to any minimum of the energy and an equilibrium, if it exists. The relaxation parameter ω is a positive small number.

For the finite waterbag model discussed above (Fig. 8), the variational equations (61) become discrete according to the local jump conditions across each surface [cf. Eq. (55)]. At the step $\mu + 1$, the waterbags are moved:

$$\mathbf{r}^{(\mu+1)} = \mathbf{r}^{(\mu)} + \omega' [\delta p + \delta (I'^2/2R^2) - \delta (1/2R^2 h^{\psi 2})]^{(\mu)} \mathbf{n}. \tag{63}$$

After the waterbags are moved according to this algorithm, the space is reorthogonalized, which in turn defines the toroidal current and h^ψ.

2. Implicit Variation of the Contours

Other variational algorithms exist which ensure the minimization of the potential energy [Eq. (58)]. By varying the flux implicitly, a method which provides a considerably faster rate of convergence is defined. We choose from Eq. (58) a variation of the flux $\delta\psi$ according to the prescription

$$\nabla^2 (\delta\psi/R) = -\omega_2 [(dp/d\psi) R + (I'/R)(dI'/d\psi) - J_\phi], \tag{64}$$

where ω_2 is again a relaxation parameter and the ∇^2 operator is the toroidal component of the curl curl operator. Thus from Eq. (58) the variation of energy is

$$\delta E = \frac{2\pi}{\omega_2} \iint_D \nabla^2 \left(\frac{\delta\psi}{R} \right) \delta\psi \, dS.$$

Using Green's theorem, the integrand may be expressed as a quadratic:

$$\delta E = -\frac{2\pi}{\omega_2} \iint_D (\nabla \delta\psi)^2 \frac{1}{R} dS + \frac{2\pi}{\omega_2} \oint_W \delta\psi \mathbf{B}_\chi \cdot \mathbf{d}l. \tag{65}$$

Again, if the flux at the wall is conserved during the variation the second term vanishes and the variational method minimizes the energy.

The variational prescription [Eq. (64)] defines an iteration in which successively improved solutions for the flux $\psi = \psi^{(\mu)}$ are obtained at each step μ:

$$\nabla^2 \frac{\psi^{(\mu)}}{R} = (1 - \omega_2) \nabla^2 \frac{\psi^{(\mu-1)}}{R} - \omega_2 \left(R \frac{dp^{(\mu-1)}}{d\psi} + \frac{I'}{R} \frac{dI'^{(\mu-1)}}{d\psi} \right). \quad (66)$$

At the step $(\mu - 1)$, the existing surfaces, j, are surfaces of constant flux $\psi^{(\mu-1)} = c_1, \ldots, c_j, \ldots, c_J$. Keeping the surfaces fixed, we shall determine the new flux $\psi^{(\mu)}(i, j)$ on the previous coordinates (i, j). Thus transforming Eq. (66) onto the existing coordinates (i, j) we obtain

$$\frac{\partial}{\partial j} \left(\frac{h^i}{Rh^j} \frac{\partial \psi^{(\mu)}}{\partial j} \right) + \frac{\partial}{\partial i} \left(\frac{h^j}{Rh^i} \frac{\partial \psi^{(\mu)}}{\partial i} \right)$$
$$= (1 - \omega_2) \frac{\partial}{\partial j} \left(\frac{h^i}{Rh^j} \frac{dc}{dj} \right) - \omega_2 \frac{h^i h^j}{dc/dj} \left[R \frac{dp}{dj} + \frac{1}{R} \frac{d}{dj} \left(\frac{I'^2}{2} \right) \right]. \quad (67)$$

This equation is differenced using a five-point scheme and the solution $\psi^{(\mu)}(i, j)$ is obtained by the alternating direction implicit (ADI) scheme (Potter, 1973). Since the solution $\psi^{(\mu)}(i, j)$ depends only weakly on the coordinates i, very few iterations of the ADI method need be taken. In the limit of convergence

$$\lim_{\mu \to \infty} \psi^{(\mu)}(i, j) = c_j.$$

The finite difference form of the equation (67) is taken to satisfy exactly the equilibrium equation (55) in the limit of this convergence.

3. Interpolation of the Contours in the Implicit Method

At each step, μ, of the implicit method an improved solution $\psi^{(\mu)}(i, j)$ for the poloidal flux is determined. The variation of ψ from the constant predefined values c_j of each contour j is a measure of the departure from equilibrium of the contours. The contours are now moved in the poloidal plane so as to lie along the constant flux lines $\psi = c_1, \ldots, c_J$. This is achieved by interpolation from the old space $\mathbf{r}^{(\mu-1)}(i, j)$, where each point of a contour is moved along a constant i-line. The values $\mathbf{r} = (R, Z)$ in the poloidal plane are varied normal to the contour, until for each contour j, the point has an appropriate flux c_j. Fourth-order interpolation is used:

$$\delta \mathbf{r} = \left(\frac{\partial \mathbf{r}}{\partial \psi}\right)_i \delta\psi + \left(\frac{\partial^2 \mathbf{r}}{\partial \psi^2}\right)_i \frac{\delta\psi^2}{2} + \left(\frac{\partial^3 \mathbf{r}}{\partial \psi^3}\right)_i \frac{\delta\psi^3}{6} + \left(\frac{\partial^4 \mathbf{r}}{\partial \psi^4}\right) \frac{\delta\psi^4}{24} + o(\delta\psi^4). \tag{68}$$

We reexpress this expansion in the coordinate j. It is useful to define the required change in j:

$$\Delta j = [\psi^{(\mu)}(i,j) - c_j]/(dc/dj),$$

then

$$\mathbf{r}^{(\mu)}(\tilde{\imath}, j) = \mathbf{r}^{(\mu-1)}(i, j) + \left[\frac{\partial \mathbf{r}^{(\mu-1)}}{\partial j}\right]_i \Delta j + \left(\frac{\partial^2 \mathbf{r}^{(\mu-1)}}{\partial j^2}\right)_i \frac{\Delta j^2}{2}$$
$$+ \left(\frac{\partial^3 \mathbf{r}^{(\mu-1)}}{\partial j^3}\right)_i \frac{\Delta j^3}{6} + \left(\frac{\partial^4 \mathbf{r}^{(\mu-1)}}{\partial j^4}\right)_i \frac{\Delta j^4}{24}. \tag{69}$$

The derivatives are determined by appropriate central differences.

Once the contours have been moved by this interpolation, the curves $\tilde{\imath}$ are no longer orthogonal to j. Thus, keeping the j contours fixed, the points $\tilde{\imath}$ are moved by the orthogonalization procedure. The iteration is then complete and a new step in which the flux is varied (Section III, B, 2) can be carried out.

C. Some Illustrative Solutions

Examples of equilibrium solutions obtained by the implicit variational method are illustrated in Figs. 9 and 10. In Fig. 9 equilibrium solutions for a torus of circular cross section and large poloidal beta are shown. Figure 10 illustrates a solution for a pear-shaped cross section.

The solutions illustrated have used eight free contours in which each contour is defined by 32 points. Though many more surfaces may be used, a much greater accuracy is achieved compared to an Eulerian method with the same number of points, since the natural coordinates are being used. The solutions shown took typically 10 iterations and 45 sec on an IBM 360/91. The rate of convergence depends on the value adopted for the relaxation parameter ω_2. Convergence was typically obtained for values

$$0.0 < \omega < 0.5,$$

and optimum values were found to be of the order 0.3. As ω_2 approaches

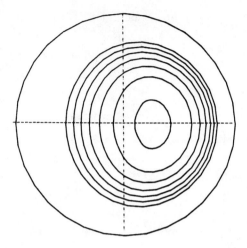

FIG. 9. The equilibrium solution for a finite number of magnetic surfaces in the Tokamak. The aspect ratio of the torus is 2 and $\beta_{pol} = 1.3$. A vacuum region exists between the circular wall and first plasma surface. The major axis of the torus is on the left.

zero its effect is to allow only small changes in the positions of the contours at each iteration. ω_2 plays the role of a time-step, and small changes are assumed in the variational procedure. Additionally, the contours are moved by explicit interpolation, so that too large a step will be inappropriate for the Taylor expansion.

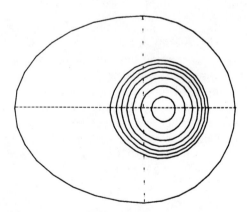

FIG. 10. The equilibrium of a finite number of magnetic surfaces in a Tokamak of pear-shaped cross section. The pressure profile is linear and a vacuum region exists outside the largest surface.

IV. Adiabatic Constraints

A. Adiabatic Changes of Equilibrium

The equilibrium between field and plasma which is described by the magnetostatic equations is not only valid in the ideal, purely static situation. Indeed, it might be questioned what is meant by static equations in describing a plasma which is intrinsically transitory. The equilibrium between plasma and field described by Eq. (4) derives from the momentum equations where the inertial forces have been neglected. This model is valid therefore if, first, equilibrium flows are small. That is, rotations within a magnetic surface are sufficiently small that the corresponding Coriolis and centrifugal forces may be ignored. The second requirement is that any transitory motion which occurs must take place sufficiently slowly that the acceleration terms may be neglected.

The term "sufficiently slowly" may be quantified by referring to the condition that the equilibrium equations impose, namely, that a magnetic surface is one of constant pressure. Suppose a perturbation of the pressure occurs at some point on a surface. In the low-beta case, the strength of the magnetic field disallows an expansion of the pressure pulse normal to the field. Equilibrium of the pressure on the surface can only be achieved by the propagation of the pulse parallel to the field line. Thus the time for the surface pressure to equilibrate is the time of flight of a sound wave along a field line. In the toroidal problem the characteristic length along a field line is $2\pi qR$, where q is the safety factor. $2\pi qR$ is the length of a field line as it moves once around the poloidal plane. This time for equilibriation is therefore

$$\tau_A = 2\pi qR/v_s,$$

where v_s is the sound speed. It follows that if the frequency of transistory phenomena is less than $2\pi/\tau_A = v_s/qR$, the momentum equilibrium between field and plasma remains valid.

Such changes may occur through diffusion, through slow instabilities (such as the tearing mode), or through the electromagnetic fields at the boundary as occurs, for example, in the adiabatic toroidal compression experiment. In each case, given an initial equilibrium state, a new equilibrium state is produced by the change. Furthermore, every intermediate state is one described by the equilibrium equations.

Diffusion changes the pressure and the poloidal and toroidal fluxes, and thereby changes the sources in the equilibrium equations. However, adiabatic changes may even occur in the absence of diffusion. For example, externally applied voltages change the boundary conditions of the elliptic problem.

Before discussing diffusion it is useful therefore to consider the mathematical form which may describe such adiabatic changes. A conceptual experiment may be imagined when the fluxes are changed at the boundary. As a result, the surfaces change and correspondingly the source functions of pressure and flux change. A self-consistent solution must therefore be sought.

The magnetic fluxes may be examined first. The electric field is expressed in terms of the vector potential **A** and electrostatic potential ϕ by

$$\partial \mathbf{A}/\partial t + \mathbf{E} = -\nabla \phi. \tag{70}$$

An ideal Ohm's law is assumed where the electric field is described by the Lorentz field

$$\partial \mathbf{A}/\partial t - \mathbf{v} \wedge \mathbf{B} = -\nabla \phi. \tag{71}$$

We integrate Eq. (71) around a surface S bounded by a closed contour l:

$$\oint_l \frac{\partial \mathbf{A}}{\partial t} \cdot \mathbf{dl} - \oint_l \mathbf{v} \wedge \mathbf{B} \cdot \mathbf{dl} = 0. \tag{72}$$

Using Stoke's theorem,

$$\iint_S \frac{\partial \mathbf{B}}{\partial t} \cdot \mathbf{dS} + \oint \mathbf{B} \cdot (\mathbf{v} \wedge \mathbf{dl}) = 0 \tag{73}$$

or

$$\frac{d}{dt} \iint_S \mathbf{B} \cdot \mathbf{dS} = 0, \tag{74}$$

namely, the magnetic flux through any surface in the absence of resistivity is conserved. Thus the magnetic fluxes through every magnetic surface during such adiabatic changes are conserved,

$$d\psi/dt = 0, \tag{75}$$

$$d\Pi/dt = 0. \tag{76}$$

Equation (75) implies that the fluid moves with the surface. We may therefore state that the mass of plasma contained within each surface, $M(\psi)$, is conserved

$$M = \iiint_{V(\psi)} \rho \, dV, \tag{77}$$

where ρ is the plasma mass density. Then

$$dM/dt = 0. \tag{78}$$

Similarly the change in the pressure may be defined. Assuming an ideal gas equation of state in which γ is the ratio of specific heats, we define the integral $e(\psi)$ as

$$e(\psi) = \iiint_{V(\psi)} p^{1/\gamma} dV. \tag{79}$$

Then

$$de/dt = 0 \tag{80}$$

for each surface ψ.

B. Adiabatic Equations Applied to Axisymmetric Toroids

It is evident that the adiabatic laws [Eqs. (75), (76), (78), (80)] apply equally in differential form across pairs of neighboring surfaces. Using the geometric notation of the local flux space (ψ, χ) in the poloidal plane, it is useful to define the differential volume $V(\psi)$ between surfaces, and a quantity $A(\psi)$ related to the differential area:

$$V(\psi) = \oint_\psi R h^\psi h^\chi \, d\chi, \tag{81}$$

$$A(\psi) = \oint_\psi (h^\psi h^\chi / R) \, d\chi. \tag{82}$$

The adiabatic constraints become

$$d\psi/dt = 0,$$
$$(d/dt)(IA) = 0,$$
$$(d/dt)(\rho V) = 0,$$
$$(d/dt)(p^{1/\gamma} V) = 0, \tag{83}$$

since I, the poloidal current, ρ, and p are surface functions.

The source function in the equilibrium equation depends only on the pressure $p(\psi)$ and poloidal current $I(\psi)$. Thus, in the absence of diffusion, only the constraints for the poloidal current I and pressure p are required to specify a new equilibrium state. Given an initial equilibrium state ψ^0 with source functions $p^{(0)}$ and $I^{(0)}$, a new equilibrium state may be determined by either of the variational procedures discussed in Section V, where the new pressure and poloidal current are defined by the constraints [Eqs. (83)]. For example, using the implicit method (Section III, B, 2), each step (μ) of the iteration employs the self-consistent pressure and poloidal current:

$$\nabla^2 \left(\frac{\psi}{R}\right)^{(\mu+1)} = (1-\omega_2)\nabla^2\left(\frac{\psi}{R}\right)^{(\mu)} - \omega_2\left(R\frac{dp^{(\mu)}}{d\psi} + \frac{I^{(\mu)}}{R}\cdot\frac{dI^{(\mu)}}{d\psi}\right)$$

$$\mathbf{r}^{(\mu+1)} = \mathbf{r}(\psi^{(\mu+1)}, \chi^{(\mu+1)}),$$

$$I^{(\mu+1)} = I^0 A^0/A^{(\mu+1)},$$

$$p^{(\mu+1)} = p^0(V^0/V^{(\mu+1)})^\gamma, \tag{84}$$

where the "area" A and volume V between each pair of surfaces is determined by an appropriate difference form of the Jacobian $h^\chi h^\psi$ in the surface integrals [Eqs. (81), (82)]. In the limit of convergence ($\mu \to \infty$), the iterative scheme [Eqs. (84)] satisfies the adiabatic constraints [Eqs. (83)].

An example of such an adiabatic change of equilibrium is illustrated in Fig. 11. Figure 11a shows the equilibrium of a Tokamak of circular cross section, aspect ratio 3, and beta-poloidal 0.7. By applying a toroidal electric field, the poloidal flux ψ_w on the wall is changed, increasing the poloidal field in the torus. It may be observed that a new equilibrium is uniquely

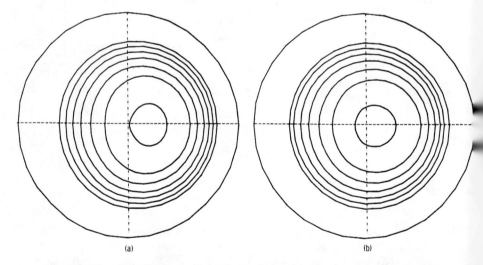

FIG. 11. The adiabatic change of equilibrium induced by the addition of poloidal flux at the wall. The initial equilibrium (a) is for a parabolic pressure profile with $\beta_{pol} = 0.7$ The addition of flux (b) compresses the toroidal column toward the major axis of symmetry on the left. The aspect ratio of the torus is 3, with the major axis of the torus on the left.

defined (Fig. 11b), in which the plasma column is compressed toward the major axis of symmetry.

It should be noted that the allowable values for the relaxation parameter ω_2, for which convergence may be achieved, are more restricted. In the low-beta case, the toroidal field produced by the plasma is small compared to the vacuum toroidal field

$$\delta I \ll I. \tag{85}$$

To ensure that large errors in the toroidal field are not produced, leading to divergence, the change of area between each pair of surfaces must be small. This may be ensured if the relaxation parameter ω_2 is sufficiently small

$$\omega_2 < \beta.$$

A particularly interesting subset of the equilibrium equations with adiabatic constraints occurs in the limit of large aspect ratio and low beta. For large aspect ratio, a constant major radius R_0, may be assumed, so that the toroidal current source function J in equilibrium is purely a function of the flux

$$R_0^2 \, dp/d\psi + I \, dI/d\psi = J(\psi).$$

Thus the equilibrium equation reduces to

$$\nabla^2 \psi = -J(\psi). \tag{86}$$

Furthermore, for large aspect ratio and low beta, the adiabatic constraint for the conservation of magnetic fluxes [Eqs. (76), (82)] reduces to

$$\delta \oint_\psi h^\chi h^\psi \, d\chi = 0. \tag{87}$$

The constraints, Eq. (87), apply to each flux surface ψ and specify the change of the current J for each surface ψ.

An example of the application of this simple model [Eqs. (86) and (87)] is to the filamentation of an equilibrium current sheet and the creation of magnetic islands (Fig. 12) (Potter and Kamimura, 1976). Figure 12a illustrates the perturbation of an initially uniform Gaussian current sheet containing a neutral line. According to linear stability theory (Furth et al., 1963), such a sheet is unstable to the resistive tearing mode as a result of the resistive cutting of field lines across the neutral line. Under conditions of high conductivity, the instability develops sufficiently slowly that the equilibrium equation (86)

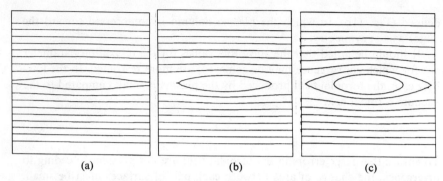

Fig. 12. The nonlinear development of magnetic islands illustrating the filamentation of an initially uniform current sheet. The system is periodic from left to right and a *Gaussian* current distribution has been taken. The contours describing field lines have been chosen to be equally spaced in "x". In (a) a small perturbation of the initially uniform distribution is shown. Further development of the "equilibrium" island states is illustrated in (b) and (c).

always applies. The constraint equations (87) apply to the high (but not perfect) conductivity case since they conserve magnetic flux but, unlike the ideal MHD equations, permit changes of topology.

The nonlinear development of magnetic islands is illustrated in Figs. 12b and 12c.

C. Relaxation of the Adiabatic Constraints by Diffusion

The adiabatic constraints, Eqs. (83), are integral relations which apply to each flux surface ψ, since in equilibrium the toroidal flux Π, the density ρ, and the pressure p are surface variables, constant on a flux surface. We define the mass density $m(\psi)$, the flux density $\pi(\psi)$, and the internal energy density $u(\psi)$ on each surface:

$$m(\psi) = \rho(\psi)V(\psi),$$
$$\pi = d\Pi/d\psi = IA, \qquad (88)$$
$$u(\psi) = p(\psi)V(\psi)/(\gamma - 1),$$

where $V(\psi)$ is the differential volume of each surface [Eq. (81)] and $A(\psi)$ is the differential "area" of each surface [Eq. (82)]. Then in the absence of diffusion, the adiabatic constraints become

$$d\pi/dt = 0,$$
$$dm/dt = 0, \qquad (89)$$
$$du/dt + p(\psi)\,dV/dt = 0,$$

where the Lagrangian time derivative refers to the moving flux space $d\psi/dt = 0$.

In the presence of particle collisions, the adiabatic constraints, Eqs. (83), are relaxed by diffusion across the surfaces. Clearly the diffusion occurs at a differential rate around each surface. However, if the diffusion rate is much less than the adiabatic frequency for the restoration of equilibrium (Section IV,A), plasma flow will always occur rapidly in the plane of the surface to reestablish the equilibrium of pressure $p(\psi)$, density $\rho(\psi)$, and poloidal current $I(\psi)$ in each surface (cf. Grad and Hogan, 1971). The diffusion from each surface is therefore an integral or surface function. No matter how arbitrary the shape of the flux surfaces, the diffusion may always be described exactly by one-dimensional diffusion equations across the flux surfaces. The diffusion coefficients are integrals around each flux surface and depend on the geometry of the surfaces.

The diffusion equations for the collisional regime are illustrated here. In this regime, an Ohm's law in the frame of the flux surfaces is used,

$$\mathbf{E} + \mathbf{v} \wedge \mathbf{B} = \mathbf{\eta} \cdot \mathbf{J}, \tag{90}$$

where $\mathbf{v}$ is the plasma velocity in the ψ-frame and $\mathbf{E}$ is the electric field in this frame. In particular, with axisymmetry, the toroidal electric field in this frame is zero,

$$E_\phi = -(1/R)(d\psi/dt) = 0.$$

By considering the component of Ohm's law parallel to the magnetic field and integrating the electric field once around the poloidal plane, a diffusion equation for the toroidal flux density is obtained:

$$\frac{d\pi}{dt} = \frac{\partial}{\partial \psi}\left[\eta_\| \frac{\partial}{\partial \psi}\left(\frac{\mathscr{I}\pi}{A}\right)\right]. \tag{91}$$

$\eta_\|$ is the parallel coefficient of resistivity and $\mathscr{I}$ is a surface integral given by

$$\mathscr{I} = \oint_\psi \frac{h^\chi}{R h^\psi} d\chi. \tag{92}$$

Applying the toroidal component of Ohm's law, an expression for the velocity of the plasma normal to the flux surface and relative to the flux surface is obtained:

$$v_\psi = (\eta_\perp - \eta_\|)\frac{B_\chi}{B^2} R \frac{\partial p}{\partial \psi} + \frac{\eta_\|}{B_\chi}\left(R \frac{\partial p}{\partial \psi} + \frac{I'}{R}\frac{\partial I'}{\partial \psi}\right),$$

where $\eta_\perp$ is the perpendicular coefficient of resistivity, and the equilibrium equation for the toroidal current [Eq. (51)] has been used in the second term. Using the perpendicular velocity v_ψ, the flux of plasma through the area of each surface may be evaluated to yield a conservative diffusion equation for the mass on each surface:

$$\frac{dm}{dt} = \frac{\partial}{\partial \psi}\left[(\eta_\perp - \eta_\parallel)\lambda \frac{\partial p}{\partial \psi} + \eta_\parallel \Gamma \frac{\partial p}{\partial \psi} + \eta_\parallel m \frac{\partial}{\partial \psi}\left(\frac{\pi^2}{2A^2}\right)\right],$$

where λ and Γ are surface integrals,

$$\lambda(\psi) = \rho(\psi) \oint_\psi \frac{Rh^\chi}{B^2 h^\psi} d\chi, \tag{93}$$

$$\Gamma(\psi) = \rho(\psi) \oint_\psi R^3 h^\chi h^\psi d\chi. \tag{94}$$

The first term corresponds to classical diffusion, while the second term is related to the Pfirsh–Schluter term. The last term results from the inward diffusion of the magnetic surfaces from a net poloidal current.

Diffusion equations for the internal energy of one or more species may be obtained in an entirely similar way (Shafranov, 1965). These equations provide a model for the self-consistent determination of the equilibrium–diffusion problem in an arbitrary axisymmetric torus. The evolution of the mass, the internal energy, and the toroidal flux specify the source terms in the equilibrium equation (52). In turn the solution of the equilibrium equation (Section III) specifies the geometric variables which define the integrals in the diffusion coefficients.

V. Further Applications

A. Remarks on "Equilibrium" and Inertial Models

In this review emphasis has been placed on the application of the waterbag method to low frequency phenomena in toroidal containment. Under these conditions the plasma and magnetic field always remain in momentum equilibrium so that the coordinates of the flux surfaces or waterbags are defined by an elliptic equation, (52). The system, of course, may evolve in time by the time-dependent equations for each surface for the internal energy, density, and magnetic flux (Section IV).

The waterbag method applies equally, however, to higher frequency phenomena, where the assumption of a balance between field and plasma no longer remains valid. In this instance the inertia of the fluid, and the velocity field, must be included explicitly. The equations become hyperbolic and more conventional in form. However, it remains true that the plasma properties are anisotropic in following the flux surfaces and the waterbag method provides a multidimensional Lagrangian method for studying such a plasma. The fluid equations are expressed in the local orthogonalized Lagrangian frame and resolved by conventional hyperbolic difference methods. From the point of view of containment, this approach is particularly powerful for studying the nonlinear development of magnetohydrodynamic instabilities in low-beta plasma. For completeness some aspects of the formulation of the low-beta equations are here included.

B. Quasi-Incompressibility in the Low-Beta Approximation

In the low-beta limit, the dynamics of the fluid may be simplified *ab initio* by noting that the motion of the fluid in the plane perpendicular to the magnetic field is not independent. The situation is entirely analogous to the incompressible motion of a slowly moving fluid in hydrodynamics. In the MHD case, the fluid does not have sufficient energy to compress the vacuum field. In the Tokamak, the vacuum field is the toroidal field B_ϕ,

$$B_\phi = I/2\pi R,$$

so that the plasma currents are small,

$$\delta I \ll I. \tag{95}$$

By considering Faraday's law for the toroidal flux for a conducting fluid:

$$dI/dt + I[R^2 \nabla \cdot (\mathbf{v}/R^2)] = 0,$$

a quasi-incompressible condition is obtained in the limit of low beta [Eq. (95) (Tuttle *et al.*, 1974)],

$$\nabla \cdot (\mathbf{v}/R^2) = 0. \tag{96}$$

The effect of assuming this constraint is to allow the Alfven speed associated with the toroidal field to become infinite.

Equation (96) may be integrated to define the velocity in terms of a pseudo-scalar stream function Φ and the parallel velocity:

$$\mathbf{v} = -\mathbf{e}_\phi \wedge R\nabla\Phi + v_\parallel (\mathbf{B}/B) + O(\beta v), \tag{97}$$

where $\mathbf{e}_\phi$ is a unit vector in the toroidal direction. The first term may be identified as the "E/B" drift since the stream function is proportional to the electrostatic potential.

The necessary elliptic equation for the stream function may be obtained by taking the appropriate curl of the perpendicular momentum equations which eliminates the magnetic forces from the toroidal field. A "pseudo-scalar" vorticity ξ is thereby defined:

$$\xi = (\nabla \wedge R^2 \rho \mathbf{v})_\phi. \tag{98}$$

The system of equations which describes the dynamics of the fluid in the axisymmetric toroidal problem reduces to

$$d\xi/dt = \partial(R^2\rho, \tfrac{1}{2}v^2) + \partial(R^2, p) + \partial(\psi, RJ_\phi), \tag{99}$$

$$d\psi/dt = \eta\nabla^2\psi, \tag{100}$$

$$\nabla^2\psi = -RJ_\phi, \tag{101}$$

$$\nabla R^3 \rho \nabla \Phi = -\xi, \tag{102}$$

where the ∇^2 operator [Eq. (101)] arises from the toroidal component of the curl curl operator, and the second elliptic equation arises from Eqs. (97) and (98). The terms on the right-hand side of the vorticity equation (99) are Jacobians with respect to the poloidal coordinates describing the centrifugal forces, the toroidal curvature drift, and the force from the poloidal magnetic field.

These equations describe the dynamics of the fluid and flux surfaces in the poloidal plane. In addition, equations for the density, internal energy, and parallel velocity are included and are simplified by the incompressible condition, Eq. (96). Sound waves parallel to the magnetic field persist but their effect perpendicular to the field is only to allow the symmetric oscillation of the toroidal column which may arise from the second term of Eq. (99).

These equations are expressed in the flux space (ψ, χ). If the topology does not change, the waterbags move with the velocity c_ψ according to Eqs. (100) and (53) (Tuttle et al., 1974) given by

$$c_\psi = R(\partial\Phi/h^\chi \, \partial\chi) - \eta(R/h^\chi)(\partial/\partial\psi)(h^\chi/Rh^\psi).$$

The elliptic equation (102) is resolved by the alternating direction implicit method (Potter, 1973). An example of the solution illustrating the toroidal oscillations of the plasma column is illustrated in Fig. 13.

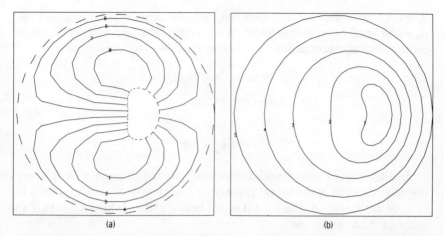

FIG. 13. Contours of the stream function (a) in a Tokamak of aspect ratio 3, illustrating the dipolar flow during the convective oscillation in the major radius of the toroidal plasma column. In (b) the magnetic surfaces are distorted during the convective oscillation.

C. Instabilities Described in the Helical Plane

One particularly important application of the low-beta inertial equations is to the description of nonlinear helical MHD instabilities (White et al., 1974). Instabilities in a toroidal chamber are generally three-dimensional. However, if the aspect ratio is large, the torus may be approximated by a periodic cylinder of length $2\pi/k$ where k is the wavenumber in the toroidal direction. Thus, if the initial equilibrium in the periodic cylinder depends only on the radial coordinate r, helical symmetry is maintained during the nonlinear development of a particular helical instability.

The instability may be denoted by the azimuthal wavenumber m and axial wavenumber k, a multiple of the fundamental $2\pi/R_0$ where R_0 is the length of the cylinder. Then helical coordinates may be defined thus:

$$\begin{aligned} r' &= r, \\ \theta' &= \theta + (k/m)z, \\ \phi' &= (1/R_0)[z - (kr^2/m)\theta]. \end{aligned} \quad (103)$$

The initial equilibrium is independent of ϕ'. The instability depends only on r' and θ' so that helical symmetry is maintained, and the two-dimensional problem is resolved in the helical plane (r', θ') which, in the discussion above (Section V, B), plays the role of the poloidal plane. The length element of the coordinate ϕ' (analogous to the toroidal curvature R) is

$$R = R_0[1 + (kr/m)^2].$$

R is assumed constant, consistent with the assumption of large aspect ratio R_0/r and small kr/m. Furthermore, White et al. (1974) assume that the plasma column is of constant density, ρ_0, so that Eqs. (99)–(102) reduce to

$$\begin{aligned} d\xi/dt &= \partial(\psi, \nabla^2 \psi), \\ d\psi/dt &= \eta \nabla^2 \psi, \\ \nabla^2 \Phi &= -\xi/(\rho_0 R_0), \\ \mathbf{v} &= \mathbf{e}_\phi \wedge \nabla \Phi. \end{aligned} \qquad (104)$$

In the vacuum region no intrinsic currents flow. However, if Ampere's law is considered in the helical plane (r', θ'), a line contour will not be closed in three dimensions due to the cut in the helical plane. Taking account of the connection across the cut,

$$\int_0^{2\pi} B_\theta, r' \, d\theta' = \int_0^{2\pi} B_\theta r \, d\theta + B_z 2\pi (kr^2/m),$$

a component of the axial field exists in the helical plane. Thus an effective current density may be ascribed to the vacuum region

$$\nabla^2 \psi = -B_z(2kR_0/m). \qquad (105)$$

The plasma region is represented by a set of nested waterbags, satisfying the equations (104). The plasma is bounded by an outermost surface C outside of which the vacuum equation (105) is satisfied. Boundary conditions are applied across the plasma–vacuum interface, namely, that the normal components of field and current vanish and that the total pressure is continuous. Consequently the parallel velocity on the vacuum–plasma interface may be defined in terms of the change of poloidal fields across the surface:

$$\rho_0 \, dv_\chi/dt = -(1/R_0^2 h^\chi)(\partial/\partial \chi)[(\nabla \psi)_v^2 - (\nabla \psi)_p^2]. \qquad (106)$$

White et al. (1974) solve the vacuum equation (105) in integral form to define the vacuum field on the surface. This permits the parallel velocity [Eq. (106)] to be determined on the surface. The parallel velocity provides Neumann boundary conditions for the solution of the elliptic equation (104) for the potential. This in turn permits the perpendicular Lagrangian motion of the waterbag surface ψ.

An example of the application of the model to the formation of "magnetic bubbles" is included in Fig. 14. In the figure shown, no resistivity and no shear are included. More generally, however, many surfaces describing a current distribution are included.

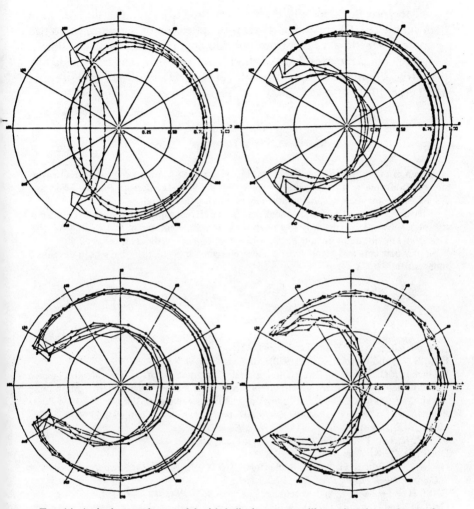

FIG. 14. A single waterbag model with helical symmetry illustrating the tendency of a periodic cylinder to form "bubbles" in the presence of a rational surface ($q_{\text{vacuum}} \sim 1$). The figures illustrate the nonlinear evolution of a single surface in time (White et al., 1974).

D. Concluding Remarks

A very wide application of the waterbag method to containment in closed magnetic surfaces has been illustrated. The method is particularly appropriate to equilibria and low-beta phenomena. Clearly in sonic or supersonic MHD phenomena, where turbulence may occur, the increasing complexity of the field structure disallows the use of the method.

At the present time only the simplest topologies for the surfaces have been considered. Of great importance is the extension of the approach to two or more magnetic axes. Provided the topology remains relatively simple, additional magnetic axes and the effect of a separatrix may be included by the introduction of additional branch cuts in the logical flux space (Potter and Tuttle, 1973).

ACKNOWLEDGMENTS

Much of this work was developed during a year's leave of absence spent at the University of California at Los Angeles. I would like to thank Professors Burton Fried and John Dawson for their hospitality and support during this period.

I have worked with a number of colleagues on different aspects of the work included in this paper. I would particularly like to mention the contributions of Dr. Andrew Wilson, Dr. John Triplett, and Graham Tuttle to the development of these ideas.

Similar concepts and algorithms to those described in Section IV have been developed independently by H. Grad.

REFERENCES

Artsimovich, L. A. (1972). *Nucl. Fusion* **12**, 215.
Berk, H. L., and Roberts, K. V. (1970). *Methods Comput. Phys.* **9**, 88.
Betancourt, O. (1974). "Three Dimensional Computation of Magneto-hydrodynamic Equilibrium of Toroidal Plasma without Axial Symmetry," Courant Inst. Math. Sci., Rep. MF-67 and COO-3077-49, New York University, New York.
Betancourt, O., and Garabedian, P. (1974). *Amer. Phys. Soc. Conf. Plasma Phys., 1974*.
Callan, J. D., and Dory, R. A. (1972). *Phys. Fluids* **15**, 1523.
Chu, C. K., and Taussig, R. T. (1967). *Phys. Fluids* **10**, 249.
Duchs, D. (1968). *Phys. Fluids* **11**, 2010.
Duchs, D, F., Furth, H. P., and Rutherford, P. H. (1972). *Proc. IAEA Conf. Plasma Phys. Control. Nucl. Fusion Res., 4th, 1971* p. 369.
Furth, H. P., Killeen, J., and Rosenbluth, M. N. (1963). *Phys. Fluids* **6**, 459.
Grad, H., and Hogan, J. (1971). *Phys. Rev. Lett.* **24**, 1337.
Hain, K., Hain, G., Roberts, K. V., Roberts, S. J., and Köppendorfer, W. (1960). *Z. Naturforsch A* **15**, 1039.
Jeffrey, A. (1966). "Magnetohydrodynamics." Wiley (Interscience), New York.
Killeen, J., and Lindemuth, I. R. (1973). *J. Comput. Phys.* **13**, 181.
Kruskal, M. D., and Kulsrud, R. M. (1958). *Phys. Fluids* **1**, 265.
McConnell, A. J. (1957). "Application of Tensor Calculus," p. 155, Dover, New York.
Navet, M., and Bertrand, P. (1971). *Phys. Lett. A* **34**, 117.
Oliphant, T. A. (1963). "Numerical Studies of the Theta Pinch," Rep. LAMS-2944, Los Alamos Sci. Lab. Los Alamos, New Mexico.
Potter, D. E. (1971). *Phys. Fluids* **14**, 1911.
Potter, D. E. (1973). "Computational Physics." Wiley (Interscience), New York.
Potter, D. E., and Kamimura, T. (1976). To be published.

Potter, D. E., and Tuttle, G. H. (1973). *J. Comput. Phys.* **13**, 483.
Roberts, K. V., and Potter, D. E. (1970). *Methods Comput. Phys.* **9**, 340.
Shafranov, V. D. (1965). *At. Energ.* **19**, 120.
Tuttle, G. H., Potter, D. E., and Haines, M. G. (1974). *IAEA Conf. Plasma Phys. Control. Nucl. Fusion Res., 5th 1974* Paper CN-33/A17-3.
White, R., Monticello, D., Rosenbluth, M. N., Strauss, H., and Kadometsev, B. B. (1974). *IAEA Conf. Plasma Phys. Control. Nucl. Fusion Res., 5th, 1974* Paper CN-33/A13-3.

Solution of Continuity Equations by the Method of Flux-Corrected Transport

J. P. BORIS AND D. L. BOOK

PLASMA PHYSICS DIVISION
NAVAL RESEARCH LABORATORY
WASHINGTON, D.C.

I. Introduction	85
A. Continuity Equations in Physics	85
B. Requirements for Finite-Difference Algorithms	86
C. Types of Errors To Be Minimized	88
II. Elements of FCT Algorithms	93
A. Development of the Basic Ideas	93
B. Some Comparisons with Standard Techniques	96
C. Three Types of Antidiffusion	97
D. The Flux-Correction Process	100
E. More FCT Algorithms	102
III. Optimization of FCT Algorithms	105
A. Errors and Optimization	105
B. Reduction of Phase and Amplitude Errors	108
C. Algorithm Comparisons and Other Computational Requirements	112
IV. Applications of Flux-Corrected Transport	115
A. CYLAZR, a Two-Dimensional Laser–Target Model	115
B. Moving Flux Coordinate Code	118
C. FCT Vlasov Solver	119
D. Other Applications	126
References	128

I. Introduction

A. CONTINUITY EQUATIONS IN PHYSICS

THIS ARTICLE DESCRIBES A method of solving the continuity equation,

$$\partial \rho / \partial t + \mathbf{V} \cdot \nabla \rho = -\rho \nabla \cdot \mathbf{V}, \tag{1}$$

by a new set of numerical techniques which we have called Flux-Corrected Transport (FCT) Equation (1), which also can be written in the conservation form

$$\partial \rho / \partial t = -\nabla \cdot \rho \mathbf{V}, \tag{2}$$

describes two phenomena taking place simultaneously, convection and compression. Compression, described by the $\rho \mathbf{V} \cdot \mathbf{V}$ term, does not really pose as many numerical difficulties as convection, i.e., propagation along the characteristics. The convective term, $\mathbf{V} \cdot \nabla \rho$ in Eq. (1), makes this equation one of the most difficult partial differential equations of continuum physics to solve stably and accurately.

The finite-difference algorithms which we develop to advance the "density" function $\rho(\mathbf{x}, t)$ in time are specifically aimed at providing versatile models of fluid problems on stationary grids, but apply equally well for Lagrangian models. The FCT techniques use a physical property of the continuity equation, namely, positivity, to construct algorithms more accurate and reliable than straightforward mathematical expansion alone could provide. From a set of seven requirements for excellent continuity equation algorithms, we derive the basic FCT concepts in one dimension and then develop several simple generalizations and variations. These FCT algorithms are developed with a variable coefficient which we then use as a free parameter to minimize the residual numerical errors. Various possible optimizations are compared with each other and with an "optimal" (but rather special purpose) algorithm to select "excellent" continuity equation algorithms. Finally, we apply these new FCT techniques to several problems in CTR and plasma physics. A one-dimensional FCT code illustrating the techniques is available from the authors. In this code, the three conservative continuity equations for mass, momentum, and energy are solved by a single, fairly general FCT module which can be easily optimized for any given computer.

Throughout this article frequent references will be made to a series of papers (Boris and Book, 1973; Book et al., 1975; Boris and Book, 1976) which have appeared in the *Journal of Computational Physics*. For convenience they will be designated FCT/I, FCT/II, and FCT/III, respectively.

B. Requirements for Finite-Difference Algorithms

Over the years many schemes have been tried in attempts to improve numerical solutions of continuity equations. These include characteristic methods (Moretti, 1972), spectral methods (Orszag and Israel, 1972), finite-element methods (Strang and Fix, 1973), splines (Ahlberg et al., 1967), and finite-difference methods (Roache, 1972). Characteristic methods and spectral methods, while giving superb results in some cases, are not generally applicable to complicated nonlinear systems of continuity equations generalized to include, for example, diffusion effects as well as convection and compression. Finite-element and spline approaches also give excellent results in most cases where they are usable but their complexity and computational cost are often prohibitively high. Thus, attention naturally turns time and again to the simple, generally applicable, computationally efficient, finite-difference formulations.

Since finite-difference algorithms are generally simple and fast, it is not too surprising that they are often less accurate for a given spatial resolution than other methods which might be applied if the programmer's patience were longer or his budget larger. Flux-corrected transport is a rather natural step to take when we systematically try to improve basic finite-difference schemes without sacrificing too much of their simplicity or speed. The basic ideas leading to FCT can be "derived" from a set of seven requirements which an "excellent" finite-difference algorithm should satisfy. These requirements, reproduced below, are discussed in FCT/III.

1. Exact conservation properties of the physical equation should be mirrored in the finite-difference approximations.
2. The algorithm should ensure stability of all the harmonics in some useful range of the grid spacing δx and the time-step δt.
3. The positivity (nonnegativity) property of $\rho(x, t)$ in Eq. (1) should be preserved. Thus if the density ρ is positive and decreasing (in time), it stops changing as it approaches zero density.
4. The algorithm should not be built around special properties such as giving exactly the correct answer when $V\delta t/\delta x = 1$.
5. The overall algorithm should be effectively second order in regions of the problem where the concept of order is related usefully to accuracy. This requirement is included to provide a minimal long-term accuracy free of at least the worst types of secularity.
6. The algorithm should leave the numerical density profile $\{\rho_j\}$ undisturbed when the flow velocity is zero.
7. The algorithm should have a single- or double-step time integration to ensure simple, fast, efficient calculations.

These requirements are listed roughly in order of decreasing importance. The last four refer to rather nebulous concepts concerning flexibility and accuracy, so one expects that they could be stretched a bit if the need arises and the payoff achieved warrants. The first three requirements governing conservation, stability, and positivity would appear to be absolutely vital to a general purpose, reliable continuity-equation algorithm. Nevertheless, it has been the custom in the past to neglect the nonnegative properties in favor of solving the equations to some specified asymptotic mathematical order. People have argued that sufficiently improved resolution will always remove nonphysical behavior such as might be obtained by a rigorous Taylor series expansion to second order. In fact, requirements 3 and 5 are mutually exclusive in regions where gradients are large or where the density goes to zero.

Only by sacrificing the second-order requirement when the gradients are very steep can we hope to obtain a nonnegative continuity-equation algorithm.

The continuity equation written in the form of Eq. (1) suggests the use of Lagrangian methods, in which the finite-difference grid moves with the local fluid velocity, to circumvent the positivity problem. The rate of change of density in a moving frame of reference is usually easy to compute because the convective term is absent in the moving frame. Such Lagrangian approaches are often useful and sometimes necessary. In contrast, Eulerian methods are needed in long calculations, where complicated flows would distort a Lagrangian grid incredibly, or in steady-state calculations, where one would like to choose a fixed grid structure and allow the flow to approach a steady state while moving through that grid. We concentrate our interest on Eulerian solutions to Eq. (1), because the Lagrangian methods often have serious computational drawbacks and because the Lagrangian approach still has unavoidable Eulerian features to it when used to solve a coupled set of fluid equations (see FCT/III).

For example, in ideal hydrodynamics the velocity field which makes the energy Lagrangian is not the same as the velocity which makes the density Lagrangian unless the pressure is identically zero. Thus, in a frame of reference moving with the fluid, energy and momentum flow across cell boundaries. In a frame of reference moving with the energy, density and momentum flow across cell boundaries. Convection of some fluid quantities across cell boundaries cannot be eliminated; thus the good convective treatments needed in the Eulerian description are still essential even in Lagrangian models.

This emphasis on the Eulerian nature of most coupled fluid problems and this argument against reliance on Lagrangian techniques as a panacea for convective flows should not be construed as implying a limitation on FCT algorithms. Flux-corrected transport has been applied to Lagrangian calculations without loss of accuracy and with the added good features of the Lagrangian grid. However, the development of the ideas behind FCT is greatly simplified by considering only the Eulerian problem at first.

C. Types of Errors To Be Minimized

Since the algorithms for solving the compressible part of the continuity equation in Eulerian form are generally much more accurate and less troublesome than solutions of the convective part, much of our analysis will be cast in terms of a one-dimensional continuity equation in which the velocity divergence is zero. The equation then becomes linear with constant coefficients. As an immediate consequence of these restrictions the analytic solution of Eq. (1) is

$$\rho(x,t) = \rho(x - V_0 t, 0), \qquad (3)$$

where we are given the initial profile $\rho(x,0)$ and V_0 is the constant velocity. We assume that our knowledge of $\rho(x,0)$ is itself limited, extending only to the initial values of ρ on a set of N discrete grid points with separation δx. These values we denote as $\{\rho_j^0\}$ ($0 \leq j \leq N$) where $\rho_N^0 = \rho_0^0$. Furthermore, we only expect our finite-difference algorithms to yield approximations to $\rho(x,n\delta t)$ at discrete times $t = 0, \delta t, 2\delta t, \ldots$ on the same discrete spatial grid; these approximate solutions of Eq. (3) we denote by $\{\rho_j^n\}$.

In this article we use the squarewave test problem introduced in FCT/I to illustrate the behavior of our various algorithms and algorithm modifications. As shown in Fig. 1a, the squarewave test has 100 grid points in periodic geometry. A square step 20 points wide is propagated at a constant velocity. Initially the 100 values $\{\rho_j^n\}$ ($j = 1, 2, \ldots, 100$) are as shown. At each cycle of a test calculation 100 new values $\{\rho_j^n\}$ have to be determined unambiguously. The error at any time-step n is given by

$$\text{A.E.} \equiv \frac{1}{100} \sum_{j=1}^{100} |\rho_j^n - \rho_j^{\text{exact}}|, \tag{4}$$

an average absolute error criterion which measures the difference between the known exact solution and the numerically computed solution.

Under the above assumptions and simplifications one can Fourier analyze $\rho(x,t)$ in space using a discrete transform.

$$\rho_j^n = \sum_{k=-N/2+1}^{N/2} \hat{\rho}_k^n e^{2\pi i k j \delta x / L}, \tag{5}$$

where $L = N\delta x$ is the length of the system. This has the inverse finite discrete transform

$$\hat{\rho}_k^n \equiv \frac{1}{N} \sum_{j=0}^{N-1} \rho_j^n e^{-2\pi i k j \delta x / L}, \tag{6}$$

where reality requires $\hat{\rho}_k^n = (\hat{\rho}_{-k}^n)^*$. Since there are only N independent real values of $\{\rho_j^n\}$ at time-step n, Eq. (6) can only define N distinct linear combinations. The real and imaginary components of the finite-discrete Fourier harmonics $\{\hat{\rho}_j^n\}$ are truly independent only for $0 \leq k \leq N/2$.

The advantage of Fourier analysis lies in the particularly simple form of the closed solution, Eq. (3), in k-space with time discretization

$$\hat{\rho}_k^n = \hat{\rho}_k^0 e^{-2\pi i k V_0 n \delta t / L}. \tag{7}$$

Of course Eq. (7) would only be true if the algorithm were "optimal." That is,

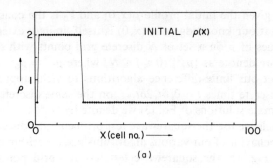

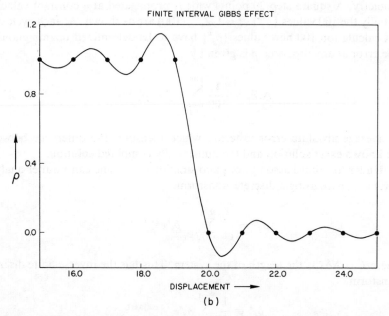

FIG. 1. (a) Initial conditions for the squarewave test problem. The velocity is constant in space and time and the boundary conditions on the 100-point mesh are periodic. The density squarewave of height 2.0 is 20 cells across. The background density is 0.5. The squarewave should propagate across the grid unchanged. (b) Plot of $\rho(x)$ versus x near the edge of a step in density. Numerical grid values are indicated by dots. The oscillations arise from the Gibbs phenomenon and can lead to negative densities as shown.

Eq. (7) holds for each harmonic after n time-steps only if the algorithm propagates the harmonic phases at exactly V_0 and leaves the harmonic amplitudes unchanged in magnitude.

In general Eq. (7) is not satisfied by a finite-difference algorithm, so it is valuable to catalog the types of numerical error which can occur. We have identified three potential linear sources of error:

1. Amplitude errors—called damping if the harmonics decay and called instability if the harmonics grow in magnitude. These errors are usually most severe for short wavelength harmonics.
2. Phase errors—generally called dispersion. The harmonics propagate at the wrong speed and the error usually increases with the wavenumber k.
3. Gibbs errors, a form of subgrid uncertainty, arise because the density profile is numerically unknown over length scales shorter than the grid spacing δx. The consequence is nonphysical oscillations in the Fourier representation.

Damping, the first type of error, dominates many numerical solutions of the continuity equation which must display a stable nonnegative solution. Numerical diffusion in continuity equation algorithms is a particular form of damping arising from unwanted error terms which appear because of spatial averaging. These errors are called numerical diffusion because the error terms can usually be written in the form of a three-point diffusion approximation,

$$(D\delta t/\delta x^2)(\rho_{j+1} - 2\rho_j + \rho_{j-1}). \tag{8}$$

The major consequence of numerical damping or diffusion is the smoothing of short wavelength features, some of which may be physically important.

The usual measure of damping for an algorithm is a quantity called the amplification factor. This is just the magnitude $|A_k|$ of the algorithm's transfer function for one time-step,

$$A_k \equiv \hat{\rho}_k^1/\hat{\rho}_k^0, \tag{9}$$

which is defined for each harmonic. When the harmonic is damped, $|A_k| < 1$. When the harmonic is unstable, $|A_k| > 1$. When the algorithm has zero residual damping (ZRD), $|A_k| = 1$ for all Fourier harmonics. When the damping is due entirely to an explicit diffusive term such as Eq. (8), the amplification factor becomes

$$A_k = 1 - 2D(1 - \cos k\delta x). \tag{10}$$

The damping gets progressively worse at short wavelength in Eq. 10.

Why is numerical damping such a serious problem in many algorithms? Instability is disastrous, and therefore to be avoided at any cost. It is usually necessary to have $|A_k|$ substantially smaller than unity for large k just to ensure that the stability requirement is satisfied for all k. The donor-cell algorithm (Courant *et al.*, 1952), also called one-sided, upstreamed, or windward differencing, was used as a test in FCT/I and FCT/III, and is a good example of excessive numerical damping. FCT derives much of its power, as we will show, by making use of this excess damping rather than just living with it.

The second source of error which can plague numerical solutions of the continuity equation is numerical dispersion. Dispersive errors occur when the phase velocity of some harmonics differs from the velocity of the flow V_0 in Eq. (7). This type of error occurs because the space and time derivatives in the continuity equation are being approximated by finite differences and hence the harmonic phase velocity is no longer correct and independent of the harmonic index k. Phase errors were discussed very briefly in FCT/I, where we noted that diffusion and antidiffusion in certain FCT algorithms has a beneficial effect on phase errors, reducing them by a factor of 4. To study the numerical phase errors further in this paper we define a relative phase error for each harmonic of a given algorithm,

$$R(k) \equiv [V_\phi(k) - V_0]/V_0, \qquad (11)$$

where $V_\phi(k)$ is the numerical phase velocity and V_0 is the correct phase velocity.

The third type of error, the Gibbs phenomenon, is displayed in Fig. 1b. If, in Eq. (5), x is written in place of $j\delta x$, the result is a continuous function $\rho''(x)$ which satisfies $\rho''(x_j) = \rho_j^n$ on the grid points. The oscillations arise because the short wavelength harmonics are truncated in the discrete expansion. Furthermore, these oscillations do not vanish as the resolution is improved. If $V_0 \delta t$ is an integer multiple of δx, propagation according to Eq. (7) yields exactly a discretized version of (3). If the profile is displaced by a fraction of a cell in one time-step, however, the values ρ_j^n on the mesh will exhibit spurious wiggles. These oscillations are errors in the sense that they can give rise to negative densities. Since they really appear in the finite representation and are not artifacts of any particular algorithm, they can be viewed as irreducible in the Eulerian, or fixed grid, representation.

Before deriving the basic concepts of FCT, we give a brief explanation of our notation. The sequence of values ρ_j^n are the values of the density profile we are calculating at time-step n and grid point j. Here ρ indicates a generic conserved quantity such as mass, momentum, or energy. Since much of our analysis considers only one time-step, ρ_j^0 and ρ_j^1 are used for the profile values at the beginning and the end of the time-step, respectively. The quantities $\tilde{\rho}_j$ and ρ_j' are used to indicate certain intermediate or provisional values of the density. Wherever k appears as a subscript or argument, it refers to the wavenumber of the particular harmonic being considered. Generally we will use $\varepsilon \approx V\delta t/\delta x$ as a nondimensional transport coefficient, ν as a nondimensional diffusion coefficient, and μ as a nondimensional antidiffusion coefficient.

II. Elements of FCT Algorithms

A. Development of the Basic Ideas

Using the seven requirements given in Section I, the basic concepts of flux-corrected transport can be derived in a rather straightforward way. Simplicity leads us to start with an explicit three-point approximation to the continuity equation,

$$\tilde{\rho}_j = a_j \rho_{j-1}^0 + b_j \rho_j^0 + c_j \rho_{j+1}^0. \tag{12}$$

Conservation of mass (assuming a uniform grid for now) requires that

$$a_{j+1} + b_j + c_{j-1} = 1, \tag{13}$$

while positivity requires that a_j, b_j, and c_j be nonnegative for all j. Equation (12) may be rewritten in a simple conservative form which automatically satisfies Eq. (13).

$$\begin{aligned}\tilde{\rho}_j = \rho_j^0 &- \tfrac{1}{2}[\varepsilon_{j+1/2}(\rho_{j+1}^0 + \rho_j^0) - \varepsilon_{j-1/2}(\rho_j^0 + \rho_{j-1}^0)] \\ &+ [v_{j+1/2}(\rho_{j+1}^0 - \rho_j^0) - v_{j-1/2}(\rho_j^0 - \rho_{j-1}^0)].\end{aligned} \tag{14}$$

For each mesh point j, Eq. (14) has only two independent parameters. Term identification between Eq. (12) and (14) gives

$$a_j = v_{j-1/2} + \tfrac{1}{2}\varepsilon_{j-1/2},$$
$$b_j = 1 - \tfrac{1}{2}\varepsilon_{j+1/2} + \tfrac{1}{2}\varepsilon_{j-1/2} - v_{j+1/2} - v_{j-1/2}, \tag{15}$$

and

$$c_j = v_{j+1/2} - \tfrac{1}{2}\varepsilon_{j+1/2}.$$

In Eq. (14), $\{v_{j+1/2}\}$ are dimensionless diffusion coefficients included to ensure positivity for the provisional new densities $\{\tilde{\rho}_j\}$, given positivity of $\{\rho_j^0\}$. The positivity condition derived from Eq. (15) is

$$v_{j+1/2} \geq \tfrac{1}{2}|\varepsilon_{j+1/2}|, \quad \text{all } j. \tag{16}$$

As long as Eq. (16) is satisfied, the provisional new values $\{\tilde{\rho}_j\}$ will be nonnegative. After the constraint (13) is enforced, two of the three coefficients in Eq. (12) remain to be determined. One of these two degrees of freedom must be fixed to ensure an accurate representation of the mass flux terms. Thus in the more transparent conservative form Eq. (14), $\varepsilon_{j+1/2} \approx V_{j+1/2} \delta t/\delta x$, where

$V_{j+1/2}$ is the flow velocity approximated halfway between the grid points j and $j+1$. The other coefficients $\{v_{j+1/2}\}$ are subject only to requirement (16) maintaining positivity.

As an immediate consequence of Eq. (16) the provisional values $\{\tilde{\rho}_j\}$ calculated from Eq. (14) must be strongly diffused to ensure positivity. Equality in Eq. (16) gives the heavily diffusive donor-cell algorithm. Since any other positivity-preserving choices for $\{v_{j+1/2}\}$ can only be more diffusive, it would appear that positivity and accuracy are mutually exclusive. An obvious correction to get around this overstrong diffusion (discussed in FCT/I) is

$$\rho_j^1 = \tilde{\rho}_j - \mu_{j+1/2}(\tilde{\rho}_{j+1} - \tilde{\rho}_j) + \mu_{j-1/2}(\tilde{\rho}_j - \tilde{\rho}_{j-1}), \quad (17)$$

where $\{\mu_{j+1/2}\}$ are positive antidiffusion coefficients. Antidiffusion reduces the strong diffusion implied by Eq. (16) but also reintroduces the possibility of negative densities in the new profile $\{\rho_j^1\}$. Instability is even possible if the values $\mu_{j+1/2}$ are too large. The only way out of this apparent bind is to modify the antidiffusive fluxes in Eq. (17). We called this flux correction (or flux limiting) because the antidiffusive fluxes

$$\phi_{j+1/2} = \mu_{j+1/2}(\tilde{\rho}_{j+1} - \tilde{\rho}_j) \quad (18)$$

appearing in Eq. (17) have to be corrected (limited) to assure positivity as well as stability.

The best linear choice of $\{\phi_{j+1/2}\}$ which still preserves positivity is $\mu_{j+1/2} \approx v_{j+1/2} - \frac{1}{2}|\varepsilon_{j+1/2}|$. This is not good enough. To reduce the residual diffusion $(v - \mu)$ further, the flux correction must be nonlinear, i.e., must depend on the actual values of the density $\tilde{\rho}_j$. The idea for the correct nonlinear flux-limiting formula arises as follows: suppose the density $\tilde{\rho}_j$ at grid point j reaches zero sooner than its neighbors. Then the second derivative is locally positive and so any finite antidiffusion would force the minimum density value $\tilde{\rho}_j = 0$ to be negative. Since this cannot be allowed physically, the antidiffusive fluxes should be limited so that minima in the profile can be made no deeper by the antidiffusive stage defined by Eq. (17). Since the continuity equation is linear, we could equally well solve for $\{-\tilde{\rho}_j^1\}$. Hence, we also must require that maxima in the profile be made no higher by the antidiffusion. Combining these two conditions, we arrive quite naturally at the prescription which forms the basis for Flux-Corrected Transport (FCT/I):

The antidiffusion stage should generate no new maxima or minima in the solution, nor should it accentuate already existing extrema.

This qualitative formulation for nonlinear filtering can be quantified easily.

The new flux-corrected transport values $\{\rho_j^1\}$ are given by

$$\rho_j^1 = \tilde{\rho}_j - \tilde{\phi}_{j+1/2} + \tilde{\phi}_{j-1/2}, \tag{19}$$

where the corrected fluxes $\tilde{\phi}_{j+1/2}$ should satisfy

$$\tilde{\phi}_{j+1/2} = S \cdot \max\{0, \min[S \cdot (\tilde{\rho}_{j+2} - \tilde{\rho}_{j+1}), |\phi_{j+1/2}|, S \cdot (\tilde{\rho}_j - \tilde{\rho}_{j-1}]\}. \tag{20}$$

Here $|S| = 1$ and sign $S \equiv \text{sign}(\tilde{\rho}_{j+1} - \tilde{\rho}_j)$.

To see what this flux-correction formula is doing, assume that $(\tilde{\rho}_{j+1} - \tilde{\rho}_j)$ is greater than zero. Then Eq. (20) gives

$$\tilde{\phi}_{j+1/2} = \min[(\tilde{\rho}_{j+2} - \tilde{\rho}_{j+1}), \mu_{j+1/2}(\tilde{\rho}_{j+1} - \tilde{\rho}_j), (\tilde{\rho}_j - \tilde{\rho}_{j-1})]$$

or zero, whichever is larger. The "raw" antidiffusive flux $\mu_{j+1/2}(\tilde{\rho}_{j+1} - \tilde{\rho}_j)$ always tends to decrease ρ_j^1 and to increase ρ_{j+1}^1. The flux-limiting formula merely ensures that the corrected flux cannot push ρ_j^1 below ρ_{j-1}^1 (which would be a new minimum) nor push ρ_{j+1}^1 above ρ_{j+2}^1 (which would give a new maximum). The general formula Eq. (20) is constructed to take care of all cases of sign and slope.

The full formulation of a flux-corrected transport algorithm consists of the following four sequential stages:

1. Compute the transported and diffused values $\{\tilde{\rho}_j\}$ from Eq. (14) where the diffusion coefficients $\{v_{j+1/2}\}$ satisfy Eq. (16). The optimum choice of the transport coefficients $\{\varepsilon_{j+1/2}\}$ and diffusion coefficients $\{v_{j+1/2}\}$ will be left for later.
2. Compute the raw antidiffusive fluxes from Eq. (18).
3. Compute the corrected fluxes using Eq. (20).
4. Perform the indicated antidiffusion via Eq. (19).

Steps 3 and 4 are the ones uniquely characteristic of an FCT algorithm. The particular versions employed in the above example are called explicit antidiffusion, Eq. (18), and strong two-sided flux correction (FCT/II), Eq. (20). There are many minor modifications which accentuate various properties of the solutions. Several of these are discussed in Sections II,D and II,E below. Historically the above reasoning was first applied in FCT/I to a three-point diffusive transport scheme called SHASTA. Subsequent work, as reported in FCT/II and FCT/III and surveyed here, generalized the concepts to many different basic transport schemes, to three types of antidiffusion, and to several variants of the flux-correction formula.

B. Some Comparisons with Standard Techniques

Before treating more advanced concepts in flux-corrected transport, we will compare the simple, explicit SHASTA algorithm with some standard algorithms which are not flux-corrected. For the squarewave test, Fig. 2 compares the usual leapfrog (Courant *et al.*, 1928) and two-step Lax–Wendroff (Lax and Wendroff, 1960) algorithms with and without added numerical smoothing after 800 steps. When $v = 0$ in both cases, the added damping is zero. The dispersive errors dominate here and nonpositive transport occurs. In both of these standard second-order algorithms somewhat smaller values of A.E. can be achieved by adding a nonphysical damping in the form of three-point diffusion. The large errors from short wavelength dispersion, reflected in the figure, are reduced when these harmonics are numerically damped. Too much damping, however, is often the bane of numerical algorithms. Figure 3a shows the donor-cell algorithm (flux-uncorrected). This algorithm has second-order relative phase errors but has strong first-order numerical damping coefficients. The value A.E. = 0.260 is larger than for the standard unsmoothed Lax–Wendroff and leapfrog algorithms (A.E. = 0.175 and A.E. = 0.245, respectively) and much larger than the value for the SHASTA explicit FCT algorithm which is shown in Fig. 3b.

The comparison of these three rather standard treatments with the SHASTA FCT solution for the same problem graphically illustrates the reason for the strong interest in FCT algorithms. The numerical error in FCT appears as a residual smoothing or diffusion near discontinuities and large gradients. Potentially disastrous dispersive ripples as in Fig. 2 or an unnaturally broadened profile as in Fig. 3a do not appear.

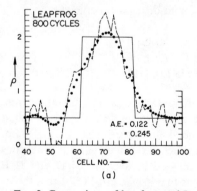
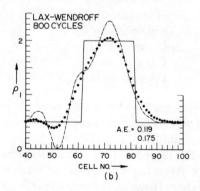

FIG. 2. Comparison of leapfrog and Lax–Wendroff algorithms on the squarewave test. (a) The reversible leapfrog algorithm with and without additional diffusive damping. Dotted line, $v = 0.01$; dashed line, $v = 0$. (b) The Lax–Wendroff algorithm with and without additional diffusive damping. The dispersion errors are so bad that added nonphysical damping actually improves the solution. Dotted line, $v = 0.01 \pm \frac{1}{2}\varepsilon^2$; dashed line, $v = 0$.

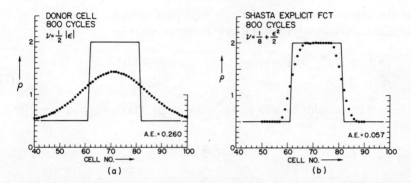

FIG. 3. Comparison of the noncorrected donor-cell algorithm, (a), with the original SHASTA calculation using FCT, (b). The improvement with FCT, a factor of almost 5 quantitatively, is qualitatively striking.

We chose this particularly simple test problem with constant velocity because the theoretical analysis of various techniques is quite easy and because the difficulties associated with Eulerian convection remain. What we have lost from the test problem are the phenomena associated with fluid compressibility, i.e., with a variable velocity field. However, as long as conservation is maintained in the general variable-velocity algorithms and no special properties of the $V = $ const. problem are invoked, the reintroduction of variable velocity below poses no great difficulty.

C. Three Types of Antidiffusion

The explicit FCT algorithm has one drawback. Even though its errors are three to five times smaller than the standard algorithms, requirement 6 is not satisfied. The numerical errors made in computing changes in ρ when the velocity is zero appear as "residual diffusion." Such errors result from the explicit form of the antidiffusion formula, Eq. (17). To minimize "residual diffusion," we first perform an amplification factor analysis on the general explicit three-point formula, Eq. (14). When the velocity is a constant V_0 and $\varepsilon \equiv V_0 \, \delta t/\delta x$, Eq. (14) simplifies to

$$\tilde{\rho}_j = \rho_j^{\,0} - \tfrac{1}{2}\varepsilon(\rho_{j+1}^0 - \rho_{j-1}^0) + \nu(\rho_{j+1}^0 - 2\rho_j^{\,0} + \rho_{j-1}^0), \tag{21}$$

where $\nu \geqslant \tfrac{1}{2}|\varepsilon|$ for positivity.

We treat each harmonic separately, choosing profiles $\rho_j^{\,0} = \rho^0 e^{ij\beta}$, where $\beta = k\delta x$. Substituting this profile in Eq. (21) gives ($\tilde{\rho}_j = \tilde{\rho} e^{ij\beta}$)

$$\tilde{\rho} = \rho^0[1 - 2\nu(1 - \cos\beta) - i\varepsilon \sin\beta] \tag{22}$$

as the explicit three-point diffusive transport formula. The square of the amplification factor for this diffusive transport stage is

$$|\tilde{A}|^2 \equiv |\tilde{\rho}/\rho^0|^2 = 1 - (4v - 2\varepsilon^2)(1 - \cos\beta) + (4v^2 - \varepsilon^2)(1 - \cos\beta)^2. \tag{23}$$

The general stability as well as positivity of Eq. (33) is assured when $|\varepsilon| < 1$ and

$$\tfrac{1}{2} > v > \tfrac{1}{2}|\varepsilon| > \tfrac{1}{2}\varepsilon^2. \tag{24}$$

The linear stability of the whole algorithm must take into account the antidiffusion of Eq. (17) as well as the diffusive transport stage. The transfer function for Eq. (17) is

$$A^1 = (\rho^1/\tilde{\rho}) = 1 + 2\mu(1 - \cos\beta). \tag{25}$$

This must be squared and multiplied by Eq. (37) to give the overall squared amplification factor for explicit three-point transport with explicit antidiffusion,

$$|A|^2 = [1 - (4v - 2\varepsilon^2)(1 - \cos\beta) + (4v^2 - \varepsilon^2)(1 - \cos\beta)^2]$$
$$\times [1 + 2\mu(1 - \cos\beta)]^2. \tag{26}$$

What is the largest value of the antidiffusion coefficient μ which can be used that still gives overall stability when v is larger than the limiting value $\tfrac{1}{2}\varepsilon^2$? When β is small but increasing, $|A|^2$ should decrease from unity for stability, rather than increase. Thus overall stability at long wavelengths requires

$$\mu \leqslant v - \varepsilon^2/2. \tag{27}$$

In addition, if $v \leqslant \tfrac{1}{2}(1 + \varepsilon^2)$, then the limiting value of μ given by Eq. (27) is stable for all wavelengths. To summarize, for explicit antidiffusion and the explicit three-point formula, the diffusion coefficient v must satisfy

$$\tfrac{1}{4}(1 + \varepsilon^2) \geqslant v \geqslant \tfrac{1}{2}|\varepsilon| \tag{28}$$

for positivity and stability, with the antidiffusion coefficient μ given by equality in Eq. (27), and the nondimensionalized velocity ε limited by $|\varepsilon| < 1$.

We have already stated that the finite-difference algorithm should not perturb $\{\rho_j{}^0\}$ when the velocities are zero. When $\varepsilon = 0$ and $\mu = v - \tfrac{1}{2}\varepsilon^2$ are substituted into Eq. (26) we find

$$|A|^2 = [1 - 2\nu(1 - \cos\beta)]^2 [1 + 2\nu(1 - \cos\beta)]^2$$
$$= [1 - 4\nu^2(1 - \cos\beta)^2]^2. \qquad (29)$$

Equation (29) states that there is a residual damping of all finite wavelength harmonics unless the diffusion coefficient ν goes to zero with ε. This residual damping is linear and appears even without the nonlinear flux-correction process which leaves still more residual damping near extrema. Although this nonphysical damping is quite small, of order β^4 for long wavelengths, we would like to remove it altogether to satisfy requirement 6.

One way to do this, as suggested in FCT/I, involves implicit antidiffusion. Equation (17) is replaced by

$$\bar{\rho}_j = \tilde{\rho}_j - \mu_{j+1/2}(\bar{\rho}_{j+1} - \bar{\rho}_j) + \mu_{j-1/2}(\bar{\rho}_j - \bar{\rho}_{j-1}), \qquad (30)$$

where a tridiagonal set of equations must be solved to find $\{\bar{\rho}_j\}$. The raw antidiffusive fluxes in the case of implicit antidiffusion are defined as

$$\phi_{j+1/2} = \mu_{j+1/2}(\bar{\rho}_{j+1} - \bar{\rho}_j) \qquad (31)$$

instead of Eq. (18). The rest of the flux-correction procedure remains the same and the analog of Eq. (26) for implicit antidiffusion is

$$|A|^2 = \frac{[1 - (4\nu - 2\varepsilon^2)(1 - \cos\beta) + (4\nu^2 - \varepsilon^2)(1 - \cos\beta)^2]}{[1 - 2\mu(1 - \cos\beta)]^2}, \qquad (32)$$

where the antidiffusive factor appears in the denominator rather than in the numerator. In Eq. (32) the upper limit (and hence best) value of μ is again $\nu - \varepsilon^2/2$. When this value is used, the algorithm is stable for all allowed β and ε and the residual damping vanishes when ε is zero.

Using implicit antidiffusion we have gained a smaller residual damping in all cases because the antidiffusive fluxes from Eqs. (31) and (30) are larger than from Eqs. (18) and (17). The zero damping requirement at $\varepsilon = 0$ is also achieved. We have had to pay extra for these benefits: the implicit tridiagonal equations (31) require extra computation and may be relatively expensive on vector and pipeline computers. Furthermore, the algorithm is now nonlocal and this might cause problems with boundaries, source terms, etc.

To develop a local form of flux correction which has zero residual damping at $\varepsilon = 0$, one must calculate raw antidiffusive fluxes as the sum of two terms, i.e.,

$$\phi_{j+1/2} = \mu_{j+1/2}(\rho_{j+1}^0 - \rho_j^0) + \mu_{j+1/2}(\Delta\tilde{\rho}_{j+1} - \Delta\tilde{\rho}_j) \qquad (33)$$

where $\Delta\tilde{\rho}_j$ is the difference between $\tilde{\rho}_j$ and the values one would get using Eq. (14) with $\{\varepsilon_{j+1/2}\}$ set to zero. That is,

$$\Delta\tilde{\rho}_j = \tilde{\rho}_j - \rho_j^{\,0} - v_{j+1/2}^0(\rho_{j+1}^0 - \rho_j^{\,0}) + v_{j-1/2}^0(\rho_j^{\,0} - \rho_{j-1}^0). \tag{34}$$

Here $\{v_{j+1/2}^0\}$ has no velocity dependence, being the limit of $\{v_{j+1/2}\}$ for zero velocity. Clearly $\Delta\tilde{\rho}_j$ vanishes when $V = 0$, so the antidiffusive fluxes remaining from Eq. (33) just cancel diffusion in Eq. (14), giving the desired result.

In the constant velocity limit the overall amplification factor for explicit three-point transport with antidiffusion given by Eq. (33) is

$$(\rho^1/\rho^0) = A = 1 - 2(v - \mu)(1 - \cos\beta) - 4\mu(v - v^0)(1 - \cos\beta)^2$$
$$- i\varepsilon \sin\beta[1 + 2\mu(1 - \cos\beta)].$$

This differs slightly from the form given in FCT/III since here we have included in v the velocity-dependent diffusion term $\varepsilon^2/2$. Choosing $\mu = v - \tfrac{1}{2}\varepsilon^2$ again gives minimum residual damping for finite ε and no linear damping for $\varepsilon = 0$.

$$(\rho^1/\rho^0) = A = 1 - \varepsilon^2(1 - \cos\beta) - 4(v - \varepsilon^2/2)(v - v^0)(1 - \cos\beta)^2$$
$$- i\varepsilon \sin\beta[1 + (2v - \varepsilon^2)(1 - \cos\beta)]. \tag{35}$$

This kind of antidiffusion has been called "phoenical" because the undamped solution "rises whole from the ashes" as did the phoenix.

Figure 2 of FCT/II shows the results of using phoenical FCT on the squarewave test problem with the original SHASTA algorithm. The phoenical version shows a slightly smaller value of A.E. (0.052) than the explicit version with A.E. = 0.057. The implicit version, shown in FCT/I, is better still with A.E. = 0.049. These average absolute error values are almost inversely proportional to the amount of time needed to perform the calculations.

D. The Flux-Correction Process

The strong flux correction obtained by using Eq. (20) is fast-running and easy to code but other prescriptions with useful properties are possible. In the previous subsection, we showed how strong flux correction guarantees that no local extrema, or "ripples," result from antidiffusion. Of course local extrema can form and increase during the diffusive transport stage as physics dictates.

Errors can arise during antidiffusion in two ways: the raw flux $\phi_{j+1/2}$

may not yield the correct antidiffusion even when the flux limiter plays no role; or the flux limiter, which is decidedly pessimistic, may overcorrect the fluxes and leave an unnecessarily large net diffusion. We have already treated the former problem. Now we examine the nonlinear errors introduced by the flux limiter itself. These errors are exhibited primarily in a phenomenon called "clipping." Clipping results from the property of Eq. (20) which prevents existing extrema from being enhanced in the antidiffusive stage (FCT/II). Clipping is shown clearly in Fig. 7 of FCT/II.

The most useful alternative to strong flux correction is one-sided flux limiting. In problems where the solution is known to be positive and to have the form of a single pulse or peak growing out of a more or less uniform background, FCT wipes out the train of dispersive ripples which would otherwise trail the peak. However, the height of the maximum can be too low by 30–40% in the case of very narrow peaks because of clipping from strong flux correction. If maxima are allowed to appear or grow while *minima* are restricted, as in strong flux correction, the algorithm retains both positivity and stability. Intuitively, this reflects the fact that shifting mass from a mesh point to its neighbor in such a way that the latter rises dramatically makes the former decrease by an equal amount, creating a local minimum. Preventing the formation of minima implies at least some measure of control on the formation of maxima. The prescription for one-sided flux limiting to replace Eq. (20) is

$$\tilde{\phi}_{j+1/2} = S \cdot \max\{0, \min[S \cdot (\rho^1_{j'+1} - \rho^1_j), |\phi_{j+1/2}|]\}, \qquad (36)$$

with $j' = j - S$, $S = \text{sign}(\tilde{\rho}_{j+1} - \tilde{\rho}_j)$, and $\phi_{j+1/2}$ as defined previously. Use of Eq. (36) results in profiles where the troughs can be clipped, but not the peaks. If ρ is strictly negative, it is natural to redefine $j' = j + S$. This has the effect of allowing the deepening of minima but not the growth of maxima. If ρ can take either sign and we are interested in accurately calculating the regions of largest excursion from $\rho = 0$, the obvious generalization is $j' = j - S \cdot \tilde{S}$, where $\tilde{S} = \text{sign}(\rho_j)$ (double one-sided flux limiting).

One-sided and double one-sided limiters have been successfully used in the construction of a finite-difference Vlasov solver (see Section IV, C) and in studies of magnetic flux compression and of the motion of barium clouds coupled electrostatically to the ionosphere (Section IV, D). These examples involve the solution of a continuity equation, or a set of continuity equations, with the velocity field derived from a potential field. Application to coupled nonlinear fluid equations in which the velocity is propagated by means of a Navier–Stokes equation tells a different story. Dispersive ripples are almost as bad as without FCT; the only difference is that the ripples are "one-sided."

Hence we conclude as a matter of experience that use of one-sided flux limiting should be restricted to the solution of convective problems with a prescribed flow field which is decoupled from the density.

E. More FCT Algorithms

The references give a broad spectrum of generalized FCT algorithms: FCT/II and FCT/III give the details for many of these. In this subsection, we consider (1) various three-point explicit transport algorithms; (2) a three-point implicit transport algorithm; and (3) a Lagrangian FCT algorithm with curvilinear coordinates.

Equation (14) for basic three-point explicit transport encompasses many of the usual finite-difference algorithms. Term identification in Eq. (14) gives the definitions of $\varepsilon_{j+1/2}$ and $v_{j+1/2}$ for the original SHASTA algorithm. Variations using more or less diffusion with SHASTA are also possible. In Section III we use this freedom to vary v to generate zero residual damping (ZRD) and low phase error (LPE) versions of three-point explicit transport.

When $v_{j+1/2} = \frac{1}{2}|\varepsilon_{j+1/2}|$, the three-point diffusive transport formula reduces to a form of the donor-cell algorithm. The correct antidiffusion for minimum residual damping is $\mu_{j+1/2} = \frac{1}{2}(|\varepsilon_{j+1/2}| - \varepsilon_{j+1/2}^2)$.

When $v = \varepsilon^2/2$, the three-point transport algorithm becomes the usual two-step Lax–Wendroff formula. Since this diffusion is not large enough to ensure positivity, the basic Lax–Wendroff method is uninteresting for FCT applications unless an additional velocity-independent diffusion is added. If $v = \varepsilon^2/2 + \frac{1}{8}$, the Lax–Wendroff and SHASTA algorithms are formally identical in the constant velocity limit. All of these algorithms arising from Eq. (14) differ in the nonlinear terms, and infinitely many other conservative variable-velocity algorithms can be constructed by modifying the higher order terms in $\varepsilon_{j+1/2}$.

The explicit three-point transport formula need not have been our starting point. Interesting algorithms arise from the implicit three-point diffusive transport equation,

$$\bar{\rho}_j + \tfrac{1}{4}[\varepsilon_{j+1/2}(\bar{\rho}_{j+1} + \bar{\rho}_j) - \varepsilon_{j-1/2}(\bar{\rho}_j + \bar{\rho}_{j-1})]$$
$$+ [v_{j+1/2}(\bar{\rho}_{j+1} - \bar{\rho}_j) - v_{j-1/2}(\bar{\rho}_j - \bar{\rho}_{j-1})]$$
$$= \rho_j^0 - \tfrac{1}{4}[\varepsilon_{j+1/2}(\rho_{j+1}^0 + \rho_j^0) - \varepsilon_{j-1/2}(\rho_j^0 + \rho_{j-1}^0)]$$
$$+ [v_{j+1/2}(\rho_{j+1}^0 - \rho_j^0) - v_{j-1/2}(\rho_j^0 - \rho_{j-1}^0)]. \tag{37}$$

In Eq. (37) $\{\rho_j^0\}$ and $\{\bar{\rho}_j\}$ appear symmetrically so the algorithm is linearly reversible. This means that there is zero residual damping (ZRD) for all $\{\varepsilon_{j+1/2}\}$ profiles when $\{v_{j+1/2}\}$ is chosen within reasonable bounds. The strong diffusion is both added and subtracted simultaneously. To find the transported, diffused solution we set

$$\tilde{\rho}_j = \bar{\rho}_j + v_{j+1/2}(\bar{\rho}_{j+1} - \bar{\rho}_j) - v_{j-1/2}(\bar{\rho}_j - \bar{\rho}_{j-1}), \qquad (38)$$

and use as raw antidiffusion fluxes

$$\phi_{j+1/2} = v_{j+1/2}(\bar{\rho}_{j+1} - \bar{\rho}_j). \qquad (39)$$

After flux correction according to Eq. (20), the final density is given by

$$\rho_j^1 = \tilde{\rho}_j - \tilde{\phi}_{j+1/2} + \tilde{\phi}_{j-1/2}. \qquad (40)$$

This implicit three-point transport formula (and the resultant FCT algorithm) is interesting because of its linear reversibility and because the choice of $\{v_{j+1/2}\}$ remains completely free (so far).

Up to this point we have considered only algorithms for a uniformly spaced one-dimensional Eulerian cartesian grid. The FCT techniques developed above can be applied directly to cylindrical and spherical forms, but, to obtain a really good algorithm, care must be taken to properly treat curvilinear scale factors. Specifically, one must be cognizant of axes in the physical system. Various radial modules have been written specifically to provide more flexible and accurate treatments at the axis. This topic was treated briefly in FCT/II and by Boris et al. (1975a).

The time integration used most often for coupled systems of equations is a simple second-order-accurate midpoint rule, where dependent or driving variables such as V and P are determined at the half-step by a forward differenced first-order algorithm. Since there is no staggering in space or time, the time-step can be varied freely from cycle to cycle as the flow field changes. Three basic realizations of the original SHASTA algorithm exist. SHASTX is a highly optimized phoenical cartesian version with a reduced calling sequence for constant-spacing fixed-zone calculations. A listing and description of SHASTX are available from the authors on request. SHASTZ is a cartesian version that has arbitrarily varying zone sizes and incorporates a sliding or moving zone, also using the phoenical form of the original SHASTA. SHASTR is a radial geometry version used in the two-dimensional cylindrical codes for the radial component. The SHASTR algorithm uses an implicit form of antidiffusion but has in addition the arbitrary sliding zone features of SHASTZ which make Lagrangian calculation possible. The finite-difference algorithm for a general form of the SHASTA algorithm will be described here. Special cases can be reduced from this and a variety of antidiffusive methods may be employed.

Suppose $\{\rho_j^0\}$ and $\{V_j\}$ are given at $\{R_j\}$, the position of the grid points at the beginning of the cycle. Suppose further that the grid points are moved to $\{R_j^1\}$ at the end of the time-step. If the velocities are time centered and $R_j^1 = R_j + V_j\delta t$, the algorithm is fully Lagrangian. Other choices of $\{R_j^1\}$ are also possible; hence the name "sliding zone." In the following, the exponent $\eta = 0, 1, 2$ for cartesian, cylindrical, or spherical geometry.

First a set of Lagrangian coordinates for each grid point and a number of geometric factors are determined:

$$
\begin{aligned}
r_j &= R_j + V_j\delta t, \\
r_{j+1/2} &= \tfrac{1}{2}(r_{j+1} + r_j), \\
R_{j+1/2} &= \tfrac{1}{2}(R_{j+1} + R_j), \\
R_{j+1/2}^1 &= \tfrac{1}{2}(R_j^1 + R_{j+1}^1), \\
C_{j+1/2} &= (R_{j+1/2}/r_{j+1/2})^\eta (R_{j+1} - R_j)/(r_{j+1} - r_j), \\
C_{j+1/2}^+ &= (R_{j+1/2}^1 - r_j)/(r_{j+1} - r_j), \quad\quad\quad (41) \\
C_{j+1/2}^- &= (r_{j+1} - R_{j+1/2}^1)/(r_{j+1} - r_j) = 1 - C_{j+1/2}^+, \\
A_{j+1/2}^+ &= r_{j+1/2}^\eta (r_{j+1} - R_{j+1/2}^1), \\
A_{j+1/2}^- &= r_{j+1/2}^\eta (R_{j+1/2}^1 - r_j), \\
A_{j+1/2} &= R_{j+1/2}^\eta (R_{j+1} - R_j), \\
A_j &= A_{j+1/2} + A_{j-1/2}.
\end{aligned}
$$

Neglecting the complications of source and driving terms, which are treated in Boris et al. (1975a), the convection and compression of the generalized trapezoid are handled as follows:

$$
\left\{\begin{array}{c}\rho_{j+1}^* \\ \rho_j^*\end{array}\right\} = C_{j+1/2}\left\{\begin{array}{c}\rho_{j+1}^0 \\ \rho_j^0\end{array}\right\}. \quad\quad\quad (42)
$$

Equation (42) gives modified density values bounding the Lagrangian element extending from grid point j to $j+1$. Next we interpolate these values back onto the new rezoned mesh $\{R_j^1\}$. Define

$$
\rho_{j+1/2}^* = C_{j+1/2}^+ \rho_{j+1}^* + C_{j+1/2}^- \rho_j^*. \quad\quad\quad (43)
$$

Then the values of the transported and diffused function $\{\tilde{\rho}_j\}$ are given by

$$
\tilde{\rho}_j = [A_{j+1/2}^-(\rho_{j+1/2}^* + \rho_j^*) + A_{j-1/2}^+(\rho_j^* + \rho_{j-1/2}^*)]/A_j. \quad\quad\quad (44)
$$

Computing the fluxes for implicit antidiffusion involves solving the tridiagonal system of equations:

$$\bar{\rho}_j + (1/4A_j)[A_{j+1/2}(\bar{\rho}_{j+1} - \bar{\rho}_j) - A_{j-1/2}(\bar{\rho}_j - \bar{\rho}_{j-1})] = \tilde{\rho}_j \qquad (45)$$

to get

$$\phi_{j+1/2} = (A_{j+1/2}/4)(\bar{\rho}_{j+1} - \bar{\rho}_j). \qquad (46)$$

Flux correction then follows in a manner exactly analogous to Eq. (20),

$$\tilde{\phi}_{j+1/2} = S \cdot \max\{0, \min[S \cdot A_{j+1}(\tilde{\rho}_{j+2} - \tilde{\rho}_{j+1}), |\phi_{j+1/2}|, S \cdot A_j(\tilde{\rho}_j - \tilde{\rho}_{j-1})]\}, \qquad (47)$$

where $S = \text{sign}(\tilde{\rho}_{j+1} - \rho_j)$.

The final flux-corrected results are found from

$$\rho_j^{\ 1} = \tilde{\rho}_j - (1/A_j)(\tilde{\phi}_{j+1/2} - \tilde{\phi}_{j-1/2}). \qquad (48)$$

In the case where $R_j^{\ 1} = R_j$, it is obvious that considerable algebraic simplification can be done to obtain simpler, more efficient formulas.

III. Optimization of FCT Algorithms

A. Errors and Optimization

This section presents an analysis of numerical errors in FCT algorithms and considers the somewhat broader problem of overall code optimization to obtain numerical accuracy, flexibility, simplicity, and computational efficiency. We first emphasize accuracy by developing improved FCT algorithms and evaluating their performance relative to an "optimal" algorithm on our squarewave test problem. The other three criteria are then invoked to help choose the best of the algorithms presented.

The FCT technique achieves its success by replacing strict asymptotic ordering expansions with the much more physical fact that positivity must be maintained in solving the continuity equation. Although clipping, the nonlinear artifact of flux correction, is mildly annoying, it is a small enough price to pay for the extraordinary stability and accuracy of FCT algorithms near strong gradients. This accuracy arises in part from an interesting side effect of the diffusion/antidiffusion process (FCT/I and FCT/III), which causes a roughly fourfold reduction of dispersion in the basic finite-difference approximation itself. Even in the absence of flux correction this improvement can be exploited systematically.

The three types of linear convective errors which occur—amplitude errors, phase or dispersion errors, and the Gibbs phenomenon—have already been described and discussed briefly. They all contribute to the overall error and collectively should be reduced to as small a level as possible in the improved algorithms we are trying to develop. This overall error also includes components arising from nonuniform grids and nonlinear interactions such as compression and feedback from other coupled equations. Because convective errors are the most basic and serious, in the analysis below we concentrate on minimizing these three essentially linear errors. In the actual test computations, however, all the nonlinearities of the flux limiter are included.

It is important to bound the numerical errors in some way from below as well as above so that our comparison calculations can tell us not only how far FCT has brought us, but also how far we have yet to go. Since the Fourier transform, when it is applicable, allows complete control over the phase and amplitude of each harmonic component of the density structure, it provides a basis for developing an apparently "optimal" algorithm for treating the convective part of the continuity equation (see FCT/III).

When the density $\rho(x, t)$ is Fourier transformed, Eq. (7) allows each harmonic to be advanced the proper phase distance and at the same time to maintain its correct amplitude. These new harmonics can then be Fourier synthesized on the finite-difference grid according to the inverse Fourier transform formula Eq. (5). If the quantity $V_0 \delta t$ is an integral multiple of δx, the density distribution at the new time is a perfectly reproduced version of the old density shifted over exactly the correct distance.

The situation is not so good when the distribution is transported some fraction of a cell. Because the function is defined only at a finite number of harmonics, short wavelength components of the distribution are not present in the representation and their absence leads to the Gibbs phenomenon described earlier. If $V_0 \delta t$ is some fraction of δx, one of the Gibbs oscillations near a sharp gradient or discontinuity will appear after one cycle and can give values of the density which are negative, as shown in Fig. 1b.

Although the Fourier transform has zero residual damping and zero phase error, the irreducible Gibbs pheomenon requires us to enforce positivity on the solution. Our "optimal" algorithm therefore combines the ideal linear phase and amplitude properties of the Fourier transform with the properties of FCT needed to maintain a nonnegative solution. The Fourier solution is diffused at each cycle and then implicitly antidiffused by equal amounts. The diffusion eliminates the nonnegativity caused by the Gibbs oscillations, and the flux-correction formula then ensures against developing negative values via minima. The effect of the flux-correction formula is to leave a residual diffusion at the edges of sharp discontinuities in order to remove the nonnegative tendencies inherent in any finite-grid representation. The solution will no longer be linear and reversible, as with the pure Fourier transform, but it will be much more reasonable and reliable in the context of physically motivated calculations.

Another way of looking at this algorithm is to consider a given profile of densities to be transported and then to seek the closest discrete approximation for which the Fourier synthesis has no extra maxima or minima between the grid points. If we were to take the original profile and to replace it with this slightly smoothed profile, the pure Fourier transform would then be a "perfect algorithm" for the modified profile. No new maxima or minima could be generated, the phase errors and amplitude errors would be nonexistent, and the algorithm would be reversible after the initial smoothing.

Clearly the smaller the diffusion and antidiffusion in our optimal algorithm, consistent with positivity, the better. However, we found in test calculations that the final solution and the asymptotic value of the average absolute error (A.E.) are quite insensitive to the level of diffusion and antidiffusion. Figure 4 shows results of the optimal algorithm computation on the standard test problem using a small value of v. The value of A.E. for these calculations is about 0.022, more than ten times better than the standard algorithms previously tested.

We certainly do not mean to imply that the Fourier transform with FCT is the best algorithm to use in general. Transformation techniques are very complicated to use when a nonuniform mesh is considered or when the solution and the equations have nonlinearities. While Fourier transforming would still be possible in a calculation with spatially varying velocity, the interaction of the nonlinear terms would certainly confuse the concepts of phase and amplitude. Furthermore, the expense of performing such transform calculations far exceeds the gains which can be realized over good finite-difference algorithms.

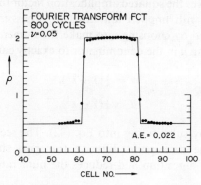

FIG. 4. Use of the optimal Fourier transform FCT algorithm on the squarewave test. With $v = 1/20$ after 800 cycles, the absolute error A.E. = 0.022 is the smallest value obtained with any positive algorithm. Since phases and amplitudes are treated exactly, the only inaccuracy arises from the finite discrete representation of physically continuous functions.

Rather, this "optimal" algorithm emphasizes the importance and the irreducible nature of the Gibbs phenomenon. The algorithm has both Zero Residual Damping (ZRD) and Zero Phase Errors (ZPE), the best that can be required of a linear algorithm. Thus we now have a realistic basis for comparison of more flexible, inexpensive finite-difference algorithms. Our Fourier algorithm gives a lower bound on the error which we usually associate with more general, but less accurate algorithms.

B. Reduction of Phase and Amplitude Errors

In Section II, one implicit and several explicit three-point transport algorithms were described. The great diversity and flexibility arose because our formalism left $\{v_{j+1/2}\}$ and $\{\varepsilon_{j+1/2}\}$, the diffusion and transport coefficients, free and virtually decoupled. In this section we use this freedom to endow our algorithms with additional desirable properties and look at the relative importance of minimizing amplitude or phase errors. Only the implicit three-point algorithm, which we have called REVFCT (REVersible), permits minimization of both (see FCT/III).

Since numerical diffusion has been of such concern in the past, we consider this source of error first by constructing ZRD algorithms in which the linear amplification factors after antidiffusion are unity for all harmonics. There are two variants of explicit three-point transport which need to be considered. One is a derivative of SHASTA (or Lax–Wendroff) where flux correction and implicit antidiffusion are applied every cycle. The second is based on the donor-cell algorithm; flux correction and implicit antidiffusion are only applied every other cycle. Phoenical algorithms cannot have ZRD in general because of their form; REVFCT already has ZRD.

Equation (32) gives the squared amplification factor for explicit three-point diffusive transport with implicit antidiffusion. The zero residual damping condition is enforced by choosing v to make the numerator a perfect square and then by choosing μ in the denominator to exactly cancel the numerator.

$$v = \tfrac{1}{4}(1 + \varepsilon^2),$$
$$\mu = \tfrac{1}{4}(1 - \varepsilon^2), \qquad (49)$$

as can be verified by substituting into Eq. (32). The second ZRD algorithm uses $v = \tfrac{1}{2}|\varepsilon|$ in the three-point diffusive transport stage. This yields the ordinary donor-cell algorithm and reduces the numerator of Eq. (32) to the expression

$$1 - 2(|\varepsilon| - \varepsilon^2)(1 - \cos\beta) \qquad (50)$$

in which the quadratic term vanishes identically. This expression gives the damping for two cycles of donor cell but has the form of a single diffusion step using the three-point formula. Therefore, a form of two-step ZRD can be obtained by implicit antidiffusion using $\mu = |\varepsilon|(1 - |\varepsilon|)$ and $\mu = 0$ on alternate cycles. Figure 5a shows this two-step ZRD algorithm based on donor cell; Fig. 5b shows the ZRD algorithm derived from SHASTA for comparison. As can be seen, the two-step algorithm is the worst of the FCT algorithms (even though almost twice as good as the flux-uncorrected algorithms), and the SHASTA ZRD algorithm is not much better. In general we do not recommend any algorithms where sequential cycles are treated differently. Clipping and nonlinear interactions from one cycle to the next are generally both more noxious when a large antidiffusion and a small antidiffusion alternate.

An even more sweeping conclusion can be drawn, since ZRD has actually made the overall errors larger. Phase errors due to numerical dispersion are generally more serious than the combination of residual amplitude errors and the Gibbs phenomenon. It should not be surprising that the phase properties of an algorithm are more important than the usual sorts of amplitude errors. Damping generally leaves the long wavelengths untouched while removing the very short wavelengths. Since these short wavelength harmonics of the solution generally suffer the most dispersion anyway, damping in conjunction with dispersion can sometimes actually reduce the overall A.E., as we have seen. The phase properties are generally more important because phase errors grow secularly when the velocity is predominantly in one direction. The difference in position between the correct phase front and the numerically computed phase front increases linearly in time when the velocity is constant.

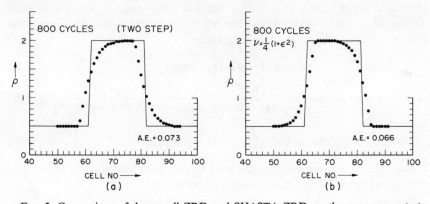

FIG. 5. Comparison of donor-cell ZRD and SHASTA ZRD on the squarewave test. Implicit antidiffusion is required for zero residual damping in each case. (a) In donor-cell ZRD, zero and finite antidiffusion are used on alternating cycles. (b) In SHASTA ZRD, ν is chosen to complete the square in the amplification factor so that two equal antidiffusion steps can exactly cancel the damping.

The phase errors are not secular when the velocity is oscillatory and the distance of oscillatory motion is small compared to spatial wavelengths of interest. In such specialized situations, the phase errors increase in one direction for half a period and then increase in the other for half a period. The net integrated phase errors go to zero on the average, leaving amplitude errors (damping) the major remaining source of numerical error. In such special situations, the ZRD algorithms discussed in the previous section might well be best.

When dispersion is secular, reductions in numerical phase errors should improve the solutions appreciably. Appendix A of FCT/III gives expansions of the relative phase error, $R = (X - V_0 \delta t)/V_0 \delta t$, for long wavelength (small β). Here X, the distance the numerical phase front with wavenumber $\beta = k\delta x$ moves in one time-step, is calculated by inverting

$$\tan kX = \text{Im}(\rho^1/\rho^0)/\text{Re}(\rho^1/\rho^0). \tag{51}$$

We expand Eq. (51) for small β and note that the correct value of X should be $V_0 \delta t$. In general, this will not be found, but at least v can be chosen to reduce the relative phase error from second order to fourth order in β. Since explicit and implicit antidiffusion in the three-point diffusive transport algorithms do not affect the phase properties, μ is still chosen to minimize the residual amplitude errors. In the phoenical and reversible algorithms, other properties have been built in, removing this additional freedom.

Thus there are three algorithms whose Low Phase Error (LPE) forms we need to investigate here: (a) explicit three-point diffusive transport with implicit antidiffusion (SHA); (b) explicit three-point diffusive transport with phoenical antidiffusion (PHO); (c) implicit three-point diffusive transport with implicit antidiffusion (REV). The relative phase error for these algorithms can be calculated from the three formulas

$$\tan kX_{\text{SHA}} = (\varepsilon \sin \beta)/[1 - 2v(1 - \cos \beta)], \tag{52a}$$

$$\tan kX_{\text{PHO}} = \frac{\varepsilon \sin \beta [1 + 2(v - \varepsilon^2/2)(1 - \cos \beta)]}{[1 - \varepsilon^2(1 - \cos \beta) - 2(v - \varepsilon^2/2)\varepsilon^2(1 - \cos \beta)^2]}, \tag{52b}$$

and

$$\tan \tfrac{1}{2}kX_{\text{REV}} = (\tfrac{1}{2}\varepsilon \sin \beta)/[1 - 2v(1 - \cos \beta)]. \tag{52c}$$

Using Eq. (11) and substituting the long wavelength expansions of Eq. (52a)–(52c) gives the following three relative phase errors:

$$R_{\text{SHA}} = \beta^2 [v - \tfrac{1}{6} - \tfrac{1}{3}\varepsilon^2]$$
$$+ \beta^4 [(1/120) - \tfrac{1}{4}v + v^2 - \varepsilon^2 v + \tfrac{1}{6}\varepsilon^2 + \tfrac{1}{3}\varepsilon^4] + O(\beta^6), \tag{53a}$$
$$R_{\text{PHO}} = \beta^2 [v - \tfrac{1}{6} - \tfrac{1}{3}\varepsilon^2]$$
$$+ \beta^4 [(1/120) - \tfrac{1}{4}v + \tfrac{1}{6}\varepsilon^2 - (\varepsilon^4/20)] + O(\beta^6), \tag{53b}$$

and

$$R_{\text{REV}} = \beta^2 [v - \tfrac{1}{6} - (\varepsilon^2/12)]$$
$$+ \beta^4 [(1/120) - \tfrac{1}{4}v + v^2 - \tfrac{1}{4}\varepsilon^2 v + (\varepsilon^2/24) + (\varepsilon^4/80)] + O(\beta^6). \tag{53c}$$

In these formulas the subscript "SHA" denotes our modified SHASTA algorithm, i.e., the three-point explicit diffusive transport algorithm. The subscript "PHO" denotes the phoenical antidiffusion version, and the subscript "REV" denotes the linearly reversible FCT algorithm with implicit antidiffusion.

For the three algorithms, Eqs. (53a)–(53c) indicate directly how to drive the relative phase error from second to fourth order. For explicit three-point diffusive transport with implicit antidiffusion we choose

$$v = \tfrac{1}{6} + \tfrac{1}{3}\varepsilon^2, \qquad \mu = \tfrac{1}{6} - \tfrac{1}{6}\varepsilon^2. \tag{54}$$

The results of this SHASTA algorithm as applied to our test problem are shown in Fig. 6a. The same choices of v and μ, Eq. (54), also minimize the phase errors of the explicit three-point diffusive transport algorithm with phoenical antidiffusion. The result is shown in Fig. 6b. Although better than previous algorithms, phoenical LPE is not quite as good as the implicit antidiffusion result (both described as SHASTA in the figure because of the genesis of the algorithm). The disadvantage of the phoenical version relative to implicit antidiffusion is clearly understood. For the implicit version, the

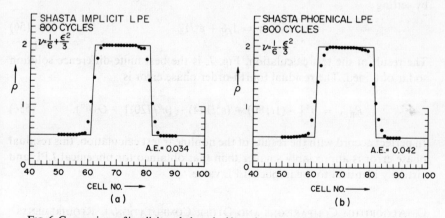

FIG. 6. Comparison of implicit and phoenical low phase error SHASTA on the squarewave test problem. Using $v = \tfrac{1}{6}(1 + 2\varepsilon^2)$ in each case reduces phase errors from second order to fourth order in $k\delta x$. Thus dispersive ripples are minimized making the work on the flux corrector much easier. Implicit antidiffusion gives almost as good a result as with the reversible FCT algorithm but phoenical antidiffusion is not quite as accurate.

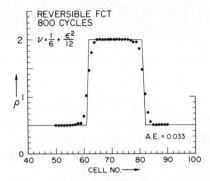

FIG. 7. Use of a reversible FCT algorithm on the squarewave test. The diffusion coefficient v is chosen to minimize phase errors and gives the lowest error of any finite-difference algorithm tested to date.

result of substituting Eq. (54) into (53a) is

$$R_{\text{SHA}} = \beta^4 [-(1/180) + (\varepsilon^2/36) + (\varepsilon^4/45)] + O(\beta^6), \tag{55a}$$

and the corresponding result for the phoenical version is

$$R_{\text{PHO}} = \beta^4 [-(1/30) + (\varepsilon^2/12) - (\varepsilon^4/20)] + O(\beta^6). \tag{55b}$$

The residual phase errors for the phoenical version are quite a bit larger.

The LPE version of REVFCT is obtained, as can be seen from Eq. (53c), by setting

$$v = 1/6 + \varepsilon^2/12. \tag{56}$$

The result of the test calculation, Fig. 7, is the best finite-difference solution so far obtained. The residual fourth-order phase error is

$$R_{\text{REV}} = \beta^4 [-(1/180) + (\varepsilon^2/144) - (\varepsilon^4/720)] + O(\beta^6). \tag{55c}$$

In perfect accord with the results of the nonlinear test calculation, this residual phase error is appreciably smaller than that obtained for phoenical LPE and virtually identical to the implicit LPE value.

C. Algorithm Comparisons and Other Computational Requirements

In the preceding two sections, many versions and types of FCT algorithms have been considered. They all stem from a general explicit three-point diffusive

TABLE I

COMPARISON OF CONTINUITY EQUATION ALGORITHMS
ON THE SQUAREWAVE TEST PROBLEM

Algorithm	Absolute error	ν	μ
Optimal Fourier FCT (implicit)	0.022	0.050	0.050
Reversible FCT (LPE)	0.033	$\frac{1}{6}(1 + \frac{1}{2}\varepsilon^2)$	$\frac{1}{6}(1 + \frac{1}{2}\varepsilon^2)$
SHASTA[a] (implicit LPE)	0.034	$\frac{1}{6} + \frac{1}{3}\varepsilon^2$	$\frac{1}{6}(1 - \varepsilon^2)$
SHASTA[a] (phoenical LPE)	0.042	$\frac{1}{6} + \frac{1}{3}\varepsilon^2$	$\frac{1}{6}(1 - \varepsilon^2)$
SHASTA[a] (implicit FCT)	0.049	$\frac{1}{8} + \frac{1}{2}\varepsilon^2$	$\frac{1}{8}$
SHASTA[a] (phoenical FCT)	0.052	$\frac{1}{8} + \frac{1}{2}\varepsilon^2$	$\frac{1}{8}$
SHASTA[a] (explicit FCT)	0.057	$\frac{1}{8} + \frac{1}{2}\varepsilon^2$	$\frac{1}{8}$
SHASTA[a] (implicit ZRD)	0.066	$\frac{1}{4}(1 + \varepsilon^2)$	$\frac{1}{4}(1 - \varepsilon^2)$
Donor cell (two-step ZRD)	0.073	$\frac{1}{2}|\varepsilon|$	$|\varepsilon|(1 - |\varepsilon|)$
Lax–Wendroff (diffused)	0.119	$0.01 + \frac{1}{2}\varepsilon^2$	0
Leapfrog (diffused)	0.122	0.01	0
Lax–Wendroff (simple)	0.175	$\frac{1}{2}\varepsilon^2$	0
Leapfrog (simple)	0.245	0	0
Donor cell (simple)	0.260	$\frac{1}{2}|\varepsilon|$	0

[a] Also applies to Lax–Wendroff and donor cell in constant velocity case.

transport formula [Eq. (14)] and a reversible implicit three-point diffusive transport formula [Eq. (37)]. Although we could have included more than the two nearest-neighbor grid points in deriving the diffusive transport algorithm (the ultimate example is the Fourier FCT algorithm developed for the constant-velocity case), the added complexity and computational inefficiency did not appear to be worth the slightly increased generality. Since the implicit three-point formula already requires the solution of a tridiagonal system, one gains nothing in simplicity or computational efficiency by considering either explicit or phoenical antidiffusion with this algorithm. The three distinct variations from the explicit three-point formula are explicit, implicit, and phoenical antidiffusion.

The value of the diffusion coefficient was left free in each of these four cases and thus could be chosen to optimize some other property of the algorithm. Table I provides a comparison of all the algorithms we have considered. When coupled with the analysis and tests of these two sections it allows us to draw the following conclusions:

1. Reducing phase errors (LPE algorithms) is generally more beneficial than reducing amplitude errors (ZRD algorithms).

2. The "optimal" finite-difference result has nonzero error because the irreducible Gibbs errors from the finite discrete representation force a modest amount of smoothing to ensure positivity.

3. The best of the generally applicable FCT algorithms (REVFCT, implicit LPE, or phoenical LPE) approach optimal to within a factor of two and average more than a factor of five better than standard flux-uncorrected transport algorithms derived from a three-point transport template.

4. Implicit antidiffusion is slightly better than phoenical antidiffusion but takes correspondingly longer to calculate. Both satisfy the requirement that the profile $\{\rho_j\}$ be unperturbed when $\{V_j\} = 0$ and both are slightly more accurate than explicit antidiffusion algorithms.

5. Algorithms in which large and small antidiffusion alternate on successive cycles seem to give poor results (donor cell two-step ZRD, for example). This is due to at least two causes: errors during the small antidiffusion cycle propagate beyond the large antidiffusion limiter, and nonlinear clipping effects are made worse.

These conclusions all point to what is good and what is bad but do not really allow us to pick a "best" algorithm. Instead, it becomes obvious that the definition of "best" has to depend on the specific problem being solved and the computer on which it is being solved. Although the algorithms are each designed to work well under all circumstances, each shows up to very best advantage in different situations. The phoenical LPE algorithm would appear best for vector computation (the flux limiter is fully vectorizable), and the reversible FCT algorithm, REVFCT, would appear best for scalar computers where the recursion relations involved in solving tridiagonal systems exact no great computational penalty. REVFCT suffers one major deficiency, however, which is revealed in equations where nonlinear interactions occur. Because both the transport and the antidiffusion are implicit, the transport causes numerical precursors which cross the mesh in one cycle. These extend far beyond the reach of the relatively local flux limiter and hence cannot be fully controlled by it. Thus REVFCT is great for passive convection but totally unsatisfactory for shocks.

Implicit antidiffusion with explicit transport does not suffer this problem at all and so the implicit LPE algorithm should really be rated overall "best buy" for scalar computers. However, when the phenomena of interest are essentially oscillatory, the one-step ZRD algorithm might actually prove better than the LPE versions because the phase errors need no longer be secular.

For the purposes of the present exposition, we have concentrated heavily on the one-dimensional continuity equation with constant coefficients on a uniform grid, and have not gone very far into the nonuniform grid and nonconstant velocity aspects of these algorithms. Even the functional form of the dependence of v on ε (for example, $v = 1/6 + \varepsilon^2/3$ for implicit LPE) does not give a clue how to evaluate $v_{j+1/2}$ on a nonuniform moving grid where compression and convection are both taking place. Fortunately the linear, constant coefficient effects which we have treated seem to be the dominant sources of error. Therefore reasonable and simple choices of the variable coefficients which reduce to one of the standard FCT forms in the constant coefficient case generally seem to work. In the next section multidimensional problems are considered, now that the one-dimensional problem is well in hand. Applications in several areas are discussed briefly, and several special purpose techniques are presented.

IV. Applications of Flux-Corrected Transport

A number of tests involving solutions of the one-dimensional ideal fluid equations were described in FCT/I. They included shocks, rarefaction waves, and contact discontinuities. In Boris (1972), Boris et al. (1975a), and FCT/II, applications to multidimensional fluids were discussed, and generalizations of the one-dimensional calculations presented. Here we describe additional applications more directly related to CTR.

A. CYLAZR, A Two-Dimensional Laser–Target Model

A number of codes have been written to describe the interaction of high-power laser pulses with a material target. To date, the most extensive one employing FCT for the hydro portion is CYLAZR (Columbant et al., 1974), developed to study the X-ray conversion efficiency in a high-atomic-number target. The model embodies cylindrical symmetry, and consists of coupled hydrodynamic, radiation transport, and chemical rate equations. Several versions of the code exist, differing chiefly in their treatments of atomic chemistry. In the more complex, separate number densities and temperatures are propagated for each ionization state and excitation level. The version described here (the original form of the model) uses a coronal model and describes the ions in terms of a single temperature T_i and ionization state Z.

FCT is particularly effective in describing the expansion of a laser-heated plasma because: (i) rapid deposition of laser energy engenders sharp discontinuities in the direction of expansion (which may develop into shocks); and (ii) transport coefficients are extremely anisotropic owing to the presence of spontaneous magnetic fields, giving rise to steep gradients transverse to **B**.

The fluid equations are solved numerically on an r–z Eulerian grid. They propagate the total ion number density $N = \sum_i N_i$, fluid velocity **V**, internal energy density $\mathscr{E}$, ion temperature T_i, and magnetic field **B**. The equations are

$$\partial N/\partial t + \nabla \cdot (N\mathbf{V}) = 0; \tag{57}$$

$$NM\left(\frac{\partial}{\partial t} + \mathbf{V} \cdot \nabla\right)\mathbf{V} + \nabla \cdot \left[\left(P + \frac{B^2}{8\pi}\right)\mathbf{I} - \frac{\mathbf{BB}}{4\pi}\right] + \nabla \cdot \mathbf{P}_R = 0; \tag{58}$$

$$\partial\mathscr{E}/\partial t + \nabla \cdot (\mathscr{E}\mathbf{V}) = -P\nabla \cdot \mathbf{V} + \mathbf{J} \cdot \mathbf{r} \cdot \mathbf{J} + P_L - R_e - \nabla \cdot \mathbf{Q}_e; \tag{59}$$

$$(\partial/\partial t + \mathbf{V} \cdot \nabla)(\tfrac{3}{2}NkT_i) + \tfrac{5}{2}NkT_i \nabla \cdot \mathbf{V} = v_{ei}\tfrac{3}{2}Nk(T_e - T_i); \tag{60}$$

$$\partial\mathbf{B}/\partial t = \nabla \times [\mathbf{V} \times \mathbf{B} - (c^2/4\pi)\,\mathbf{r} \cdot (\nabla \times \mathbf{B})]$$
$$- (c/4\pi e)\nabla \times [(1/N_e)(\nabla \times \mathbf{B}) \times \mathbf{B}] - (ck/eN_e)\nabla N_e \times \nabla T_e \tag{61}$$
$$+ \text{ other spontaneous } \mathbf{B}\text{-field terms}.$$

Here all quantities are independent of θ; $\mathbf{B} = \mathbf{e}_\theta B$ and all other vectors are of the form $\mathbf{V} = \mathbf{e}_r V_r + \mathbf{e}_z V_z$; electron density is $N_e \equiv NZ = \sum N_i Z_i$ (summed over ion species); $M = N^{-1}\sum_i N_i M_i$ is the mean ion mass; P_L is the laser energy deposition power; $P = k(N_i T_i + N_e T_e)$ is the total pressure; $\mathscr{E}$ is the thermal plus ionization energy density; R_e is the rate of change of $\mathscr{E}$ due to radiation losses; and **I** is the unit dyad. The transport coefficients **r** (resistivity), Q_e (heat flux), and v_{ei} (momentum transfer collision rate) are those given by Braginskii (1965).

Equations (57)–(59) are solved using the FCT modules SHASTZ and SHASTR (see Section II,E above) in time-step-split fashion. Equations (60) and (61) are further split as follows: the convective terms are solved using FCT, while the remaining (diffusive and source) terms are differenced implicitly. The radiation and atomic chemistry calculations are performed on a fast time-scale by integrating ordinary differential rate equations. The results are used to redefine N, T_i, and $\mathscr{E}$ for the subsequent fluid transport step.

As an example of the capabilities of this code, consider Figs. 8 and 9 taken from Ripin et al. (1975). Figure 8 compares X-ray intensity obtained by numerical calculations with that measured experimentally. It shows the importance of including spontaneous **B** fields in the calculations, especially for the short wavelength portion of the spectrum.

Figure 9 shows axial variations of the fluid qualities, X-ray production rate and collision time. Note the very steep density gradient in the absorption region, produced by pressure balance of the cold dense plasma with the heated underdense plasma and laser light. Other versions of the code are being used to study the dynamics and X-ray production of exploding wires, and a variety of problems involving plasma streaming.

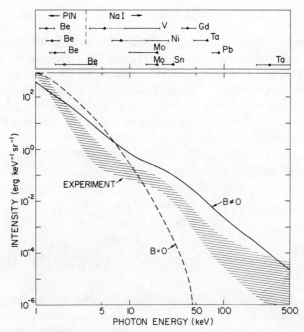

FIG. 8. X-Ray intensity versus energy from experiment and numerical calculations. The cross-hatched area spans the range between high and low results observed for six shots. Note the improvement in agreement between experiment and numerical results when spontaneous magnetic fields are included. The detector ranges (bars) and peak sensitivity energies (points) are indicated across the top.

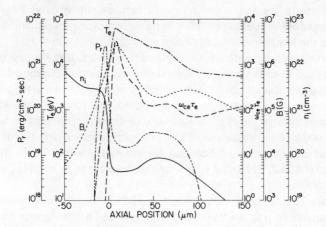

FIG. 9. Computed axial variation of P_v, T_e, n_i, B and the product of electron angular frequency ω_{ce} and collision time τ_e. The profiles are shown 2 psec after the peak intensity of the laser pulse (1.2×10^{10} W/cm²).

B. Moving Flux Coordinate Code

In order to perform simulations of high-beta flows in the presence of large transport anisotropies, Anderson (1975) has developed an FCT algorithm called SLIDE, based on a continuously rezoned grid which follows the motion of magnetic flux surfaces. The method replaces an axisymmetric (r,z) coordinate system with a flux tube (ψ,χ) system. Here ψ and χ are the usual azimuthal and longitudinal flux functions. As in the previous example, the third coordinate, θ, is ignorable.

Under this transformation the heat flow term, for example, appears without mixed derivatives:

$$\mathbf{\nabla}\cdot(\mathbf{K}\cdot\mathbf{\nabla}T) = \frac{1}{h_\psi h_\chi h_\theta}\left\{\frac{\partial}{\partial\psi}\left(\frac{h_\chi h_\theta}{h_\psi}K_\perp\frac{\partial T}{\partial\psi}\right)\right.$$
$$\left. + \frac{\partial}{\partial\chi}\left[\frac{h_\psi h_\theta}{h_\chi}\left(K_\perp + \left(\frac{K_\parallel - K_\perp}{B^2}\right)B_\chi^{\,2}\right)\right]\right\}. \quad (62)$$

Here K in the thermal conductivity tensor, and the h's are the scale factors defined for curvilinear coordinates. Since time-step-splitting is used in differencing the fluid equations, extreme field-oriented anisotropies in K can be modeled without difficulty. A straightforward differencing of the above term in r–z coordinates, on the other hand, leads to spurious transport along the coordinate directions. As a consequence, anisotropies of more than $\sim 4:1$ cannot be accurately described in the cartesian representation, but present no problem in the flux coordinates. In fact, anisotropies as large as $10^5:1$ can be handled. Anderson (1975) has published examples illustrating this and showing the limitations of a cartesian mesh.

Central to the algorithm is the method for constructing the new orthogonal grid required at each time level. One set of grid lines is determined by the flux contours (lines of constant ψ). The second set is defined by an orthogonalization procedure. As a consequence of the method, the metric of the moving coordinate system is an explicit function of time, introducing a new source term in some of the equations. This is lumped together with the usual source terms present in standard Eulerian formulations, using the FCT transport module SHASTZ. SHASTZ is appropriate because of its continuous rezoning capability.

Figure 10 illustrates the application of SLIDE to a fully ionized plasma pellet expanding into a 15 kG magnetic field and a low density background. The plasma density is 10^{19} cm^{-3} (central) with an exponential dropoff outside $r = 1$ mm; its temperature is $T = 20$ eV and initial expansion velocity is 3.5×10^7 cm/sec. At $t = 3$ nsec (Fig. 10a,b) field retardation of the radial motion is quite noticeable. Later at 6 nsec (Fig. 10c, d) the density maximum is seen to be off-axis where the plasma has been snowplowed by the magnetic field. Also at this time the axial profile is seen to be constricted by pinching forces (sausage instability).

C. FCT Vlasov Solver

Here we describe a novel application made possible by the nonlinear properties of FCT. The Vlasov equation from plasma theory describes the evolution of the single particle distribution function $f(\mathbf{x}, \mathbf{v}, t)$,

$$\partial f/\partial t + \mathbf{v} \cdot \nabla_x f + (q/m)\,\mathbf{E} \cdot \nabla_v f = 0. \tag{63}$$

It can be viewed as a continuity equation propagating a fluid in an $N_x + N_v$ dimensional space, where N_x and N_v are the dimensions of the configuration and velocity subspaces, respectively. This phase fluid density is incompressible because $\mathbf{v}$ is independent of position and $\mathbf{E}$ is independent of $\mathbf{v}$. Hence Eq. (63) is in exactly the form of Eq. (1) with the compression term omitted.

When Eq. (63) and a self-consistent equation for $\mathbf{E}$ are solved by finite differences using conventional linear algorithms, multistreaming instabilities develop due to the discrete approximation. Each row of points with a common velocity v_j behaves like a beam. Any two such beams have a finite relative velocity and hence are capable of exhibiting charge-bunching instabilities. These instability growth rates are of the order of the ratio of the velocity to displacement mesh spacing and can distort or completely swamp physical processes. Still worse, they eventually drive f negative.

Fortunately FCT affords a degree of control over these errors which is lacking in conventional difference schemes. Let us suppose for simplicity that we have a single mobile charge species, electrons, and that the electric field $\mathbf{E}$ satisfies Poisson's equation,

$$\nabla \cdot \mathbf{E} = -4\pi e [n_e(\mathbf{x}, t) - n_0]. \tag{64}$$

Here n_0 is a uniform background ion number density and $n_e(\mathbf{x}, t)$ is the number density of electrons in configuration space, given by

$$n_e = \int d^3v\, f(\mathbf{v}). \tag{65}$$

Let us further restrict ourselves to a one-dimensional spatially periodic system. Then Eqs. (63)–(65) can be rewritten as

$$\frac{\partial f}{\partial t} + \frac{\partial}{\partial x}(vf) - \frac{\partial}{\partial v}\left(\frac{e}{m} Ef\right) = 0, \tag{66}$$

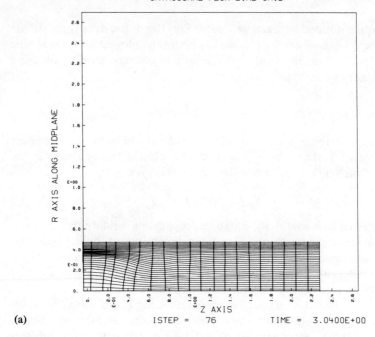

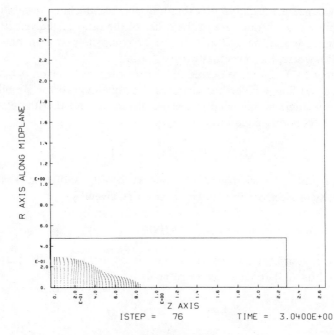

FIG. 10. The orthogonal flux tube grid and marker particle positions are shown at two early times in the simulation of the expansion of the pellet plasma into the magnetic field. Times

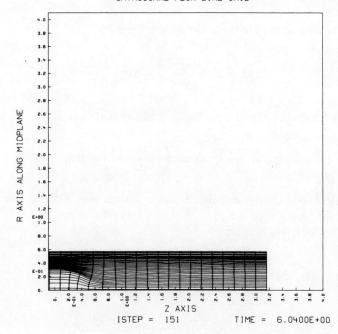

(c)

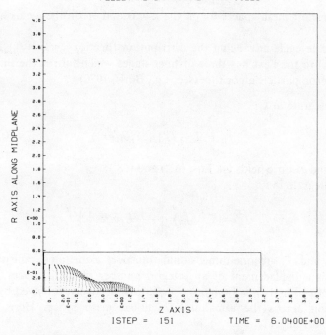

(d)

are given in nanoseconds while distances are in cm. Note the change in scale for the later time necessitated by the pellet domain expanding beyond the earlier "movie window."

where the electric field in the x direction is given by

$$E(x) = 2ieL \sum_{k=-\infty}^{\infty}{}' \frac{1}{k} \rho_k \exp\left(\frac{2\pi i k x}{L}\right). \tag{67}$$

The $k = 0$ term is omitted from the summation in (67) and

$$\rho_k = \frac{-e}{L} \int_0^L dx \exp\left(\frac{-2\pi i k x}{L}\right) \int dv f(x,v) \tag{68}$$

is the electronic charge density; L is the system size. On a discrete mesh, x takes the values x_j, $j = 1, 2, ..., N$, and v takes the values v_l, $l = 1, 2, ..., M$. Equations (61)–(68) become

$$E_j \equiv E(x_j) = 2ieL \sum_{-N/2+1}^{N/2}{}' \frac{\rho_k}{k} \exp\left(\frac{2\pi i k j \delta x}{L}\right), \tag{67'}$$

$$\rho_k = \frac{e}{L} \sum_{j=1}^{N} \delta x \sum_{l=1}^{M} \delta v f_{jl} \exp\left(\frac{-2\pi i k j \delta x}{L}\right). \tag{68'}$$

Here constant mesh spaces δx on the x-axis and δv on the v-axis have been specified.

A single cycle advancing the distribution function array $\{f_{jl}\}$ from one time level to the next has three distinct stages which mirror the three stages in a leapfrog particle algorithm (see, e.g., Boris, 1970).

1. Integrate in x,

$$\tilde{f}_{jl} = T_x(\{f_{jl}^0\}), \{v_l \delta t / \delta x\}); \tag{69}$$

2. Find electric fields via Eqs. (67') and (68');
3. Integrate in v,

$$f_{jl}^1 = T_v\left(\{\tilde{f}_{jl}\}, \left\{-\frac{e}{m} E_j \delta t / \delta v\right\}\right). \tag{70}$$

Here T_x and T_v are one-dimensional transport algorithms, consisting of a Lagrangian displacement of an integral number of mesh spaces, given by $J_l = [v_l \delta t / \delta x]$ and $L_j = [-(e/m) E_j \delta t / \delta v]$, respectively, followed by an FCT integration across the remaining fractional mesh space. Here $[z] \equiv$ the integral part of z.

This scheme is automatically time centered when the x and v integrations are alternated. Accuracy is then fully second order when the transport algorithms are independently second-order accurate. For the best results, the two FCT algorithms should in general be chosen differently for x and v. In the x direction, f seldom fluctuates drastically. Hence the emphasis is on good linear phase properties, and one of the LPE algorithms from Section III is called for. By contrast, velocity profiles frequently vary abruptly—e.g., one commonly takes a delta function for the initial velocity dependence. Further, velocity excursions are seldom large so phase errors are not secular. Here the principal error is nonlinear damping due to clipping, especially at early times, before instabilities can produce large values of the dimensionless impulse, $(-e/m) E \delta t/\delta v$.

For this application a "minimum diffusion" FCT algorithm is most appropriate. In its simplest form, this means taking

$$v_j = \tfrac{1}{2}|\varepsilon_j| = \mu_j + \tfrac{1}{2}\varepsilon_j{}^2 \tag{71}$$

for the diffusion/antidiffusion coefficients to be used at position x_j. Here ε_j is the fractional part of the impulse in the velocity direction. This value of v_j is the smallest which maintains positivity in the transport stage.

Figure 11 shows the evolution of a two-stream instability (Book *et al.*, 1973). The two initially cold beams broaden, slow down, and eventually become trapped. The wave energy as a function of time (not shown) oscillates with the bounce frequency of the trapped electrons; these oscillations eventually damp when phase mixing is complete.

A second example is shown in Fig. 12 (Book and Sprangle, 1974). A two-species plasma (mass ratio $m_i/m_e = 1836$) is driven by a large amplitude pump wave, whose wavenumber and frequency satisfy the ion-sound dispersion relation. The equations are solved in the wave frame but are transformed back to the lab frame for display purposes. Figure 12 shows that a small electron population is quickly trapped, while the remaining electrons transit. On the much slower ion time-scale, the bulk of the ions are accelerated, and the ion waves show a tendency to brake.

In both examples the runs were terminated when unphysical effects began to propagate in from the $\pm V_{\max}$ edges of the mesh. This behavior can be prevented by tinkering with the velocity boundary conditions or by making the velocity mesh nonuniform near the edges.

"Gitterbewegung" is another kind of fine tuning. The action of T_x, the spatial transport operator defined earlier, is to treat rows with $v_l \delta t/\delta x$ close to an integer more accurately than those for which the fractional part of a cell crossed is large. This inequity can be avoided by using successive Galilean transformations which change the values of the fractional parts from cycle to

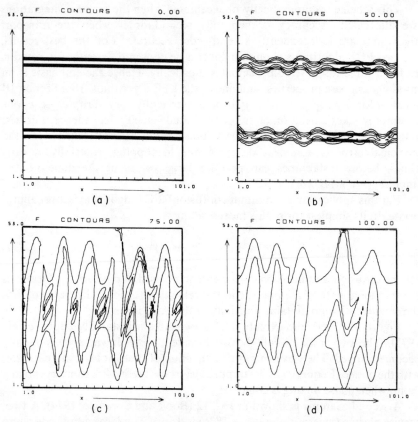

FIG. 11. Evolution of a two-stream instability using SHASOV. Phase contours show the linear growth (b), trapping (c), and saturation (d) of the counterstreaming beams. Time is measured in units $\delta t = 0.2\omega_p^{-1}$, where ω_p is the plasma frequency associated with a single beam.

cycle. Each row of grid points at constant velocity runs the gamut from most to least favored status and receives the same average treatment. (Displays should always be carried out in one stationary frame.)

The advantages of the FCT Vlasov solver are those usually associated with finite-difference techniques—speed, flexibility, and programming convenience. The one-dimensional examples cited above run almost as fast as a particle simulation code (Boris, 1970), having as many particles as there are grid points here. These times are extremely good, since one generally needs fewer mesh points than particles. The relative speed and resolution quickly dissipate when we go to phase spaces of higher dimension, however. Going from the two-dimensional phase space (x–v) of a one-dimensional problem to the four-

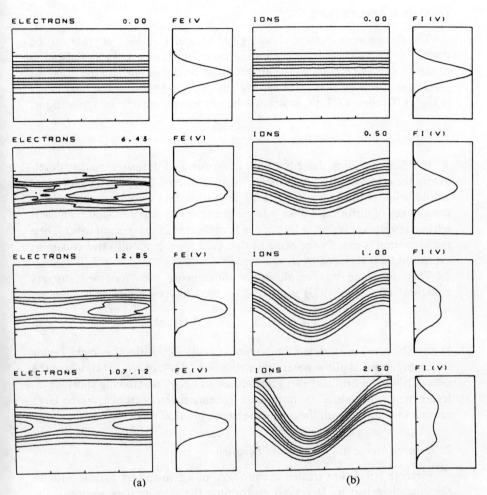

FIG. 12. Phase contours and velocity profiles for electron and ion distribution driven by an external wave. The mass ratio $m_i/m_e = 1836$; time is given in units of ω_{pe}^{-1} and ω_{pi}^{-1} above the respective phase space plots; the time-step $\delta t = 10^{-3} \omega_{pi}^{-1}$. The system length equals one driver wavelength $2\pi/k$. The driver frequency $\omega_o = kc_s$, where $c_s \approx (\kappa T_e)^{1/2} m_i$ is the ion sound speed; the driver amplitude E_o satisfies $eE_o k/\kappa T_e = .25$; $T_e/T_i = 10^3$.

dimensional phase space (x, y, v_x, v_y) of a two-dimensional problem nominally requires us to square the number of mesh points. Since the degrees of freedom for particles in a particle code operate at least semi-independently, it is usually not necessary to square the number of particles going from one to two dimensions. Therefore, even the best Vlasov code is outstripped by particle simulation techniques when the phase space exceeds three dimensions.

D. Other Applications

The foregoing discussion should give the reader an idea of what can be done with FCT. There have been many other applications. As an indication of their scope and a guide to the literature, we briefly summarize the principal ones here. Some of these will receive detailed discussion in the final paper in the FCT series, FCT/IV, which will be addressed primarily to applications.

1. Diffusion Plus Convection

If a fluid equation describes both advection and diffusion, neither dominating the other, the FCT treatment is slightly altered. Suppose the numerical antidiffusion coefficient μ in the FCT algorithm that would be employed in the absence of diffusion is always larger than the physical diffusion coefficient η (the prescription is only a little more complicated in the general case). Then the algorithm is exactly the same as that used for the nondiffusive equation, except that μ is replaced by $\mu' = \mu - \eta$.

This technique was employed in differencing the "modified Burgers equation" (Ott et al., 1973) as a model for certain types of plasma turbulence,

$$\partial V/\partial t + V \, \partial V/\partial X = \alpha V + \beta \, \partial^2 V/\partial X^2. \tag{72}$$

It was also used in studying the nonlinear evolution of the Rayleigh–Taylor instability in the shallow water approximation (Book et al., 1974), where the diffusion was an artificial effect introduced to keep fast-growing short wavelength modes initialized by numerical truncation errors from growing large enough to compete with the long wavelength physical mode.

2. Hermite Moments of the Vlasov Equation

Eltgroth (1974) has treated the problem of a two-species plasma with an electrostatic driver by Hermite-transforming the velocity dependence of the Vlasov equation. The resulting set of moment equations (analogous to the usual fluid equations) is then solved using a form of SHASTA.

3. Theta Pinch and Other Cross-Field Streaming Problems

Liewer and Krall (1973) have used FCT to study the structure of collisionless shocks in theta pinches. The code is one-dimensional radial and incorporates anomalous transport coefficients derived from consideration of microinstabilities. Because the behavior of the imploded magnetic pulse is essentially laminar (ballistic) for the early part of the calculation, minimum diffusion FCT is appropriate.

Similar codes have been applied to astrophysical plasma problems (Chevalier and Gardner, 1974; Papadopoulos *et al.*, 1974) and to the coupling between streaming plasma and an ionized ambient background (Wagner and Manheimer, 1973).

4. Z-Pinch Geometry

Mosher *et al.* (1975) have written a one-dimensional radial FCT code for treating exploding wires with radiation and chemistry determined from a coronal model. Anderson *et al.* (1972) have modeled a Z-pinch imbedded in a stabilizing high density gas, again using a one-dimensional code.

5. Compressed Magnetic Field (CMF) Generators

Freeman and Thompson (1975) have written a two-dimensional MHD code to study the dynamics of CMF generators, in which a metallic armature filled with high explosive expands outward, compressing magnetic flux established between it and an outer helical coil. The expansion decreases the inductance in the winding, amplifying the electrical current. Thus chemical energy in the explosive is converted to kinetic energy of the armature, which in turn is transformed into electrical energy delivered to an external load. CSQ is an Eulerian code which propagates both the solid material fluid quantities and, using a form of SHASTA, the magnetic flux function.

6. Studies of Barium Clouds

Book and Scannapieco (Goldman *et al.*, 1974; Scannapieco *et al.*, 1974) have applied FCT to solve the continuity equation for field-line integrated ion Pederson conductivity in studying barium releases and F-region irregularities. The flow velocity in these problems is basically an $E \times B$ drift derived from the charge neutrality condition, $\nabla \cdot \mathbf{J} = 0$. In this case, one-sided flux limiting (see FCT/II) is very successful.

7. Magnetic Compression

FCT/II described a method employed by Book and Clark (1975) to compute the buildup of magnetic fields occurring when a plasma streams transverse to a second, magnetized plasma. This method, applicable in a variety of situations, involves direct differencing of the inductive law, with flux limiting invoked only in the direction of streaming.

8. Kelvin-Helmholtz Instability of a Compressible Medium

SHAS2D (Boris, 1972) has been used to study the stability of two-dimensional compressible shear flow. Solutions of the linear dispersion relation for the Kelvin–Helmholtz instability in a compressible medium are in good agreement with the onset and initial growth phase (Boris *et al.*, 1975b). A Karman vortex street appears, and the instability saturates, whereupon a nonlinear mixing phase sets in. Averages along the original streamlines show an apparent homogenization, even though mixing on the fine scale is nearly totally absent.

Many other successful applications have not yet been published. In addition, the reader, like the authors, will probably be able to add to this list in the future as new problems are encountered for which FCT is the numerical treatment of choice.

ACKNOWLEDGMENT

The authors would like to acknowledge the help and contributions of J. H. Gardner, K. Hain, E. S. Oran, and S. Zalesak in the editing and preparation of this article and in the development and application of these FCT techniques.

REFERENCES

Ahlberg, J. H., Nilson, E. N., and Walsh, J. L. (1967). "The Theory of Splines and Their Applications." Academic Press, New York.
Anderson, D. V. (1975). *J. Comput. Phys.* **17**, 246.
Anderson, D. V., Lampe, M., and Manheimer, W. M. (1972). "Numerical Study of the Gas-Enclosed Z-Pinch," NRL Memo. Rep. No. 2486. Naval Res. Lab., Washington, D.C.
Book, D. L., and Clark, R. W. (1973). "Effects of a Finite Debris Density Profile on the Development of the Longmire Shell," NRL Memo. Rep. No. 3066. Naval Res. Lab., Washington, D.C.
Book, D. L., and Sprangle, P. (1974). *Bull. Amer. Phys. Soc.* [2] **19**, 882.
Book, D. L., Boris, J. P., Ossakow, S. L., and Pierre, J. M. (1973). *Proc. Conf. Numer. Simul. Plasmas, 6th, 1973* p. 6.
Book, D. L., Ott, E., and Sulton, A. L. (1974). *Phys. Fluids* **17**, 676.
Book, D. L., Boris, J. P., and Hain, K. H. (1975). *J. Comput. Phys.* **18**, 248.
Boris, J. P. (1970). *Proc. Conf. Numer. Simul. Plasmas, 4th, 1970* p. 3.
Boris, J. P. (1972). "SHAS2D, A Fully Compressible Hydrodynamics Code in Two Dimensions," NRL Memo. Rep. No. 2542. Naval Res. Lab., Washington, D.C.
Boris, J. P., and Book, D. L. (1973). *J. Comput. Phys.* **11**, 38.
Boris, J. P., and Book, D. L. (1976). *J. Comput. Phys.* (to be published).
Boris, J. P., Gardner, J. H., and Zalesak, S. (1975a). "Atmospheric Hydrocodes Using FCT Algorithms," NRL Memo. Rep. No. 3081. Naval Res. Lab., Washington, D.C.
Boris, J. P., Coffey, T. C., and Fisher, S. (1975b). "The Kelvin-Helmhotz Instability and Turbulent Mixing," NRL Memo. Rep. No. 3124. Naval Res. Lab., Washington, D.C.

Braginskii, S. I. (1965). *Rev. Plasma Phys.* **1**, 205.
Chevalier, R. A., and Gardner, J. H. (1974). *Astrophys. J.* **192**, 457.
Columbant, D. G., Whitney, K. G., Tidman, D. A., Winsor, N. K., and Davis, J. (1974). "Laser Target Model," NRL Memo. Rep. No. 2954. Naval Res. Lab., Washington, D.C.
Courant, R., Friedrichs, K. O., and Lewy, H. (1928). *Math. Ann.* **100**, 32; transl.: *IBM J. Res. Dev.* **11**, 215 (1967).
Courant, R., Isaacson, E., and Rees, M. (1952). *Commun. Pure Appl. Math.* **5**, 243.
Eltgroth, P. G. (1974). *Phys. Fluids* **17**, 1602.
Freeman, J. R., and Thompson, S. L. (1975). "Two-Dimensional MHD Modeling of Compressed Magnetic Field Generators I: The Magnetic Field Solver." Rep. SAND-75-375. Sandia.
Goldman, S. R., Ossakow, S. L., and Book, D. L. (1974). *J. Geophys. Res.* **79**, 1471.
Lax, P. D., and Wendroff, B. (1960). *Commun. Pure Appl. Math.* **13**, 217.
Liewer, P. C., and Krall, N. A. (1973). *Proc. 1973 Sherwood Theory Meet.* p. 129.
Moretti, G. (1972). "Thoughts and Afterthoughts about Shock Computations," Rep. PIBAL 72-37. Polytechnic Institute of Brooklyn, New York.
Mosher, D., Stephanakis, S. J., Hain, K., Dozier, C. M., and Young, F. C. (1975). *Ann. N.Y. Acad. Sci.* **251**, 632.
Orszag, S. A., and Israel, M. (1972). "Numerical Flow Simulation by Spectral Methods," *Proc. Symp. Numer. Models Ocean Circ.* Nat. Acad. Sci., Washington, D.C.
Ott, E., Manheimer, W. M., Book, D. L., and Boris, J. P. (1973). *Phys. Fluids* **16**, 855.
Papadopoulos, K., Clark, R. W., and Wagner, C. E. (1974). *Proc. Solar Wind Conf., 1974* p. 343.
Ripin, B. H., Burkhalter, P. G., Young, F. C., McMahon, J. M., Columbant, D. G., Bodner, S. E., Whitlock, R. R., Nagel, D. J., Johnson, D. J., Winsor, N. K., Dozier, C. M., Bleach, R. D., Stamper, J. A., and McLean, F. A. (1975). *Phys. Rev. Lett.* **34**, 1313.
Roache, P. (1972). "Computational Fluid Dynamics." Hermosa Publ., Albuquerque, New Mexico.
Scannapieco, A. J., Ossakow, S. L., Book, D. L., McDonald, B. E., and Goldman, S. R. (1974). *J. Geophys. Res.* **79**, 2913.
Strang, G., and Fix, G. (1973). "An Analysis of the Finite Element Method." Prentice-Hall, Englewood Cliffs, New Jersey.
Wagner, C. E., and Manheimer, W. M. (1973). *Proc. Conf. Numer. Simul. Plasmas, 6th, 1973* p. 126.

Multifluid Tokamak Transport Models[*]

JOHN T. HOGAN

OAK RIDGE NATIONAL LABORATORY
OAK RIDGE, TENNESSEE

I. General Remarks 131
   A. The Development of One-Dimensional Tokamak Codes 131
   B. A Brief Description of the Tokamak 133
   C. A Survey of Codes 136
   D. Commonly Used Moment Equations 136
   E. Some Relevant Physical Processes 141
II. Plasma Models 142
   A. Transport by Collisional Diffusion and Turbulence 143
   B. Methods Used To Solve the Diffusion Problem 147
III. Suprathermal Plasma: Injected Ions and Alpha Particles 150
   A. Beam Trapping and Thermalization 151
   B. Implementation 152
IV. Neutral Gas 153
   A. Models for Neutral Transport 154
   B. Implementation of Neutral Models 156
V. Impurities 158
   Physical Models 159
VI. Summary 161
   References 162
   Appendix: Bibliography 164

I. General Remarks

A. THE DEVELOPMENT OF ONE-DIMENSIONAL TOKAMAK CODES

ALONG WITH SUCCESS IN confining plasmas in the Tokamak geometry (Artsimovich, 1972) there has arisen a need for a more detailed understanding of transport processes. The desire to do work relevant to Tokamak confinement physics suggests calculations that incorporate most significant physical features of the experiment, each perhaps somewhat roughly described. It is this quality of "maximal ordering" which has been useful in analytic asymptotic analysis (Kruskal, 1963) and which characterizes Tokamak computer models.

---

[*] Work supported in part by the U. S. Energy Research and Development Administration under contract with the Union Carbide Corporation.

This article is a review of numerical work carried out to model the behavior of Tokamak plasmas. It is an outline of the physical processes which have been found to be important in reproducing experimental trends, along with a discussion of the numerical techniques which have been used to effect solutions. While the numerical techniques described here are by no means unusual, there is an expectation of concentrated work in this field which may employ more sophisticated numerical techniques to solve more complex problems. Thus, while the true significance of the results lies outside the sphere of computational physics, these one-dimensional codes may be seen as a collective "feasibility experiment," proving the worth of proceeding with more elaborate development, with the assurance that the information so obtained will be useful in controlled fusion work.

The historical stimulus for Tokamak modeling was the measurement of spatial profiles of electron density and temperature in the T-3 experiment at the Kurchatov Institute, by a team of scientists from the Culham Laboratory (Robinson *et al.*, 1969). Successful modeling of other controlled fusion experiments had been done earlier (Roberts and Potter, 1970), but there had been a lack of detailed spatial information in low-beta toroidal research and so there was no convincing check for computational models. Early numerical work on Stellerator confinement (Hinnov *et al.*, 1968) was zero-dimensional, and proceeded from *a priori* assumptions about plasma transport to match the data by varying the magnitude of unmeasured coefficients describing the interaction of the plasma with the wall. Since theories of thermal conduction and convection are at issue, and since these processes have a strong effect on spatial profiles, the new measurements made it possible to compare theory and experiment more closely.

In a conference on the topic of low-beta toroidal research held at Dubna, USSR, in 1969, the first Tokamak transport codes were introduced. The papers by Dnestrovskii and Kostamarov (Dnestrovskii *et al.*, 1969) and by Mercier and Soubbaramayer (1969) presented models for the plasma energy balance and for the establishment of the discharge current.

Following the Thomson scattering measurements and the Dubna conference, there was renewed international interest in the Tokamak. The U.S. controlled-fusion program was revised to include several new Tokamaks, and these developments, in turn, stimulated the development of new codes. The Fourth International Conference on Controlled Fusion, sponsored by the International Atomic Energy Agency in Madison, Wisconsin in 1971, saw the presentation of results of new codes from Düchs *et al.* (1971) and Widner *et al.* (1971b).

The first results from this new generation of Tokamak experiments and fresh theoretical stimulus from the working out of neoclassical transport theory (Rosenbluth *et al.*, 1972; Hinton and Rosenbluth, 1973) were responsible for new work in developing codes which could properly treat the neoclassical theory. These new versions were discussed at the Fifth European Conference of Controlled Fusion and Plasma Physics at Grenoble (Hogan and Dory, 1972; Killeen, 1972) and at a subsequent specialist's meeting at Fontenay-aux-Roses. These papers and that of Wiley et al. (1972) contain results of a "benchmark" comparison of a standard case, run to facilitate evaluation of the different numerical techniques used in each code, as well as to ensure uniformity of notation in transport coefficients and the like. In this regard, steady-state solutions of the neoclassical equations found by Wiley and Hinton (1972) were valuable as a diagnostic tool. Mercier (1972) reviewed the status of Tokamak transport codes at the Grenoble conference.

Interest in neutral beam injection heating of Tokamaks to reactor parameters stimulated computation of the effects of suprathermal particles on the Tokamak plasma. Models for neutral injection into Tokamaks were given at the Grenoble conference by Hogan and Dory (1972) and by Girard et al. (1972). Studies of critical processes affecting the thermal plasma (such as the provision of large ohmic heating currents) and of conditions in beam-dominated D–T plasma have since been carried out. Düchs reviewed transport code work at the Sixth European Conference on Controlled Fusion and Plasma Physics held in Moscow in 1973 (Düchs et al., 1973a). A second specialists workshop to survey European work was held at Frascati, in September 1974. Tokamaks with noncircular cross section have been constructed, and one-dimensional codes written to model them (Krall, Rawls, and Helton, 1974). Proposals to achieve thermonuclear break-even by beam–plasma reactions stimulated further numerical work. (Results from a one-dimensional code are given by Meade et al., 1974.)

There has existed a continuing experimental stimulus to improve the treatment of physical processes associated with neutral gas and impurity evolution. As these models were developed a number of different plasma transport models resulted, which gave a reasonable account of experimental behavior. Many models have been proposed to fit the data, and prompt the observation that these codes have reached maturity. Their adequacy can be tested against new experiments in new regimes, and further improvements may require a deeper treatment of the underlying physics.

B. A Brief Description of the Tokamak

1. *Fields and Geometry*

The geometry of a Tokamak is shown in Fig. 1. An electric field is induced in the toroidal direction, causing a toroidal current to flow. The poloidal

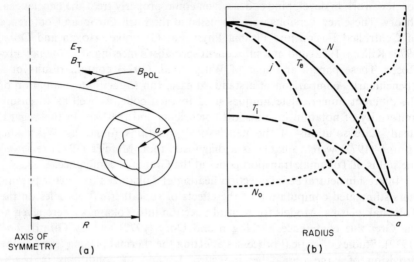

FIG. 1. (a) Arrangement of electromagnetic fields in a Tokamak and (b) representative computed profiles of plasma variables.

field and an externally applied toroidal magnetic field combine to give satisfactory plasma confinement. To a good approximation, surfaces of constant poloidal flux are nested toroids of circular cross section, whose major axes are slightly shifted from the major axis of the confining vessel. Thus, the radial profiles are roughly symmetric in the minor cross section, and they are symmetric longitudinally as well on the transport time-scales of interest. This, and the fact that the experimental profiles are roughly symmetric (Fig. 2), justifies the one-dimensional treatment adopted in the codes to be discussed.

There has been some preliminary work on multidimensional transport (Dnestrovskii et al., 1972). In cases with nonvarying total plasma current (Fowler et al., 1972), the magnetic flux surfaces can be regarded as steady and the transport equations can be solved with poloidal flux and time as variables instead of r, t (J. Rawls, J. Helton, and R. Miller, private communication, 1974).

2. Constituent Species

The Tokamak plasma constituents are:

1. Thermal distributions of electrons and protons (deuterons and tritons in reactor plasmas).
2. Suprathermal hydrogenic ions entering from injection at high energy, or alpha particles formed as reaction products (at 3.52 MeV) from the D–T

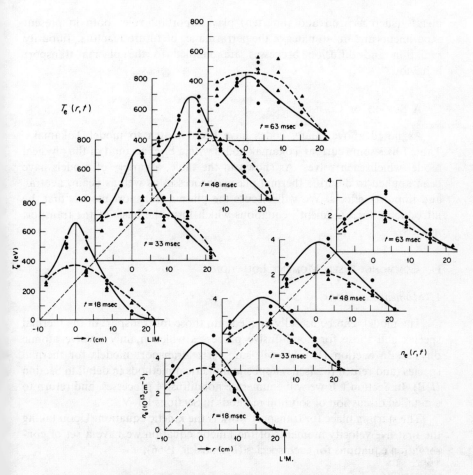

FIG. 2. Typical experimental profiles from Thomson scattering measurement of electron density and temperature (ORMAK data courtesy of M. Murakami). —●— Type A: $p = 4.8 \times 10^{-4}$ Torr. --▲-- Type B: $p = 4.4 \times 10^{-4}$ Torr. $I_p \simeq 95$ kA; $B_T \simeq 18$ kG.

reaction. These thermalizing distributions are most accurately treated by a Fokker–Planck equation (Cordey and Core, 1974; Killeen and Marx, 1970; Callen et al., 1974), but for application in fluid codes approximate expressions for number, momentum, and energy moments of the Fokker–Planck equation are also used.

3. *Neutral gas.* The plasma is created by ionization of neutral hydrogen, and the balance between transport loss of plasma ions and recycling of neutrals determines the plasma parameters.

4. *Impurities.* Low-Z contaminants (such as carbon and oxygen) and heavy

metals (such as iron and tungsten) play important roles both in present experiments and in estimates of the performance of future reactors. Impurity radiation and diffusion processes are coupled to the plasma transport behavior.

C. A Survey of Codes

As noted above, many codes have been written to model Tokamaks. Table I lists some current Tokamak codes with a brief account of the physical model which each solves. As shown in the table, a variety of models have been applied to describe thermal plasma processes as well as beam, neutral, and impurity effects. We will discuss the plasma transport models first, by introducing the "moment" equations which form the basis of the transport models.

D. Commonly Used Moment Equations

1. *Moment Equations*

The models can be usefully divided into those for transport of the thermal species and those for nondiffusive processes which mainly involve atomic physics. In Section II,A we will summarize transport models for thermal species (and return to discuss apposite numerical methods in detail in Section II,B). In Section I,E we will enumerate nondiffusive processes, and return to a detailed discussion of solution methods in Sections III–V.

The starting place for transport work is the kinetic equation. Upon taking the first five velocity moments of the kinetic equation we have a set of conservation equations for each species (Braginskii, 1966),

$$\partial n_i/\partial t + \nabla \cdot (n_i \mathbf{V}_i) = \Sigma^{(1)}, \tag{1}$$

$$p_i(d\mathbf{V}_i/dt + \mathbf{V}_i \cdot \nabla V_i^\alpha) = -\nabla \cdot \mathbf{P}_i + \int dv\, m\mathbf{V}_i C + \Sigma^{(2)}, \tag{2}$$

$$\frac{3}{2}\frac{\partial}{\partial t}(n_i T_i) + \nabla \cdot (n_i T_i \mathbf{V}_i) = \int dv\, m\frac{V_i^2}{2} C + \Sigma^{(3)}, \tag{3}$$

where n_i is the number density, $\mathbf{P}_i$ the pressure tensor, V_i^α the α-component of velocity, $\mathbf{q}_i$ the heat flow vector, C the scattering operator resulting from collisions or turbulence, and $\Sigma^{(k)}$ are the sources of mass, momentum, and energy from suprathermal or atomic phenomena.

There are serious conceptual questions regarding the foundation of toroidal equilibrium and transport calculations (Grad, 1967). This fact is reflected in the lack of detailed agreement between self-consistent theoretical predictions and experiment. In the absence of a fundamental basis for the calculations, numerical modeling must have a strong empirical component, and may only serve to develop scaling relationships over modest parameter ranges. The models discussed here all proceed in a semiempirical way from a common set of equations, which may be viewed as moments of a generalized kinetic equation. (The moments of the Fokker–Planck drift-kinetic equation give the "neoclassical" model.) The scattering integral C determines the form of the macroscopic equations, and also whether or not they are closed. For a number of physical models the set closes with no need of moments higher than $q_{e,i}$. These may be discussed by first examining the "neoclassical" equations, solved for the toroidal geometry of Fig. 1. For this case, C is given by the Landau, or Fokker–Planck, collision term (Rosenbluth et al., 1957). These equations are expressed in terms of averages over a magnetic flux surface, and these averages are, in turn, reduced to a one-dimensional variation in space:

$$\frac{\partial}{\partial t} n(r,t) + \frac{1}{r} \frac{\partial}{\partial r}(rnV) = \Sigma_n, \tag{4}$$

$$\frac{3}{2} \frac{\partial}{\partial t}(nT_e) = -\frac{1}{r} \frac{\partial}{\partial r}(rq_e) + \Sigma_e, \tag{5}$$

$$\frac{3}{2} \frac{\partial}{\partial t}(nT_i) = -\frac{1}{r} \frac{\partial}{\partial r}(rq_i) + \Sigma_i, \tag{6}$$

$$j = \sigma E + \sum_k j_k \tag{7}$$

where

$$nV = -\sum_{12}(L_{11} \, \partial n/\partial r + L_{12} \, \partial T_e/\partial r + L_{13} \, \partial T_i/\partial r + L_{14} \mathscr{E}), \tag{8}$$

$$q_e = \tfrac{3}{2} nVT_e + (L_{12} \, \partial n/\partial r + L_{22} \, \partial T_e/\partial r + L_{23} \, \partial T_i/\partial r + L_{24} \mathscr{E}), \tag{9}$$

$$q_i = (\tfrac{3}{2} - y)nVT_i + (L_{13} \, \partial n/\partial r + L_{23} \, \partial T_e/\partial r + L_{33} \, \partial T_i/\partial r + L_{34} \mathscr{E}), \tag{10}$$

$$j_n = L_{14} \, \partial n/\partial r + L_{24} \, \partial T_e/\partial r + L_{34} \, \partial T_i/\partial r, \tag{11}$$

where L_{ij} and y are given in Table II. (The notation follows Hinton and Rosenbluth, 1973). These equations describe the irreducible lower limit to transport rates in toroidal systems, and they have been used in many applications. They do not adequately describe the experiments, however, and so alternative models have been proposed. These "semiempirical" models

TABLE I

TOKAMAK TRANSPORT CODES[a]

Authors	Thermal plasma models	Neutrals	Impurities	Beams	Alpha particles	Geometry
Dnestrovskii, Kostamarov, Pavlova	Galeev-Sagdeev	Analytical model: integral equation	Effective Z^b	None	None	1-D, 2-D
Mercier, Soubaramayer, Boujot	Pfirsch-Schluter (enhanced), Galeev-Sagdeev plateau	Integral equation	Effective Z^b	None	None	
Düchs, Furth, Rutherford, Seidl	"6-regime model"	Particle simulation	Low Z: Multifluid rate equations High Z: Corona equilibria	Analytical models	Analytical models	1-D

Author	Pseudo/neoclassical, trapped particle modes, MHD turbulence	Integral equation	Low Z: Multifluid rate equations High Z: Corona equilibria	$H(r)$, G_e, G_i, k_e, k_i	Analytical models	1-D, 1-DM[c]
Hogan, Widner, Dory, Howe						
Killeen, Keeping, Grimm	Neoclassical	None	Effective Z[b]	None	None	1-D
Wiley, Hinton	Neoclassical	None	Ionic Z	None	None	1-D
Girard, Marty, Khelladi, Moriette	Pseudo/neoclassical	Monte Carlo	Effective Z[b]	Monte Carlo	Analytical models	1-D
Helton, Rawls	Neoclassical	None	Effective Z[b]	None	None	1-DM[c]
Krall, Thomson Bryne	Neoclassical turbulence	None	Effective Z[b]	None	None	1-DM[c]
Tajima, Takeda, Itoh	Neoclassical	Analytical	Effective Z[b]	None	None	1-D

[a] A summary of one-dimensional Tokamak transport codes known to the author.
[b] Impurities are treated by increasing the ionic charge assigned to plasma ions.
[c] One-dimensional code with a variable metric to include two-dimensional effects.

involve taking a subset of Eqs. (8)–(11), usually the diagonal terms L_{jj}, and then modifying the diffusion coefficients. These models will be described in Section II, A.

TABLE II

NEOCLASSICAL TRANSPORT COEFFICIENTS

These coefficients are expressed as functions of $v_{e,i}^*$ where $v_{e,i}^*$ is the ratio of effective electron (ion) collision frequency to electron (ion) bounce frequency.

$$v_e^* = \frac{(B_t/B_\theta)r}{(r/R)^{3/2}\tau_e(T_e/m_e)^{1/2}} \qquad v_i^* = \frac{(B_t/B_\theta)r}{(r/R)^{3/2}\tau_i(T_i/m_i)^{3/2}}$$

Figure 1 shows the geometry

B_Z: toroidal magnetic field
B_θ: poloidal magnetic field
T_e: electron temperature
n: plasma density
T_i: ion temperature
$m_{e,i}$: electron (ion) mass

$$\tau_e \equiv \frac{3m_e^{1/2}}{4(2\pi)^{1/2}\ln\Lambda e^4 n}T_e^{3/2} \qquad \tau_i = \frac{3m_i^{1/2}T_i^{3/2}}{4\pi^{1/2}Ne^4\ln\Lambda}$$

where

$$L_{ij} = \begin{bmatrix} a_{11}nD & a_{12}nD & a_{13}n(r/R)^{1/2}B_\theta^{-1} \\ a_{12}nT_eD & a_{22}nT_eD & a_{23}nT_e(r/R)^{1/2}B_\theta^{-1} \\ a_{13}n(r/R)^{1/2}B_\theta^{-1} & a_{23}nT_e(r/R^{1/2})B_\theta^{-1} & a_{33}(r/R)^{1/2}(1/\eta_{11}T_e) \end{bmatrix}$$

$$[a] = \begin{bmatrix} \dfrac{1.12}{1+1.78v_e^*} & \dfrac{1.25}{1+1.78v_e^*} & 0. & \dfrac{2.44}{1+0.85v_e^*} \\ \dfrac{1.25}{1+0.66v_e^*} & \dfrac{2.64}{1+0.35v_e^*} & 0. & \dfrac{4.35}{1+0.4v_e^*} \\ 0. & 0. & (A) & 0. \\ \dfrac{2.44}{1+0.85v^*} & \dfrac{4.35}{1+0.4v_e^*} & 0. & \dfrac{1.9}{1+v_e^*} \end{bmatrix}$$

$$(A): \frac{0.68}{1+0.35v_i^*}\left[\left(\frac{m_i}{m_e}\right)^{1/2}\left(\frac{T_e}{T_i}\right)^{1/2}\frac{1}{2^{1/2}}\right]$$

In addition to these transport equations, the plasma flow is coupled to electromagnetic fields. The Maxwell equations which are used in the models we discuss couple the poloidal magnetic field (B_θ) to the toroidal electric field $(\mathscr{E})$ and current density (j).

$$\partial B_\theta/\partial t = \partial\mathscr{E}/\partial r, \tag{12}$$

$$\mu_0 j = (1/r)(\partial/\partial r)(rB_\theta). \tag{13}$$

2. Boundary Conditions

a. Symmetry at the Magnetic Axis. The cylindrical approximation (and the data from experimental applications) suggest symmetry about $r = 0$ (Fig. 1). This means that the radial derivatives are zero for all scalars and polar vector components and that all axial vector components are zero at $r = 0$. The second-order accuracy of the numerical schemes requires that we integrate the differential equations over a cylindrical volume centered about $r = 0$ (Fig. 1b).

b. The Outer Boundary Conditions. Boundary conditions applied at $r = a$, the plasma edge, are more complex. The simplest model specifies a fixed (or preprogrammed) level for the transport variables, i.e., $n_{e,i}(a,t) = \alpha_{e,i}$; $T_{e,i}(a,t) = \beta_{e,i}$; $B(a,t) = I(t)/5a$, where I is the total plasma current and a the minor radius. However, the applications may require, in general, an impedance boundary condition of the type

$$\alpha n + \beta \, dn/dr = \gamma. \tag{14}$$

Such a condition is required by the treatment of recycling at the edge where outflowing charged particles are scraped off. Here the level of density at the boundary is determined by the rate of efflux and by the lifetime of a charged particle against recombination. Boundary values for the poloidal magnetic field are found by solving the external circuit equations, treating the resistance and inductance of the plasma self-consistently. These equations vary according to the nature of the external electrotechnical apparatus of each Tokamak, with some programmed to run at constant current, and others at constant voltage.

The transport equations and boundary conditions are common to the codes discussed here. The treatment of other physical processes differs widely from code to code, and we next turn to an enumeration of some other processes which are important in Tokamaks.

E. SOME RELEVANT PHYSICAL PROCESSES

We will discuss nondiffusion losses for each constituent in turn.

1. Thermal Distribution

There are sources of electrons and hydrogenic ions in these processes: ionization of cold neutral hydrogenic atoms by electron impact; and stripping of impurity ions. The momentum equations must account for: enhanced resistivity from impurities and alpha particles; and modification to toroidal flows caused by injected beams. The energy equation must account for: (a) transfer of energy from electrons and hydrogenic ions to impurities and subsequent loss by line radiation, bremsstrahlung, and recombination radiation; (b) charge exchange between hydrogenic ions and colder neutral atoms; (c) collision frequency enhancement in diffusion terms by nonhydrogenic impurities; (d) energy input by thermalization of injected beams; (e) energy input by thermalization of alpha particle reaction products; (f) energy loss by synchrotron radiation; and (g) energy loss by electron bremsstrahlung caused by scattering from hydrogenic ions.

2. *Suprathermal Ions*

The sources of particles are charge exchange and ionization of energetic injected neutrals, and D–T reactions producing 3.5 MeV alphas. The momentum equations account for the net current introduced by injected beams. The energy equation accounts for thermalization with electrons, hydrogenic ions, and impurities; and acceleration by the electric field.

3. *Impurities*

The sources of particles are: impurities introduced by hydrogenic neutral sputtering evaporation, neutron sputtering, and impurity self-sputtering. The momentum equations account for the friction between hydrogenic ions and impurities which drives radial diffusion. The energy equation accounts for impurities equilibriating with hydrogenic ions, and their energy loss by radiation.

4. *Neutrals*

The sources of particles are the external filling gas and cold gas desorbed on the vacuum chamber wall. The momentum equations account for the plasma rotation which imparts a net momentum to the neutral distribution. The energy equation accounts for the charge exchange with plasma ions.

II. Plasma Models

In this section we review the transport models which have been applied to Tokamaks, and discuss numerical questions pertaining to their solution.

A. Transport by Collisional Diffusion and Turbulence

1. Neoclassical

a. The neoclassical formulation (without atomic physics complications) has been given in Eqs. (4)–(13) and represents the simplest model for plasma transport. Following completion of the neoclassical theory for large aspect ratio Tokamaks several independent numerical calculations of the full set of neoclassical transport equations were carried out. A standard case was solved and plasma profiles are shown in Fig. 3 (Hogan and Dory, 1972; Wiley et al., 1972; Keeping et al., 1972).

The standard case is an ST discharge with these parameters: Toroidal field 40 kG, plasma current 60,000 A, minor radius 14 cm, major radius 109 cm, no neutrals or impurities. The various numerical treatments are described in Section II, B. We note here that this is the most thoroughly developed theoretical model, and since attempts to match the data with it have been unsuccessful, alternatives have been proposed.

b. Earlier treatments of neoclassical models kept only the diagonal terms in Eqs. (8)–(11) and adopted a three-regime treatment, with coefficients derived by Galeev and Sagdeev (1968), by Pfirsch and Schlüter (1962), and by Shafranov (1966).

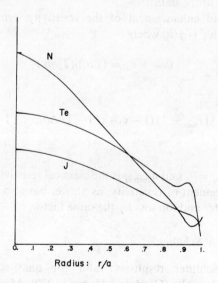

Fig. 3. Computed profiles for a standard comparison case: ST Tokamak. Maxima: T_e: 710, 319; N: 1.23 10^{13}; J: 146; t: 10 msec.

Banana: $(\nu_e^*, \nu_i^* \ll 1)$

$$\chi_e = 1.81\hat{\rho}_e^2 (r/R)^{1/2}/\tau_e, \qquad \chi_i = 0.68\hat{\rho}_i^2 (r/R)^{1/2}/\tau_i.$$

Intermediate:

$$\chi_e = 1.66 r^2 V_{Th_e}^3 / \Omega_t^e \Omega_{pd}^e R^2, \qquad \chi_i = 1.33 r^2 V_{Th_i}^3 / \Omega_t^i \Omega_{pa}^i R^2. \tag{15}$$

Classical: $(\nu_e^*, \nu_i^* \gg 1)$

$$\chi_{PS} = \chi_e = 2.33\hat{\rho}_e^2 (1 + 1.6 r^2 B_t^2 / R^2 B_{pol}^2)/\tau_e,$$

$$\chi_i = \hat{\rho}_i^2 (1 + 1.6 r^2 B_t^2 / R^2 B_{pol}^2)/\tau_i,$$

where

$$V_{Th_{e,i}}^2 = 2T_{e,i}/m_{e,i};$$

$$\Omega_{t,pol}^{e,i} = eB_{t,pol}/m_{e,i}c.$$

2. Modified Neoclassical

In a discussion of T-3 data, Dnestrovskii et al. (1971) have taken the diagonal terms of Eqs. (8)–(11), but have also included the effect of electron temperature gradient in the particle diffusion term (they have added L_{12}). In addition, they employ transport coefficients valid in the banana regime $(\nu_i^* \ll 1)$ which have been derived by Kovrizhnick (1969). Effectively, these rates are larger than the Hinton–Rosenbluth results for $\nu_i \sim 1$. They propose two models which fit the data:

(a) A sevenfold enhancement of the resistivity, employing the model $\eta = \gamma \cdot \eta_{Spitzer}$, with $\gamma = \gamma(\theta)$ where

$$\theta = u/v_d = (j/n_e)/(T_e/m_i)$$

and

$$\gamma(\theta) = \begin{cases} 1, & 0 < 1 \\ 1 + \tfrac{1}{2}(\gamma_{max} - 1)(1 - \cos\pi(\theta-1)/\Delta\theta), & 1 < \theta < 1 + \Delta\theta \\ \gamma_{max}, & \theta > +\Delta\theta \end{cases}$$

where $\Delta\theta = 2$, $\gamma_{max} = 7$, and $\eta_{Spitzer}$ is the classical resistivity.

(b) An enhancement of resistivity, as above, but also an enhancement of electron ion transfer and ion loss by the same factor.

3. Modified Pfirsch–Schlüter

The Pfirsch–Schlüter result is valid for quasi-stationary processes $(\tau_{pulse} \gg \tau_p)$ in Tokamaks (Grad and Hogan, 1970). Mercier and Soubbaramayer (1972) have obtained a fit to T-3 and ST electron temperature profiles

by employing
$$\chi_e = 2 \times 10^3 \chi_{PS}. \tag{16}$$

The T-3 temperature profiles are broad, while the ST profiles are sharply peaked. They have reproduced this trend with a single coefficient, while neglecting impurity and neutral effects.

4. *Pseudoclassical (Drift-Wave Turbulence)*

Artsimovich (1972) has inferred an electron diffusion coefficient χ_e from the laser-scattering data. He finds empirically that the *ansatz*

$$\chi_e = (5\text{--}10) n \hat{\rho}_e^2 / \tau_e \tag{17}$$

($\hat{\rho}_e$ is the electron's poloidal gyroradius) will reproduce the observed temperature profiles. Yoshikawa (1970) has presented a derivation of this formula from considerations of drift-wave turbulence and finds

$$\chi_e \sim C^2 n \hat{\rho}_e^2 / \tau_e \quad \text{where} \quad C^2 \sim 1. \tag{18}$$

This model has enjoyed wide popularity since it gives a rough fit to data from many machines (Düchs *et al.*, 1971; Kelley *et al.*, 1973).

5. *Trapped Particle Instabilities*

While there is considerable theoretical interest in the possibility of new collisionless instabilities arising from the presence of many trapped particles with bounce frequencies much greater than collision frequencies, modeling interest has centered on those described by Dean *et al.* (1974). The models reflect a sequence of transport processes, shown compared with neo- and pseudoclassical rates in Fig. 4. These estimates are made from calculations presented by Kadomtsev and Pogutse (1971). This model also seems to give rough agreement with the data.

6. *MHD Turbulence*

Robinson (1973) has proposed a turbulence model to explain the low energy lifetimes observed in the current-rise phase of Tokamaks. In this model the neoclassical rates are used unless the diffusion coefficient describing enhanced loss from poloidal field fluctuations is greater.

$$D = \begin{cases} D_{\text{neo}}, & D_{\text{neo}} < D_{\text{MHD}} \quad \text{(see Table II)}, \\ D_{\text{MHD}} = \gamma \Lambda V_\alpha, & D_{\text{MHD}} > D_{\text{neo}}. \end{cases} \tag{19}$$

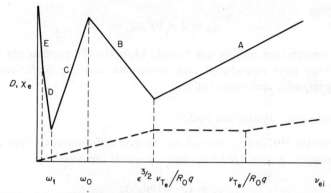

FIG. 4. Behavior of diffusion, conductivity coefficients with collision frequency for (A) pseudoclassical, (B) collisional dissipative trapped electron mode, (C) collisionless dissipative trapped electron mode, and (D), (E) trapped ion modes.

here Λ is the radial eigenfunction of the linear MHD stability problem, V_α is the Alfven velocity with respect to the poloidal field, and γ is the fluctuation amplitude.

Kadomtsev (1973) has noted that nonlinear MHD stability theory predicts the existence of macroscopic "bubble"-like solutions, with well-defined plasma vacuum regions in shear-free cases. In this model resistive diffusion fills the bubbles and this diffusion provides the plasma transport mechanism.

Tests with both these models give rough agreement with ORMAK data (Hogan, 1974).

7. Ion Acoustic Turbulence

Results from the "standard" density regime of Tokamaks ($n_e \gtrsim 2 \times 10^{13}$ cm^3) indicate that the observed resistance may be explained by the presence of impurity ions (discussed in Section III, A, 1). Nevertheless, an alternative has been proposed by Dnestrovskii et al. (1970) who employ a resistivity which varies in a way suggested by theories of ion-sound turbulence, which has been described in Section II, A, 1, a.

8. Other MHD Models

Mercier and Soubbaramayer (1970) have proposed a model based on the average magnetic shear. They state that

$$\begin{aligned} \chi_e &= \chi_1, & \theta < \theta_{\text{crit}}, \\ \chi_e &= \chi_2, & \theta \geq \theta_{\text{crit}}, \end{aligned} \qquad (20)$$

where θ_{crit} is a predetermined value of the shear, θ. Düchs et al., (1971) have proposed a model suggested by ideal interchange theory. They invoke a skin-limiter rule which enhances electron thermal conductivity by a factor $10^4 \, s(r/a)^2$ at all points outside a radius at which $d/dr(B_{\text{pol}}/r)$ is found to be positive. We now turn to an account of the numerical methods employed to implement these models.

B. Methods Used To Solve the Diffusion Problem

1. Difference Schemes

The conservation equations for the thermal plasma components [Eqs. (1)–(3), (12), (13)] are coupled, nonlinear diffusion equations. Their solutions are experimental in that proofs of convergence and accuracy are hard to find (in fact, for some coefficients it can be shown that the neoclassical equations do not have bounded solutions). The methods employed for each of the codes we have discussed are summarized in Table III.

Since most of the codes employ Crank–Nicholson schemes, we will illustrate two prominently used implicit schemes: linearization and predictor–corrector. As a model for discussion we employ the nonlinear initial boundary-value problem

$$\frac{\partial}{\partial t} u(r,t) = \frac{1}{r} \frac{\partial}{\partial r}\left[rk(u) \frac{\partial u}{\partial r} \right], \quad 0 \leqslant t \leqslant t_{\max}, \quad 0 \leqslant r \leqslant 1, \quad (21)$$

TABLE III

NUMERICAL METHODS

Code	Method for solving diffusion equations
Dnestrovskii et al. (1969)	Fully implicit integration with iteration
Mercier et al. (1969)	Implicit, variable phase, $(0 < \theta < 1)$ (Temam, 1968; Janenko, 1970)
Düchs et al. (1971)	Implicit, time centered Crank–Nicholson nonlinear diffusion coefficients linearized in time
Widner et al. (1971a)	Implicit, time centered, Crank–Nicholson nonlinear diffusion coefficients linearized in time
Wiley and Hinton (1972)	Implicit, Crank–Nicholson scheme used with predictor and iterated corrector
Killeen et al. (1972)	Implicit, backward difference. Equations solved by iteration
Girard et al. (1972)	Iterated explicit integration
Krall et al. (1975)	Central (Crank–Nicholson) or backward differencing. Velocity equation explicit. Linearized
Helton and Rawls (1975)	Predictor–corrector

with $u(x, 0)$, $\partial u/\partial r(0, t)$, and $u(1, t)$ prescribed. The Crank–Nicholson representation is

$$(u_j^{N+1} - u_j^N)/\Delta t = [r_{j+1/2} k_{j+1/2}^{N+1/2}/(\Delta r)^2 r_j][(u_{j+1}^{N+1} + u_{j+1}^N)/2]$$
$$- (r_{j+1/2} k_{j+1}^{N+1/2} + r_{j-1/2} K_{j-1/2}^{N+1/2})[(u_j^{N+1} + u_j^N/2)]$$
$$+ (r_{j-1/2} k_{j-1/2}^{N+1/2}/(\Delta r)^2 r_j)[(u_{j-1}^{N+1} + u_j^N)/2], \qquad (22)$$

where

$$k_{j+1/2}^{N+1/2} \equiv k[(u_{j+1}^{N+1/2} + u_j^{N+1/2})/2]$$

with the interval $[0, 1]$ discretized as

$$r_j = (j-1)\Delta r \quad \text{for} \quad j = 1, n$$

$$u_j^{N+1} \qquad t^{\text{new}} = t + \Delta t$$

$$t \uparrow \qquad 0$$
$$u_{j-1}^N \quad u_j^N \quad u_{j+1}^N$$
$$0 \quad 0 \quad 0 \qquad t$$

$$\xrightarrow{\quad\quad r \quad\quad}$$

a. Linearization. Düchs (1972) and Widner *et al.* (1971a) have assumed that the transport coefficients [$k(u)$ in this model] have a power-law dependence. (This assumption is valid for the transport models employing matched values in three regimes; see Section II, A, 1, b.) Here this would imply $k(u) \propto u^\alpha$. With this assumption, then $k_{i+1/2}^{n+1/2}$, which is required to construct the solution, is given by

$$k_{i+1/2}^{n+1/2} = k_{i+1/2}^n + \alpha (k_{i+1/2}^n/u_{i+1/2}^n)(u_{i+1/2}^{n+1} - u_{i+1/2}^n)/\Delta t.$$

Then the model equation becomes

$$\frac{u_i^{n+1} - u_i^n}{\Delta t} = \left\{ \frac{k_i^n}{2(\Delta r)^2} + \frac{1}{8r_i(\Delta r)^2} \left[r_{i+1} k_{i+1}^n \left(H\alpha - \alpha \frac{u_{j-1}^n}{u_{j+1}^n} \right) - r_{i-1} k_{i-1}^n \right] \right\} u_{i+1}^{n+1}$$
$$+ \left[\frac{k_j^n}{2} \left(\delta^2 u \frac{d}{u_j^n} - \frac{2}{\Delta r^2} \right) \right] u_j^{n+1} + \left\{ \frac{k_j^n}{2(\Delta r)^2} - \frac{1}{8_j{}^r(\Delta r)^2} \right.$$
$$\left. * \left[r_{j+1} k_{j+1}^n - r_{j-1} k_{j-1}^n \left(H\alpha - \alpha \frac{u_{j+1}^n}{u_{j-1}^n} \right) \right] \right\} u_{j-1}^{n+1}$$

$$+ \frac{k_j^n}{2} \delta^2 u^n (1 - \alpha) + \frac{1}{8r_j(\Delta r)^2}$$

$$\times \{(u_{j+1}^n - u_{j-1}^n)[r_{j+1} k_{j+1}^n (1 - \alpha) - r_{j-1} k_{j-1}^n (1 - \alpha)]\} \tag{23}$$

where

$$\delta^2 u = (u_{j+1}^n - 2u_j^n + u_{j+1}^n)/(\Delta r)^2, \qquad k_j^n = k(u_j^n).$$

As this equation is tridiagonal and linear in u, it can be solved by standard techniques which we discuss later.

Since theoretical work usually specifies diffusion coefficients which vary in a more complicated way than as u^α, this method is not literally applied since it would involve a reworking of the code for each different transport model. Instead a general analytic dependence $k(A(u))$ is prescribed, where $A(u)$ describes the variation of the transport coefficient with such transport variables as v_i^* (the ratio of collision to bounce frequency), ω_b, or ω^* (ω_b is the bounce frequency, ω^* the diamagnetic drift frequency). Then

$$k(u)_j^{n+1/2} = k_n^j + \frac{dk}{dA}\bigg|_j^k \frac{dA}{du}\bigg|_j^k \frac{u_j^{n+1} - u^n}{\Delta t} \cdot \frac{\Delta t}{2} \tag{24}$$

where dk/dA and dA/du are given analytically.

b. Predictor–Corrector. The predictor–corrector method is used by J. C. Wiley (unpublished thesis, 1973) with the Crank–Nicholson representation. The predictor step is accomplished by solving for $u_j^{n+1/2}$ with $t^{\text{new}} + \Delta t/2$. He assumes for the predictor step that

$$k_{i+1/2}^{n+1/2} \simeq k(\tfrac{1}{2} u_{i+1}^n + u_i^n). \tag{25}$$

This approximation to u is then used in the first corrector step to find u_j^{n+1}. The corrector step can be repeated, and he assumes

$$u_i^{n+1/2} = \tfrac{1}{2}(u_i^n + u_i^{n+1(m)}), \tag{26}$$

where $u_i^{n+1(m)}$ is the most recent approximation. He finds that one corrector step is usually sufficient for the problem. This method has been compared to the method of linearization by Wiley. In place of Eq. (23) one finds

$$(u_i^{n+1} - u_i^n)/\Delta t = (1/2(\Delta r)^2 r_j)[(r_{j+1/2} k_{j+1/2}^{t+\Delta t/2}) u_{j+1}^{t+\Delta t/2}$$
$$- (r_{j+1/2} k_{j+1/2}^{t+\Delta t/2} + r_{j-1/2} k_{j-1/2}^{t+\Delta t/2}) u_j^{t+\Delta t}$$
$$+ (r_{j-1/2} k_{j-1/2}^{t+\Delta t/2}) u_{j-1}^{t+\Delta t} + (r_{j+1/2} k_{j+1/2}^{t+\Delta t/2}) u_{j+1}^t$$
$$- (r_{j+1/2} k_{j+1/2}^{t+\Delta t/2} + r_{j-1/2} k_{j-1/2}^{t+\Delta t/2}) u_j^t + (r_{j-1/2} k_{j-1/2}^{t+\Delta t/2}) u_{j-1}^t], \tag{27}$$

where

$$k_{j+1/2}^{t+\Delta t/2} = \tfrac{1}{2}k(u_{j+1}^{t+\Delta t/2} + u_j^{t+\Delta t/2}).$$

The obvious advantage, as noted, is that coding is insensitive to the detailed form of the transport coefficients.

c. *Fully Implicit Crank–Nicholson.* Killeen (1972) has used a generalization of these procedures with successive iteration of the equations. The function k is evaluated by solving a complete forward difference for u, then using the solution to evaluate k.

All the implicit methods produce a tridiagonal system of difference equations of the form

$$-\alpha_j u_{j+1}^{N+1} + \beta_j u_j^{N+1} - \gamma_j u_{j-1}^{N+1} = \delta_j. \tag{28}$$

2. Solution Schemes

When the coupled transport equations are treated, the model equation is replaced by a matrix equation. The Gaussian elimination procedure is employed and we assume the solution (vector) has the form

$$u_j = E_j u_{j+1} + F_j \quad (1 < j \leqslant n). \tag{29}$$

Substituting u_{j-1} [as in (29)] into (28) we find

$$u_j = -(\beta_j + \gamma_j E_{j-1})^{-1}\alpha_j u_{j+1} + (\beta_j + \gamma_j E_{j-1})^{-1}(\beta_j - \gamma_j E_{j-1}). \tag{30}$$

Equating coefficients in (30) we find

$$\begin{aligned} E_j &= -(\beta_j + \gamma_j E_{j-1})^{-1}\alpha_j, \\ F_j &= (\beta_j + \gamma_j E_{j-1})^{-1}(\gamma_j - \gamma_j E_{j-1}). \end{aligned} \tag{31}$$

Thus, if E_1 and F_1 are known we may (recursively) obtain all E_j's and F_j's. If all E_j's and F_j are known, and if we have a boundary condition linking u_j to u_{j-1}, we may find all u_j. [This standard procedure is also discussed by Potter (1973)].

III. Suprathermal Plasma: Injected Ions and Alpha Particles

In a way the more interesting species in the Tokamak are the high energy particles provided by injection of neutrals and the alpha particle reaction products. The emphasis in Tokamak research is shifting toward greater reliance upon beam contributions to fusion energy output, so that a good model is required. Energetic alpha particles, which must be described to model D–T processes, will be treated as a subcase.

We will first discuss models for these processes, and then their implementation in transport codes.

A. BEAM TRAPPING AND THERMALIZATION

1. *Source*

Figure 5 shows the typical experimental arrangement for neutral injection. A beam of particles with energies W_0, $\frac{1}{2}W_0$, and $\frac{1}{3}W_0$ emerges from the neutralizing cell and passes through the plasma. These neutrals are subject to ionization by impact with electrons and hydrogenic and impurity ions, and to charge exchange with plasma protons. Neutrals are converted into ions (or lost by charge exchange) all along the path shown in Fig. 5. Rome *et al.*

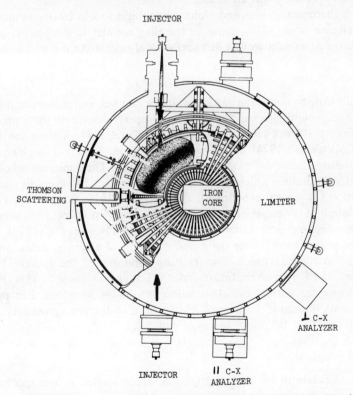

FIG. 5. Experimental arrangement for neutral injection into ORMAK. Two injectors with beam lines roughly parallel to the toroidal magnetic field.

(1974) have calculated the resulting source of energetic ions averaged over a magnetic surface (as happens physically with the motion along magnetic field lines). The relative fraction of the beam located at radius r is determined by first calculating the attenuation of the beam along the beam line, then using the particle orbits of the newly born fast ions to construct the average over a flux surface. This gives the radial distribution of injected particles as a function of the plasma variables calculated in the transport code. All treatments to date incorporate the assumption that the thermalization occurs on the original ("birth") flux surface.

For alpha particles the source is

$$\Sigma = n_D n_T \langle \sigma v \rangle_{DT} + \Sigma_{beam}$$

where $n_{D,T}$ is the number density of D(T) and Σ_{beam} are alpha particles produced during thermalization (Dawson et al., 1971). The same considerations about spatial deviation from birth flux surfaces apply to alphas as well, and the same approximations are made.

Both thermalizing ions and alphas are subject to loss by unconfined drift orbits (Rome et al., 1975). Particles traveling counter to the plasma current suffer large deviations from a flux surface and may strike the wall or limiter.

2. Thermalization

While slowing down, the particles transfer number, momentum, and energy to the background plasma. The equation properly describing these processes is the Fokker–Planck equation (Cordey and Core, 1974; Killeen and Marx, 1970; Callen et al., 1974), and work in coupling the Fokker–Planck equation to the thermal transport model is now under way. The construction of a set of moment equations which are then solved approximately gives an adequate description for present experiments (Kelley et al., 1973; Berry et al., 1974b). The solution of the equations of motion of a "slug" of injected beam current has been used to model the two-component torus (Meade et al., 1974). Work is under way to incorporate the radial variation of magnetic fields, and the effects of particle trapping, into existing solutions of the Fokker–Planck equation. Then, with larger future codes which employ these Fokker–Planck calculations, the correct radial variation can be incorporated. For present experiments the model is in error by the ratio of the ion gyroradius in the poloidal field to a, the minor radius (about 20%).

B. Implementation

The calculations of beam trapping as incorporated in transport codes proceed directly along the lines discussed in Rome et al. (1974). The solution of a first-order ordinary differential equation for spatial attenuation of the beam is followed by averaging the fast ion distribution over the appropriate drift surface. The moment calculations of beam contributions of energy to the thermal plasma (for example) use an analytic approximation to the steady-state distribution:

$$f_b(v) = [\tau_s/2\pi(v^3 + v_c^3)][(v_0^3 + v_c^3)/(v^3 + v_c^3)]^{-\tau_s/3\tau_{cx}}.$$

Then the heat input to the electrons and protons of the background is

$$Q_f^e = n_f m_f \int_0^{v_0} dv \frac{v^4}{v_c^3 + v^3} p_{cx}(v),$$

$$Q_f^p = n_f m_f \int_0^{v_0} dv \frac{v v_c^3}{v^3 + v_c^3} p_{cx}(v),$$

where v_0 is the initial speed of the fast ion, n_f the fast ion input rate, and m_f is the mass of fast ions.

$$v_c = \{(29T_e/m_f)(m_f/m_H)^{1/3}(m_f[Z]/m_i)^{2/3}\}^{1/2},$$

$$[Z] \equiv \sum_j n_j Z j^2 (m_H/m_j) / \sum_i n_i Z_i,$$

$$p_{cx}(v) = \exp\left\{-\int_v^{v_0} \frac{x^2 dx}{x^3 + v_c^3} \frac{\tau_s}{\tau_{cx}(x)}\right\},$$

where τ_s and τ_{cx} are the thermalization and charge exchange lifetimes.

IV. Neutral Gas

There can be no successful attempt to model present or proposed Tokamaks which does not include an accurate treatment of neutral gas. The issues which codes have clarified in confinement experiments are closely coupled to the role of neutrals: (a) the reasons for the existence of a density limit in Tokamaks (Düchs et al., 1972b); (b) the role of charge exchange in the energy balance at low proton densities (Dnestrovskii et al., 1973; Kelley et al., 1973); (c) the prospects for enhanced impurity evolution in D–T experiments and reactors (Meade et al., 1974); and (d) the scaling of plasma transport with density, current, and temperature (Düchs et al., 1972c; Hogan, 1974).

Charge exchange can be a significant contribution to losses in present experiments, while questions of the gas level required to achieve long pulses and adequate fueling will be of concern for reactors.

The physical processes involved in the neutrals' interaction with the plasma are straightforward. The wall and limiter contain layers of neutral hydrogenic atoms which enter the plasma by a variety of mechanisms. One must account for ionization of the neutrals by electron impact and by charge exchange with protons.

A. Models for Neutral Transport

1. Transport Equation

The starting point is the Boltzmann equation, since the problem is quite similar to neutron transport problems which have been extensively considered (Greenspan, 1974). Existing neutron transport codes are generally not well adapted to serve as subroutines in the relatively small Tokamak transport codes, so several groups have constructed neutral gas models independently. Some characteristics of the models are described briefly in Table IV.

The physical model for neutral behavior was proposed by Sakharov (1961). Cold neutral gas absorbed on the vacuum chamber walls may be knocked back into the plasma by plasma efflux. Entering the plasma at the wall energy, the atom (or molecule) has a high probability of charge exchange (or ionization). If charge exchange occurs with a hotter plasma ion, and if the particle happens to be directed inward, then it acquires a much higher mean free path until the next event. This upward cascade in energy can produce a significant number of neutrals even in the core of the plasma (Berry et al., 1974a). (In present experiments there is about 1 neutral/10^5 protons, but since the charge exchange cross section is large, even this low level contributes to the net energy balance.)

TABLE IV

Solutions for Neutral Transport

Code	Geometry	Ion distribution	Numerical method
Dnestrovskii et al. (1971)	Slab	Maxwellian	Functional iteration (integral equation for N_0)
Düchs et al. (1972)	Cylinder	Particles born at local T_i	Follow particles incident at boundary for n (≤ 10) generations of charge exchange events
Mercier et al. (1972)	3-D	Maxwellian and delta-function	Solution of integral equation for N_0 (3-dimensional treatment)
Girard et al. (1972)	Cylinder	Maxwellian	Particle following
Hogan and Dory et al. (1972)	Slab	Maxwellian	Functional iteration (integral equation for N_0)

The plasma parameters determine the lifetime for ionization and charge exchange, of course, but the effect of the exterior is important in this problem. Sources of neutral atoms are: (1) dissociative ionization of H_2^+ produced by ionization of H_2^0 from the walls (Franck–Condon transitions); (2) neutralization of outflowing plasma ions leading to an inward flux of more energetic neutrals (They are isotropic in direction, since the phase angle of the gyromotion of the ions is arbitrary at collision.); and (3) reflection of charge exchange particles from the wall. This is expected to become a more severe problem as Tokamaks achieve higher ion temperatures.

We shall briefly discuss the extant models for treating neutrals via the transport equation as has been done by Dnestrovskii et al. (1973), Hogan and Clarke (1974), and Mercier and Soubbaramayer (1972).

a. Analytic. Konstantinov and Perel (1961) gave an analytic approximation to solving the kinetic equation which is useful for checking numerical results. They assume flat plasma density and temperature profiles, and a delta-function ion energy distribution. The resulting solution is a sum of exponentials.

b. Delta-Function Model. Dnestrovskii et al. (1973) assume that with each charge exchange, the new neutral acquires the local ion temperature, with equal probability of traveling in the $\pm x$-direction. (They treat a one-dimensional, slab geometry.) Thus, they compute

$$V_x \, \partial f/\partial x = Sf = (FN/2v_0)[\delta((V-v_i)/v_u) + \delta((V+v_i)/v_u)], \quad (32)$$

where $F(r,v)$ is the ion distribution.

$$f(\pm a, v) = (N_0/V_0)(v \pm V_0/a).$$

where V_0 is the incident neutral velocity. This leads to an integral equation for density

$$N_0(x) = \sum_{k=0}^{\infty} N_k(x); \quad N_{k+1}(x) = \int_0^a d\xi \, K(x,\xi) N_k(\xi), \quad (33)$$

where

$$K(x,\xi) = \tfrac{1}{2}[F(\xi)/V_i(\xi)] \left[\Phi(\xi, x, V_i(\xi)) S(x-\xi) + \Phi(0,\xi,V_i(\xi)) \Phi(0,x,V_i(x)) \right]$$

$$\Phi(\xi, x, V) = \exp[-(1/V) S_\xi^x S dx^1], \quad S = \text{sign}.$$

c. Slab Model. Hogan and Clarke (1974) have solved the kinetic equation with two-dimensional velocity space for a slab model

$$V_x \, \partial F/\partial x = (\langle \sigma v \rangle_i + \overline{\sigma v}_{cx}) f_i + (n_0 + n_1) \overline{\sigma v}_{cx}, \quad (34)$$

where f_0 is the solution to the equation

$$V_x \, \partial f_0/\partial x \equiv (\langle \sigma v \rangle_{\text{ion}} + \overline{\sigma v}_{\text{cx}}) N_0$$

$$f_0(\pm a, V) = \varphi_{\text{incident}}(V); \qquad N_0 = \int d^3 v f_0.$$

$\langle \sigma v \rangle_i$, $\overline{\sigma v}_{\text{cx}}$ are the rate coefficients for electron impact ionization and charge exchange. After solving for n_1, the distribution function of the neutrals is computed and used for calculating the mean energy of the neutrals: to assess charge exchange loss, to compare the predicted charge exchange energy spectrum with the measured one, and to compute the sputtering induced by this charge exchange flux.

2. *Particle Models*

The codes of Düchs et al. (1972b) and of M. Khelladi (unpublished thesis, 1972) include a somewhat different treatment. In each the code follows incident particles in a cylindrical geometry, accounting for successive generations of particles in an explicit fashion.

B. IMPLEMENTATION OF NEUTRAL MODELS

The physical models for processes involving neutral gas show the great similarity which exists between this problem and neutron transport. Ionization of a neutral is analogous to neutron capture, while charge exchange represents a neutron collision. Existing neutron transport codes have been surveyed by Greenspan (1974), and are capable of treating a full three-dimensional problem by Monte Carlo techniques, as in the application to Tokamak problems by Parsons and Medley (1974). (Nevertheless, common experience with transport codes incorporating neutrals seems to be that the calculations of the neutrals employs much of the execution time.) The models have been grouped as analytic and particle-following, and we shall now turn to a discussion of these.

1. *Analytic*

Treating the slab models, the distribution function for the neutrals may conveniently be split so that one solves two linear integral differential equations—one a homogeneous equation with nonhomogeneous boundary conditions, and vice versa. That is,

$$f(x, v, \theta) = f_c(x, v, \theta) + f_h(x, v, \theta), \tag{35}$$

where

$$f_h(a,v,\theta) = 0, \qquad f_c(a,v,\theta) = f_{ext}(v,\theta), \qquad \pi/2 \leq \theta \leq \pi,$$
$$f_h(-a,v,\theta) = 0, \qquad f_c(-a,v,\theta) = f_{ext}(v,\theta), \qquad 0 \leq \theta \leq \pi/2.$$

Integrating under these assumptions we may reduce the problem by first solving for f_h, and then we find

$$n_h(y) = \alpha \int_{-1}^{1} dz \, \frac{(n_0 + n_c)}{T^{3/2}} G\left[\frac{\alpha}{T^{1/2}} \left| \int_z^y dz' e(z') \right| \right], \tag{36}$$

where

$$q(x,z) \equiv \frac{\alpha}{[T(x)]^{1/2}} \left| \int_z^x dz' e(z') \right|,$$

$$e(x) = [n_e(x)\langle\sigma v\rangle_{\text{ionization}} + n_p(x)\langle\sigma v\rangle_{\text{cx}}]/n_p(x=0)\langle\sigma v\rangle_{\text{cx}}(x=0),$$

$$G(q) \equiv \int_c^\infty dy \cdot y \, e^{-y^2} E_1(q/y).$$

E_1, E_2 are exponential integrals,

$$n_0(x) = E_1(\xi_1) + E_2(\xi_2),$$

$$\xi_1 \equiv \gamma \int_{-1}^z dz' e(z') \qquad \xi_2 \equiv \gamma \int_z^1 dz' e(z'),$$

$$\alpha = (2/\pi^{1/2})[n_p(0)\langle\sigma v\rangle_{\text{cx}}(x=0)]/[2kT_p(0)/M_p]^{3/2},$$

$$\beta = [E_0/T_p(0)]^{1/2}\alpha.$$

In Eq. (36) we have assumed a proton distribution function

$$f_p(v) = (m/2\pi kT)^{3/2} N_p e^{-mv^2/2kT}.$$

Dnestrovskii et al. (1973) represent the charge exchange event by the creation of a neutral with an energy equal to the local ion temperature, and assume an equal probability for traveling in the $\pm x$-directions (with no y component). This leads to the simpler integral equation given in Eq. (33). The difference between the two models is insignificant for calculating the energy balance, but can be important in a calculation of the sputtering of wall materials caused by charge exchange bombardment, since the single-particle sputtering yield is a rapidly varying function of energy. Each of these equations is solved by a functional iteration, with each iteration corresponding to a better approximation of the integral. A straightforward quadrature is inadequate since the function G possesses a logarithmic singularity at $q = 0$. For each j

the sequence $\{n_1^{(i)}(x_j)\}$ converges monotonically from below provided that

$$\tilde{\gamma} \equiv \underset{j}{\text{Max}} \left\{ \alpha \int_{-1}^{1} dz \frac{n_i(z)}{[T(z)]^{1/2}} G[q(z, x_j)] \right\} < 1. \tag{37}$$

Since the integral equation describes a continuous probability of charge exchange, it is hard to describe the iteration results in terms of a "number of generations" of charge exchange events.

2. Particle Simulation

The Düchs code employs a different technique. A particle source is assumed to exist on the boundary, and following a straight line trajectory the particle density is attenuated by ionization and charge exchange. The particle is assumed to undergo a charge exchange event at a point, and is assumed to acquire the incident ion energy. (Only velocities in the plane are assumed.) There is an arbitrary choice to be made of the number of charge exchange generations to be followed. Usually four generations suffice to reproduce an answer with 10% accuracy. Then

$$n_c(r) = n_{\text{cold}}(r) + \sum_{j=1}^{4} n_{\text{hot}}(r), \tag{38}$$

with

$$n_c(r) = \frac{n_c(\text{edge})}{\pi} \int_0^\pi d\theta \exp\left[-1/v_0 \int_0^{lAP} e(l) \, dl\right],$$

and, for example,

$$n_{\text{hot}}^{(1)}(r) = \frac{1}{\pi} \sum_{n=0}^{7} \int_0^{lAP(\theta_n)} NN_{\text{cold}} \, \sigma v_{\text{cx}} \frac{dl}{V_{\text{hot}}^{(1)}(l)}$$

$$* \exp\left[-1/V_{\text{hot}}^{(1)}(l) \int_0^l e(l) \, dl^1\right], \tag{39}$$

where lAP is the distance from source to field point, and θ_n are the eight angles used to approximate the θ variation.

V. Impurities

The role of impurity species may be divided into three categories according to the time period of relevance.

1. Contemporary experiments: a large variety of low-Z contaminants are present such as carbon and oxygen arising from an imperfect vacuum, pump accidents, and a host of other causes.

2. Present experiments/reactors. There is evidence that heavy metal impurities accumulate in discharges and this may be an inevitable concomitant of high temperature operation.

3. Reactors. The alpha particle reaction product in the D–T reaction is (after thermalization with the plasma) an unwelcome $Z = 2$ impurity ion. The performance of reactors may hinge on the rate of accumulation of alphas, since they contribute to the $\beta(\equiv nkT/B_{\text{pol}}^2)$ which is, in turn, limited to low values by stability requirements.

We will discuss the physical models for both low- and high-Z ($Z \geqslant 30$) impurities which have been employed, and at the same time discuss the means to implement solutions.

Physical Models

1. Low-Z

These are contaminants from hydrocarbons, ceramic insulators, and other sources. They must be described accurately since they seem to play a dominant role in transport. We will briefly discuss the charge state, radiation, and diffusion of these low-Z impurities as they affect transport codes.

The range of laboratory plasma density and temperatures in CTR magnetic confinement devices allows the use of the "corona" model to describe the equilibrium charge state of these ions. In this model the dominant processes are excitation by electron impact and recombination by radiative emission. This collisional radiative (CR) model has been widely used in Tokamak calculations. Using the CR model, the change in ionization state is given by

$$dy_k/dt = n_e[y_{k-1} S_{k-1} + y_{k+1}(\alpha_k + \gamma_k n_e) - n_k(S_k + \alpha_{h-1} + \gamma_k)], \quad (39)$$

where S_k is the ionization coefficient; α_k the radiative recombination coefficient; γ_k the three-body recombination coefficient; $y_k = N^{(k)}/N$; and $N^{(k)}$ is the density of ions in kth state.

As the rate coefficients depend on N_e, T_e, these equations must be solved simultaneously with the plasma transport equations. One cycle of this calculation will give the change in y_k from a change in the rates and this must be superimposed on the change in $N^{(k)}(x,t)$ caused by spatial diffusion. The work of Düchs et al. (1973b) described a treatment of oxygen impurities which are important to a description of the ST Tokamak (iron has been treated by Meade et al., 1974). The Oak Ridge code employs this calculation for all ionization states of carbon and oxygen. This calculation is done explicitly and adds about 20% to the execution time. Both use explicit techniques to integrate the rate equations.

The long-time equilibrium of these equations can be useful for some problems. These "corona equilibria" can be formulated (Jordan, 1969) as look-up tables, and data are available for all the low-Z elements carbon–nickel. These are much faster to run, but they cannot be used in the start-up phase in the Tokamak since the product of density and time is too low to reach these equilibria.

The contribution of these impurities to the transport calculations can come through enhancement of collision frequencies and through additional energy loss through stripping, line, and recombination radiation. Each electron taken from an impurity must be heated up to the prevailing electron temperature and the radiation from resonance fluorescence lines escapes the plasma. The expressions for line radiation loss have been given by Düchs *et al.* (1974).

These radiative losses are most important during the breakdown and current-penetration phase and near the edge, if fresh impurities are being produced by sputtering.

Besides the ionization level and radiative loss, it is important to model the diffusion of these impurity species. There is an important question bearing on the maximum allowable pulsed length for a reactor which is being addressed by the present generation of experiments. If impurity species diffuse inward and at the rate predicted by neoclassical theory, then the reactor pulse length will be shortened.

The relevant impurity fluxes are given by Moore and Hinton (1974) in a form similar to those in Table II. Results valid in the banana regime ($v_i^* \ll 1$) have been given by Sigmar *et al.* (1975). The form of the fluxes is identical with that of Eq. (8), except that now we are dealing with, say, eight species for oxygen (one for each charge state). The complexity of the difference equations multiplies as well, and the linearization technique discussed in Section II,2 becomes impractical for this problem.

Düchs *et al.* (1974) have approximated the problem by computing the $\sum Q_k k^2$ and treating this term explicitly. In studies underway at Oak Ridge the L_{kk} impurity cross terms are not differenced, while the L_{pk} terms are included and the solution is computed using a predictor–corrector method.

In view of proposals to construct the first wall of fusion reactors out of some low-Z material (such as carbon) impurity diffusion will play an increasingly important role (Kulcinski *et al.*, 1975).

2. *High-Z Metals*

The role of high-Z impurities in recent experiments is important but ill-treated by existing models. Experimental estimates of the concentration of heavy metals relative to protons range from 0.02% to 0.2%. If the former, the radiation loss would likely be small. Absolute measurement of line spectra yield the upper estimate for ST, with the result that the radiative power loss is 20% the input power.

Stripping and charge levels have been calculated from corona-equilibrium models, but as yet there is no time-dependent nonequilibrium model for these ions as there is for low Z. The metals of interest are Mo, W, V, Nb, and Au.

3. *Relation to the Calculation*

We summarize the relation of impurity ionization, radiation, and diffusion to the rest of the transport calculation in Table V.

TABLE V

IMPURITIES

Conservation	Term	Effect
Number	Σ_e	Source of electrons as impurities are stripped
	Γ	Classical models which employ collisional rates must account for enhancement by $Z_{\text{eff}} = \Sigma n_k Z_k^2 / \Sigma n_k Z_k$
	$\dot{N}$	$N_e \neq N_p$. Proton defect in present experiments, defect in reacting fuel in D–T systems
Energy: T_e	NT_e	Proton defect
T_i	$\text{div}(\Gamma T_e)$	Enhanced flow rate both by scattering and by impurity flux
	$\text{div}(\chi T_e')$	Resistivity enhanced
	ηj^2	
	P_{LR}, P_B, P_R	Line, recombination radiation, and bremsstrahlung
Momentum (Ohm's law)	ηj	Resistivity enhanced

VI. Summary

Besides the work referred to in describing the transport models, there is an extensive literature dealing with applications of these codes to CTR problems. The bibliography gives some indication of the report literature, which has been significant because of the provisional and exploratory nature of these codes.

As noted earlier the future prospects for transport code development seem to be along the lines of multidimensional calculations in physical and velocity space. When considering the upgrading of geometry (two-dimensional), beams (Fokker–Planck equation), impurities (many data files for all species), neutrals (neutron transport codes), and plasma transport (instability simulation), it can be seen that complexity can grow easily.

REFERENCES

Artsimovich, L. A. (1972). *Nucl. Fusion* **12**, 215.
Berry, L. A., Clark, J. F., and Hogan, J. T. (1974a). *Phys. Rev. Lett.* **32**, 362.
Berry, L. A., et al. (1974b). *Proc. Plasma Phys. Control. Nucl. Fusion, 5th, 1974.* Paper IAEA-CN-33/A5-1.
Braginskii, S. I. (1966). *Rev. Plasma Phys.* **1**, 205.
Callen, J. D., et al. (1974). *Proc. Conf. Plasma Phys. Control. Nucl. Fusion, 5th, 1974.* Paper IAEA-CN-33/A16-3.
Cordey, J. G., and Core, W. G. F. (1974). *Proc. Conf. Plasma Phys. Control. Nucl. Fusion, 5th, 1974.* Paper IAEA-CN-33/A16-1.
Dawson, J. M., Furth, H. P., and Tenney, F. (1971). *Phys. Rev. Lett.* **26**, 1156.
Dean, S. O., et al. (1974). "Status and Objectives of Tokamak Systems for Fusion Research," USAEC Rep. Wash-1295. U.S. At. Energy Comm., Washington, D.C.
Dnestrovskii, Y. N., Kostamarov, D. P., and Pavlova, N. L. (1969). *Proc. Int. Symp. Closed Confinement Syst., 1969.*
Dnestrovskii, Y. N., Kostamarov, D. P., and Pavlova, N. L. (1970). *Proc. Eur. Conf. Control. Fusion Plasma Phys. 7th, 1970* p. 17.
Dnestrovskii, Y. N., Kostamarov, D. P., and Pavlova, N. L. (1971). *At. Energ.* **29**, 1205.
Dnestrovskii, Y. N., Kostamarov, D. P., and Pavlova, N. L. (1972). *Proc. Eur. Conf. Control. Fusion Plasma Phys., 5th, 1972* p. 36.
Dnestrovskii, Y. N., Kostamarov, D. P., and Pavlova, N. L. (1973). *At. Energ.* **32**, 337 (1973).
Düchs, D. F. (1972), "Fluid Models for Tokamak Plasmas," NRL Rep. 7340. Naval Res. Lab., Washington, D.C.
Düchs, D. F., Furth, H. P., and Rutherford, P. H. (1971). *Proc. Conf. Plasma Phys. Control. Nucl. Fusion, 4th, 1971* p. 369.
Düchs, D. F., Furth, H. P., and Rutherford, P. H. (1972a). *Nucl. Fusion* **12**, 341.
Düchs, D. F., Furth, H. P., and Rutherford, P. H. (1972b). *Proc. Eur. Conf. Control. Fusion Plasma Phys., 5th, 1972* p. 14.
Düchs, D. F., Furth, H. P., and Rutherford, P. H. (1972c). *Bull. Amer. Phys. Soc.* [2] **17**, 984.
Düchs, D. F., Furth, H. P., and Rutherford, P. H. (1973a). *Proc. Eur. Conf. Control. Fusion Plasma Phys., 6th, 1973* p. 29.
Düchs, D. F., Furth, H. P., and Rutherford, P. H. (1973b). *Bull. Amer. Phys. Soc.* [2] **18**, 1338.
Düchs, D., Englehardt, W., and Koppendorfer, W. (1974). *Nucl. Fusion* **14**, 73.
Fowler, R. H., Hogan, J. T., and Dory, R. A. (1972). *Bull. Amer. Phys. Soc.* [2] **17**, 1055.
Galeev, A. A., and Sagdeev, R. F. (1968). *Sov. Phys.—JETP* **26**, 233.
Girard, J. P., Khelladi, M., and Marty, D. (1972). *Eur. Proc. Conf. Control. Fusion Plasma Phys., 5th, 1972* p. 105.
Grad, H. (1967). *Phys. Fluids* **10**, 137.
Grad, H., and Hogan, J. (1970). *Phys. Rev. Lett.* **24**, 1337.
Greenspan, E. (1974). *Nucl. Fusion* **14**, 771.
Hinnov, E., Bishop, A. S., and Fallon, H. P., Jr. (1968). *Plasma Phys.* **10**, 291.

Hinton, F. L., and Rosenbluth, M. N. (1973). *Phys. Fluids* **16**, 836.
Hogan, J. T. (1974). *Bull. Amer. Phys. Soc.* [2] **19**, 444.
Hogan, J. T., and Clarke, J. F. (1974). *J. Nucl. Mater.* **53**, 1.
Hogan, J. T., and Dory, R. A. (1972). *Eur. Proc. Conf. Control. Fusion Plasma Phys.*, *5th, 1972* p. 40.
Janenko, N. N. (1970). "Méthode des pas fractionnaires." Armand Colin.
Jordan, C. (1969). *Mon. Not. Roy. Astron. Soc.* **142**, 501.
Kadomtsev, B. B. (1973). *Proc. Eur. Conf. Control. Fusion Plasma Phys.*, *6th, 1973* p. 59.
Kadomtsev, B. B., and Pogutse, O. P. (1971). *Nucl. Fusion* **11**, 67.
Keeping, P., Killeen, J., and Grimm, R. (1972). *Proc. Eur. Conf. Control. Fusion Plasma Phys.*, *5th, 1972* p. 38.
Kelley, G. G., *et al.* (1973). *Proc. Int. Symp. Closed Confinement Syst., Garching, 1973* p. B-31.
Killeen, J. (1972). Culham Lab. Rep. PPN No. 2071.
Killeen, J., and Marx, K. (1970). *Methods Comput. Phys.* **9**, 422.
Konstantinov, O. V., and Perel, V. I. (1961). *Sov. Phys.—Tech. Phys.* **5**, 1403.
Kovrizhnick, L. M. (1969). *Sov. Phys.—JETP* **56**, 877.
Kruskal, M. (1963). *In* "Mathematical Models in the Physical Sciences" (S. Drobot, ed.), p. 373. Prentice-Hall, Englewood Cliffs, New Jersey.
Meade, D. *et al.* (1974). *Proc. Conf. Plasma Phys. Control. Nucl Fusion, 5th, 1974* p. 605.
Mercier, C., and Soubbaramayer. (1969). *Proc. Int. Conf. Closed Confinement Syst., 1969.*
Mercier, C., and Soubbaramayer. (1970). *Proc. Eur. Conf. Control. Fusion Plasma Phys., 4th, 1970* p. 16.
Mercier, C., and Soubbaramayer. (1972). *Proc. Eur. Conf. Control. Fusion Plasma Phys., 5th, 1972* p. 157.
Moore, T. B., and Hinton, F. L. (1974). *Nucl. Fusion* **14**, 639.
Parsons, C., and Medley, S. (1974). *Plasma Phys.* **16**, 267.
Pfirsch, D., and Schluter, A. (1962). Report No. MPI/PA/7/62, Max Planck Institute, 1962.
Potter, D. (1973). "Computational Physics." Wiley, New York.
Roberts, K. V., and Potter, D. (1970). *Methods Comput. Phys.* **9**, 340.
Robinson, D. C., *et al.* (1969). *Nature (London)* **224**, 400.
Robinson, D. C. (1973). *Notes. Working Group Joint Eur. Torus, 1973.*
Rome, J. A., *et al.* (1974). *Nucl. Fusion* **14**, 141.
Rome, J. A., *et al.* (1975). Report ORNL-TM-4855. Oak Ridge Nat. Lab. Oak Ridge, Tennessee (to be published).
Rosenbluth, M. N., MacDonald, W., and Judd, D. (1957). *Phys. Rev.* **107**, 1.
Rosenbluth, M. N., Hazeltine, R., and Hinton, F. (1972). *Phys. Fluids* **15**, 116.
Sakharov, A. (1961). "Plasma Physics and the Problem of Controlled Thermonuclear Reactions," Vol. 1, p. 21. Pergamon, Oxford.
Shafranov, V. D. (1966). *Plasma Phys.* **8**, 314.
Sigmar, D. J., Hirschman, S., and Clarke, J. F. (1975). Report ORNL-RM-4839. Oak Ridge Nat. Lab., Oak Ridge, Tennessee (to be published).
Temam, R. (1968). Sur la stabilité de la convergence de la méthode de pas fractionnaires." Ann. Math.
Widner, M., and Dory, R. (1971). Report ORNL-TM-3498. Oak Ridge Nat. Lab., Oak Ridge, Tennessee (unpublished).
Widner, M., Dory, R., and Hogan, J. (1971a). *Phys. Lett. A* **36**, 217.
Widner, M., *et al.* (1971b). *Proc. Conf. Plasma Phys. Control. Nucl. Fusion, 7th, 1971* p. 347.
Wiley, J., and Hinton, F. (1972). *Bull. Amer. Phys. Soc.* [2] **17**, 1008.
Wiley, J. C., *et al.* (1972). *Phys. Rev. Lett.* **29**, 698.
Yoshikawa, S. (1970). *Phys. Rev. Lett.* **25**, 353.

APPENDIX: BIBLIOGRAPHY

In addition to material cited in the text there are other papers, reports, and abstracts which contain information about the codes or about applications.

Boujot, J. P., Mercier, C., and Soubbaramayer (1972). *Comput. Phys. Commun.* **4**, 89.
Boujot, J. P., Mercier, C., Morera, J. P., and Soubbaramayer (1973). Fontenay-Aux-Roses Report EUR-CEA-FC683.
Dory, R. A., and Widner, M. (1970). *Bull. Amer. Phys. Soc.* [2] **15**, 1418.
Düchs, D. F. (1970). *Bull. Amer. Phys. Soc.* [2] **15**, 1488 (1970).
Düchs, D. F., Furth, H. P., and Rutherford, P. H. (1971). *Bull. Amer. Phys. Soc.* [2] **16**, 1223.
Düchs, D. F., Furth, H. P., and Rutherford, P. H. (1973). Princeton Plasma Physics Laboratory Report TM-265. Princeton University, Princeton, New Jersey.
Girard, J. P., Khelladi, M., and Marty, D. (1973). *Nucl. Fusion* **13**, 685.
Girard, J. P., Marty, D., and Moriette, P. (1974). *Proc. Conf. Plasma Phys. Control. Nucl. Fusion, 5th, 1974* Paper IAEA-CN-33.
Graf Finck von Finckenstein, F., and Düchs, D. F. (1974). Lecture Notes in Mathematics," Vol. 395, p. 3. Springer-Verlag, Berlin and New York.
Graf Finck von Finckenstein, F., and Düchs, D. F. (1974). "Methoden und Verfahren der mathematischen Physik," Vol. 11. B. I. Wischenschaftsverlag, Mannheim/Wien/Zurich.
Krall, N., *et al.* (1974). Report No. LAPS-2. Science Applications, Inc.
McBride, J., *et al.* (1974). Report No. LAPS-1. Science Applications, Inc.
Mercier, C., and Soubbaramayer (1973). "Spectrèdes vitesses des neutres, ···", Fontenay-Aux-Roses Report EUR-CEA-DPH-PFC/STGI, Note Interne No. 1107.
Mercier, C., and Soubbaramayer (1974a). Fontenay-Aux-Roses Report EUR-CEA-FC-742.
Mercier, C., and Soubbaramayer (1974b). Fontenay-Aux-Roses Report EUR-CEA-FC-744.
Mercier, C., Adam, J., Soubbaramayer, and Soule, J. L., (1973). Fontenay-Aux-Roses Report EUR-CEA-FC-708.
Takeda, T., and Itoh, S. (1974). *Jap. At. Energy Res. Inst.*, *Rep.* **JAERI-M-5697**.

ICARUS—A One-Dimensional Plasma Diffusion Code

M. L. Watkins, M. H. Hughes, K. V. Roberts, and
P. M. Keeping

CULHAM LABORATORY
UKAEA RESEARCH GROUP*
ABINGDON, OXFORDSHIRE, ENGLAND

and

J. Killeen

LAWRENCE LIVERMORE LABORATORY
UNIVERSITY OF CALIFORNIA
LIVERMORE, CALIFORNIA

I. Introduction 166
II. The Physical Model 169
   A. Introduction 169
   B. The Neoclassical Transport Model 169
   C. Boundary and Initial Conditions 175
III. The Numerical Model 176
   A. Introduction to One-Dimensional Finite-Difference Methods . . . 176
   B. The Method of Solution Employed in ICARUS 177
IV. Programming Techniques 179
   A. The OLYMPUS System 179
   B. Program Structure 182
   C. Architecture 183
   D. Initialization 185
   E. Control 186
   F. Diagnostics 189
   G. Documentation 190
V. Applications 190
   A. Introduction 190
   B. Present Generation Tokamaks 192
   C. Next Generation Tokamaks 201
VI. Summary 206
   References 207

* UKAEA/Euratom Fusion Association.

I. Introduction

ONE-DIMENSIONAL PLASMA DIFFUSION codes (Dnestrovskii et al., 1970; Dory and Widner, 1970; Düchs, 1970; Mercier and Soubbaramayer, 1970; Düchs et al., 1971; Hogan et al., 1971; Keeping et al., 1972; Watkins et al., 1976) are used to study the temporal evolution of plasma and poloidal magnetic field in axisymmetric, toroidal, plasma containment devices of the Tokamak type (Artsimovich et al., 1969; Dimock et al., 1971; Bol et al., 1972; Drummond et al., 1973; Gibson et al., 1973; Itoh et al., 1973; Kelley et al., 1973; Rebut et al., 1973). The motion of the plasma is considered to be sufficiently slow for dynamical effects to be neglected, so that the configuration evolves through a series of quasi-equilibrium states satisfying the equation

$$\nabla p = \mathbf{J} \times \mathbf{B}, \tag{1}$$

where p is the plasma pressure, $\mathbf{B}$ is the magnetic field, and $\mathbf{J}$ is the current density.[1] The magnetic surfaces are surfaces of constant pressure and form a nested toroidal set centered on the magnetic axis. An appropriate set of coordinates for the problem is (r, θ, φ) where r and θ are polar coordinates referred to the magnetic axis of the torus which has major radius, R, and φ is the azimuthal angle measured along the magnetic axis (Fig. 1). The intersections of the magnetic surfaces with a cross-sectional plane, $\varphi = $ constant, are the field lines of the poloidal magnetic field, $\mathbf{B}_p = (B_r, B_\theta)$. If the minor radius of the torus is denoted by a and the inverse aspect ratio is defined to be $\varepsilon(r=a) = a/R$, the limit of small inverse aspect ratio, $\varepsilon(r=a) \to 0$, corresponds to the assumption of cylindrical symmetry. However, even for present-day experiments, usually $\varepsilon(r=a) \gtrsim 0.1$, so that toroidal effects are important.

Tokamak calculations of current interest are concerned with the equilibrium, stability, and evolution of a magnetically confined plasma. Given a two-dimensional equilibrium configuration which satisfies Eq. (1) at any given time, diffusion, heating, and other entropy-generating processes will modify the plasma-field parameters and lead, by adiabatic readjustment, to a new equilibrium configuration at a slightly later time. At each stage the configuration should be tested for stability.

At present, however, a complete two-dimensional calculation which would follow the time evolution of the plasma-field configuration has not been undertaken. Although two-dimensional equilibrium calculations are considered routine (Feneberg and Lackner, 1973; von Hagenow and Lackner,

---

[1] The S. I. system of units is used throughout this article.

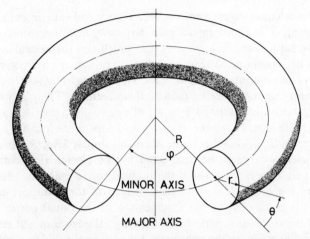

FIG. 1. Schematic diagram of the torus indicating the coordinate system (r, θ, φ).

1973) and magnetohydrodynamic stability calculations, which were first developed in one dimension (Newcomb, 1960; Friedberg, 1970; Shafranov, 1970; Goedbloed and Hagebeuk, 1972), are now being extended to two dimensions (Bateman et al., 1974; Sykes and Wesson, 1974) with some promise of realistic three-dimensional calculations in the near future (Wooten et al., 1974), most of the evolutionary calculations for Tokamaks are still carried out in one-dimensional cylindrical geometry with the toroidal corrections modifying the transport coefficients. There are several reasons for retaining the one-dimensional model for the diffusion. First, it is not yet clear which two-dimensional set of equations should be used, since the time-scale for establishing pressure and temperature equilibrium on a magnetic surface is some orders of magnitude shorter than that for transport across the surfaces. Almost certainly the standard two-dimensional, magnetohydrodynamic equations do not apply. The physical processes taking place in the Tokamak plasma are unclear, but are likely to be dominated by effects such as wall interaction, turbulent transport, and the presence of impurities and neutral gas. Second, implicit, two-dimensional, numerical methods capable of handling efficiently the high Alfvén speed and long evolutionary time-scale characteristic of Tokamaks are not yet available.

It seems preferable to establish a semiquantitative agreement between computation and experiment in one-dimensional geometry before moving on to the more complex and expensive two-dimensional case. Nevertheless there exists some pioneering work on two-dimensional evolutionary calculations for Tokamaks (Winsor et al., 1970; Potter and Tuttle, 1973; Sharp and Taylor, 1973).

Since there are so many processes that may be important in Tokamaks, and since an adequate agreement between theory and experiment has not yet been achieved, it is expedient to plan any new, one-dimensional diffusion code as a versatile research apparatus into which new theoretical or empirical effects can be incorporated readily and assumptions or parameters changed with ease. The computer code, ICARUS (Watkins et al., 1976), has been designed with due regard being paid to these features. The plasma magnetic field configuration is described by a set of essentially conservative equations for the particle density, n, the electron and ion pressures, p_e, p_i, and the poloidal magnetic field component, B_θ. The toroidal field, B_φ, is considered to be sufficiently large to be unaffected by the plasma (low-beta approximation, where beta is the ratio of the thermal to the magnetic pressure) and only enters the calculation through its effect on the transport coefficients. The model is completed by the specification of the radial particle and heat fluxes and the toroidal electric field. Although the basic model employed in the code is that given by the neoclassical theory applicable to a collision-free plasma (Hazeltine et al., 1973), small changes in the input data for the code allow the use of other scaling laws such as those associated with collision-dominated plasmas (Pfirsch and Schlüter, 1962) or nonclassical effects (Yoshikawa, 1970; Artsimovich, 1971).

The results of diffusion calculations are extremely dependent on the details of the *ad hoc* model, so it is necessary to publish together the model and the results if the literature is to maintain acceptably high standards. It is preferable that the code be available for use by other research groups so that the complex details of the physical and numerical models and the programming can be checked and the calculations confirmed and extended.

ICARUS uses the OLYMPUS programming system (Roberts, 1974; Christiansen and Roberts, 1974) which is intended to meet these requirements. Although it can be used for a wide variety of problems, OLYMPUS was originally designed for initial-value calculations of the form

$$\partial \mathbf{u}/\partial t + \mathbf{G}(\mathbf{u}) = \mathbf{0}, \qquad (2)$$

where $\mathbf{u}$ is the solution vector and $\mathbf{G}$ is a linear or nonlinear operator. Many calculations in classical physics are of this kind, including those for which one-dimensional Tokamak diffusion codes were developed. OLYMPUS codes are written in Standard FORTRAN and can be run on any computer system provided that the appropriate version of a Standard Control and Utility Package has been loaded first into the library. Packages have been developed for seven types of system so far and two have been published (Christiansen and Roberts, 1974; Hughes et al., 1975). Each code has the same underlying modular structure and is well documented so that a listing of the code is easily understood. Provision is made for *ad hoc* modifications for specific calculations without compromising the basic version. The first published example is the one-dimensional laser fusion code MEDUSA (Christiansen *et al.*, 1974). This code has been used by a number of research groups who have checked and extended the test calculations and recommended a number of minor corrections (Christiansen *et al.*, 1975).

The purpose of this article is to describe the basic physical model (Section II) and numerical scheme (Section III) used in ICARUS, together with the OLYMPUS programming techniques (Section IV), and to indicate how these can be applied to a range of specific Tokamak calculations (Section V).

II. The Physical Model

A. Introduction

The computer code ICARUS (Watkins *et al.*, 1976) provides the solution to a set of N partial differential equations which describe the temporal evolution of N dependent variables, $\mathbf{u}_n$, as functions of time (t) and a single spatial coordinate (r). The form of the equations to be solved is represented by the conservation equation in an infinitely long, cylindrically symmetric device

$$(\partial/\partial t)\mathbf{u}_n(r,t) + (1/r)(\partial/\partial r)r\mathbf{F}_n(r,t) = \mathbf{S}_n(r,t), \qquad n = 1,\ldots,N, \qquad (3)$$

where $\mathbf{F}_n(r,t)$ and $\mathbf{S}_n(r,t)$ represent, respectively, the flux and source of $\mathbf{u}_n$ at each position (r,t). These functions have to be supplied explicitly in terms of $\mathbf{u}_n(r,t)$ and their spatial gradients. The particular applications of the code (Section V) involve the representation of $\mathbf{F}_n$ in the diffusion approximation so that, typically,

$$\mathbf{F}_n \propto \partial \mathbf{u}_n/\partial r. \qquad (4)$$

When the equations represent predominantly diffusion processes, it is natural to solve these equations by an implicit technique (Section III).

B. The Neoclassical Transport Model

The basic physical model employed in the computer code ICARUS is the neoclassical transport model of Hazeltine *et al.* (1973). The fluxes obtained represent averages over both the toroidal, φ, and poloidal, θ, directions in an axisymmetric torus (Fig. 1). The velocity moments of the Boltzmann equation (Chapman and Cowling, 1953) with a Fokker–Planck collision operator (Rutherford, 1970) give a series of conservation equations in which the dependent variables are functions of the single spatial coordinate, r, the distance measured from the geometrical minor axis of the torus.

The model describes a hot, axisymmetrically-confined, low-beta plasma consisting of hydrogen ions (mass, m_i) and electrons (mass, m_e). A strong magnetic field exists in the toroidal direction, and is specified for all time to be the Knorr field (Knorr, 1965)

$$B_\varphi = B_0/(1 - \varepsilon \cos\theta), \qquad \varepsilon = r/R.$$

B_0 is a reference magnetic field, independent of position and time, and ε is the local inverse aspect ratio. An expansion of the Boltzmann equation to $O(\varepsilon^{1/2})$ indicates the radial diffusion is ambipolar (Rosenbluth *et al.*, 1972): the electron and ion particle fluxes, Γ_e and Γ_i, are equal

$$\Gamma_e = \Gamma_i = \Gamma.$$

For scale lengths sufficiently large that the plasma may be considered to be electrically neutral, the electrons and ions are present everywhere in equal numbers and the temporal variation of the particle number density, n, is described by the equation of continuity

$$(\partial n/\partial t) + (1/r)(\partial/\partial r) r\Gamma = 0. \tag{5}$$

The electron and ion heat balance equations may be written in conservative form for the electron and ion energy densities, ε_e and ε_i

$$\frac{\partial}{\partial t}\varepsilon_e + \frac{1}{r}\frac{\partial}{\partial r} rQ_e = E_\varphi J_\varphi - \left(\frac{\varepsilon_e - \varepsilon_i}{\tau_{eq}}\right) - \Gamma T_i\left(\frac{1}{n}\frac{\partial n}{\partial r} - \frac{0.172}{T_i}\frac{\partial T_i}{\partial r}\right), \tag{6}$$

$$\frac{\partial}{\partial t}\varepsilon_i + \frac{1}{r}\frac{\partial}{\partial r} rQ_i = \left(\frac{\varepsilon_e - \varepsilon_i}{\tau_{eq}}\right) + \Gamma T_i\left(\frac{1}{n}\frac{\partial n}{\partial r} - \frac{0.172}{T_i}\frac{\partial T_i}{\partial r}\right). \tag{7}$$

The heat fluxes, Q_e and Q_i, represent both thermal and particle transport of energy. Energy input to the plasma is by virtue of the ohmic heating term, $\mathbf{E}\cdot\mathbf{J}$, which involves only the toroidal electric field, E_φ, for the Tokamak ordering procedure in which the poloidal magnetic field, B_θ, is assumed to be small in comparison with the externally applied toroidal magnetic field, B_φ. The energy is transferred from electrons to ions by virtue of the equipartition term characterized by the time

$$\tau_{eq} = \tfrac{1}{2}(m_i/m_e)\tau_{ei},$$

where τ_{ei} is the electron–ion collision time given by Braginskii (1965)

$$\tau_{ei} = 3m_e^{1/2}(kT_e)^{3/2}(4\pi\varepsilon_0)^2/4(2\pi)^{1/2}ne^4 \ln \Lambda,$$

where k is the Boltzmann constant, ε_0 the permittivity of free space, e the charge on the electron, and T_e is the electron temperature. The Coulomb logarithmic function, $\ln \Lambda$, is given by (Braginskii, 1965)

$$\ln \Lambda = \begin{cases} 30.3 - 1.15 \log n + 3.45 \log(kT_e/e), & kT_e/e < 50 \text{ eV}, \\ 32.2 - 1.15 \log n + 2.3 \log(kT_e/e), & kT_e/e > 50 \text{ eV}. \end{cases}$$

The electron and ion energy equations are coupled also by the last term on the right-hand sides of Eqs. (6) and (7) representing the effects of ion motion along the magnetic field and the radial electric field.

Equations (6) and (7) may be used to describe the temporal evolution of the electron and ion pressures, p_e and p_i, by means of equations of state of the form applicable to a perfect gas

$$\begin{aligned} \varepsilon_e &= p_e/(\gamma - 1) = nkT_e/(\gamma - 1), \\ \varepsilon_i &= p_i/(\gamma - 1) = nkT_i/(\gamma - 1), \end{aligned} \quad (8)$$

where γ is the ratio of the specific heats, assumed to be 5/3, and T_i is the ion temperature.

The toroidal current density, J_φ, may be determined from Maxwell's form of Ampere's law

$$\mu_0 J_\varphi = (1/r)(\partial/\partial r)rB_\theta, \qquad (9)$$

where μ_0 is the permeability of free space, with the poloidal magnetic field being given by Faraday's Law

$$\partial B_\theta/\partial t = (\partial/\partial r)E_\varphi. \qquad (10)$$

Consistent with the Tokamak ordering procedure it is valid to consider changes only in the poloidal magnetic field, B_θ.

The plasma is described by the three dependent variables, n, p_e, p_i, representing, respectively, the particle number density and electron and ion pressures. The conservation equations for these quantities are closed by the specification of the radial particle flux, the radial electron and ion heat fluxes, and the toroidal electric field applicable to the diffusion of plasma in a torus (Galeev and Sagdeev, 1968; Kadomtsev and Pogutse, 1971). In such a system a charged particle may execute three types of nearly periodic motions. These are gyration about a magnetic field line, motion along the field line resulting in transit around the system or bouncing between mirror points, and drifting across the field lines in response to electric fields and the gradient and curvature of the magnetic field. The relation between the frequencies associated with these motions and the frequency for electron–ion collisions, $v_{ei} = \tau_{ei}^{-1}$, determines the exact form of the diffusion coefficient, which is given, in general, by an equation of the form

$$D \sim vl^2, \qquad (11)$$

where v is the effective collision frequency and l is a characteristic step size for a random walk process. The relative forms of the diffusion coefficients in the high frequency ("classical"), intermediate frequency ("plateau"), and low frequency ("banana") regimes are summarized in Fig. 2. The frequencies

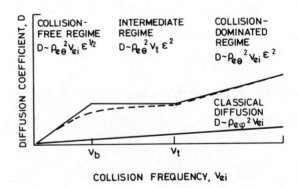

FIG. 2. Schematic representation of the neoclassical diffusion coefficient, D, as a function of the electron–ion collision frequency, v_{ei}, in the collision-free ($v_{ei} < v_b$), intermediate ($v_b < v_{ei} < v_t$), and collision-dominated ($v_{ei} > v_t$) regimes. The "smoothed" approximation for the diffusion coefficient used in the computer code, ICARUS, is also indicated together with the coefficient for classical diffusion in a cylinder.

that separate these regimes are representative of the bounce frequency, v_b, of trapped particles and the transit frequency, v_t, of untrapped particles.

The expressions for the fluxes in the low frequency regime are given most concisely by Hazeltine et al. (1973). Corresponding to the forces, A_m ($m = 1, 2, 3$), which result from gradients in the particle density, electron and ion temperatures, and the electric field there exist three fluxes, J_l

$(l = 1, 2, 3)$,

$$A_1 \equiv A1/p_e = p_e^{-1} \partial(p_e + p_i)/\partial r - \tfrac{5}{2} T_e^{-1} \partial T_e/\partial r + (y - \tfrac{5}{2}) T_e^{-1} \partial T_i/\partial r,$$
$$A_2 \equiv A2/kT_e p_e = k^{-1} T_e^{-2} \partial T_e/\partial r,$$
$$A_3 \equiv A3 = E_\varphi,$$
$$J_1 \equiv \Gamma,$$
$$J_2 \equiv Q_e,$$
$$J_3 \equiv (kT_e)^{-1}(J_\varphi - \sigma_s E_\varphi).$$
(12)

The fluxes and forces are related through the transport coefficients, L_{lm}

$$J_l = -\sum_{m=1}^{3} L_{lm} A_m, \quad l = 1, 2, 3.$$

In addition, the total ion heat flux is given by

$$J_i = y\Gamma k T_i - L_i k^{-1} T_i^{-2} \partial T_i/\partial r. \tag{13}$$

The parameter, y, has been introduced so that A_2 is dependent only on the gradient of the electron temperature. To within a reasonable degree of accuracy the transport coefficients, L_{lm}, are given by the simple *ansatz* (Hazeltine *et al.*, 1973)

$$L_{lm} = \alpha_{lm} \varepsilon^{1/2} + \beta_{lm} \varepsilon. \tag{14}$$

This gives exact results in the limits $\varepsilon \to 0$, $\varepsilon \to 1$. In the code ICARUS it has been found most useful to provide the forces, Am ($m = 1, 2, 3$), and the corresponding transport coefficients, $RLlm$ ($l, m = 1, 2, 3$), in terms of the classical diffusion coefficient in the poloidal magnetic field, D_θ^c, and the Spitzer conductivity, σ_s, defined according to

$$D_\theta^c = \rho_{e\theta}^2 \nu_{ei}, \quad \sigma_s = ne^2/m_e \nu_{ei}, \tag{15}$$

where the electron gyro radius, $\rho_{e\theta}$, in the poloidal magnetic field is given by

$$\rho_{e\theta}^2 = 2m_e k T_e/e^2 B_\theta^2. \tag{16}$$

The following definitions for the transport coefficients apply:

$$RL11 = (D_\theta^c/kT_e)(CA11\varepsilon^{1/2} + CB11\varepsilon + CC11),$$

$$RL12 = RL21/kT_e = (D_\theta^c/kT_e)(CA12\varepsilon^{1/2} + CB12\varepsilon + CC12),$$

$$RL13 = -nRL31 = (n/B_\theta)(CA13\varepsilon^{1/2} + CB13\varepsilon + CC13),$$

$$RL22 = D_\theta^c(CA22\varepsilon^{1/2} + CB22\varepsilon + CC22),$$

$$RL23 = -p_e RL32 = (p_e/B_\theta)(CA23\varepsilon^{1/2} + CB23\varepsilon + CC23),\quad (17)$$

$$RL33 = \sigma_s(CA33\varepsilon^{1/2} + CB33\varepsilon + CC33),$$

$$RLI = (m_i T_e/m_e T_i)^{1/2} D_\theta^c (CAI\varepsilon^{1/2} + CBI\varepsilon + CCI),$$

$$y = CAY\varepsilon^{1/2} + CBY\varepsilon + CCY.$$

The numerical values of the constants (CA11, ..., CCY) employed in the definitions of the transport coefficients are summarized in Table I. The range

TABLE I

THE NUMERICAL VALUES OF THE CONSTANTS IN THE NEOCLASSICAL TRANSPORT COEFFICIENTS FOR A COLLISION-FREE PLASMA

CA11 = 1.12	CB11 = −0.62	CC11 = 0.0
CA12 = 1.27	CB12 = −0.77	CC12 = 0.0
CA13 = 2.44	CB13 = −1.44	CC13 = 0.0
CA22 = 2.64	CB22 = −0.93	CC22 = 0.0
CA23 = 4.35	CB23 = −1.85	CC23 = 0.0
CA33 = −1.95	CB33 = 0.95	CC33 = 1.0
CAI = 0.48	CBI = 0.23	CCI = 0.0
CAY = 0.0	CBY = 1.17	CCY = 1.33

of validity of the transport expressions may be extended to the intermediate frequency ("plateau") regime with the aid of "smoothing functions" (Table II). These functions effect a smooth transition of the diffusion coefficients as the collision frequency increases to a value greater than the bounce frequency, v_b (Fig. 2).

The transport model applicable to low and intermediate values of the collision frequency forms the basic physical model employed in ICARUS.

TABLE II

The "Smoothing Functions" Used to Extend the Range of Validity of the Neoclassical Transport Model to More Collisional Plasmas[a]

$$\text{ZMTH11} = 1.0/(1.0 + 1.78 v_{ei}^*)$$
$$\text{ZMTH12} = 1.0/(1.0 + 0.66 v_{ei}^*)$$
$$\text{ZMTH13} = 1.0/(1.0 + 0.85 v_{ei}^*)$$
$$\text{ZMTH22} = 1.0/(1.0 + 0.35 v_{ei}^*)$$
$$\text{ZMTH23} = 1.0/(1.0 + 0.40 v_{ei}^*)$$
$$\text{ZMTH33} = 1.0/(1.0 + 1.00 v_{ei}^*)$$
$$\text{ZMTHI} = 1.0/(1.0 + 0.36 v_{ii}^*)$$

[a] "Smoothing functions," ZMTH$\alpha\beta$, act to transform $(CA\alpha\beta\varepsilon^{1/2} + CB\alpha\beta\varepsilon + CC\alpha\beta)$ into $(CA\alpha\beta\varepsilon^{1/2} + CB\alpha\beta\varepsilon) \times \text{ZMTH}\alpha\beta + CC\alpha\beta$. The normalized collision frequencies, v_{ei}^* and v_{ii}^*, are defined as

$$v_{ei}^* = v_{ei}/v_b, \qquad v_{ii}^* = v_{ii}/v_b,$$

where v_b is the bounce frequency of trapped particles.

C. Boundary and Initial Conditions

The code employs a finite-difference mesh that is mapped onto a continuous domain (Section III) in such a way that the physical boundaries of a wall (or limiter), at radius $r = a$, and the minor axis at radius $r = 0$, coincide with mesh cell boundaries rather than mesh cell centers at which the dependent variables are defined. The physical boundary conditions apply to the fluxes, Γ, J_φ, Q_e, Q_i. On the minor axis of the torus at radius, $r = 0$, all radial components of vector quantities must be zero, with the result that the gradients of the dependent variables, n, p_e, p_i, must also be zero. On the other hand the toroidal current density, J_φ, should exist and can be related in general to the poloidal magnetic field, B_θ, at a position r_0 (near $r = 0$) by Ampere's Law applied to a cylinder about the axis

$$J_\varphi = 2B_\theta/\mu_0 r_0.$$

The boundary conditions at the outer radius, $r = a$, are more difficult to treat. The simplest method is to employ the use of "pedestal" values for the dependent variables at the first mesh point with a radius, $r > a$: the boundary values of the dependent variables are set to a fixed fraction of the central values at time, $t = 0$, by specifying the functional forms

$$f = f_0(1 - (1 - P)x^l)^m, \quad f \equiv \{n, T_e, T_i\},$$
$$J_\varphi = J_0(1 - x^l)^m, \quad x = r/a.$$

The indices $\{l, m\}$ are preset to values applicable to parabolic distributions and the fractional pedestals, P, are set to 10%. These may be modified through data input, as also may the maximum values, f_0, and the total toroidal current

$$I_\varphi = 2\pi \int_0^a J_\varphi r\, dr.$$

The pedestal values are maintained at the first mesh point with $r > a$ for all time.

III. The Numerical Model

A. Introduction to One-Dimensional Finite-Difference Methods

The equations to be solved may be written concisely in the form of Eq. (2) with $\mathbf{u}(r, t)$ being the column vector $\{n, p_e, p_i, B_\theta\}$ and $\mathbf{G}$ being an operator dependent on $\mathbf{u}$ and its spatial derivatives. The problem formulated by Eq. (2) is an initial-value problem so that $\mathbf{u}$ is determined for all time given its initial value, $\mathbf{u}(r, 0)$, together with the specification at all times of the spatial boundary conditions.

In general, it is not possible to express the solutions to Eq. (2) in terms of known functions. Numerical methods rather than analytical methods have to be employed. The vector, $\mathbf{u}$, and the operator, $\mathbf{G}$, are defined on the mesh points of a grid (Fig. 3) mapped onto a continuous domain bounded by the

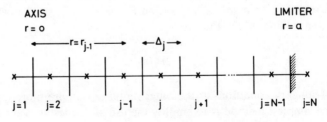

FIG. 3. The finite-difference mesh.

positions, $r = 0$ and $r = a$. Temporal differencing assumes a discrete set of lattice points along the time axis.

The general difference analog of Eq. (2) is obtained by integration

$$\mathbf{u}^{n+1} - \mathbf{u}^n = -\int_{t^n}^{t^{n+1} = t^n + \Delta t^n} \mathbf{G}(\mathbf{u}) \, dt. \tag{18}$$

The solution to this equation may be approximated by replacing the right-hand side of Eq. (18) by some temporal average

$$\mathbf{u}^{n+1} - \mathbf{u}^n = -[\zeta \mathbf{G}(\mathbf{u}^{n+1}) + (1 - \zeta)\mathbf{G}(\mathbf{u}^n)] \Delta t^n, \tag{19}$$

where

$$0 \leq \zeta \leq 1.$$

When $\mathbf{G}$ is of such a form that Eq. (2) is parabolic in (r, t), fully implicit methods of solution ($\zeta \equiv 1$) are particularly effective (Richtmyer and Morton, 1967).

B. The Method of Solution Employed in ICARUS

In the diffusion approximation the model equations which can be written to conform with Eq. (3) contain fluxes, $\mathbf{F}$, and source terms, $\mathbf{S}$, that can be functions of the dependent variable $\mathbf{u}$

$$\mathbf{F}(\mathbf{u}) = f_\mathbf{u}(\mathbf{u})(\partial/\partial r)\mathbf{u},$$

$$\mathbf{S}(\mathbf{u}) = s_\mathbf{u}(\mathbf{u})\mathbf{u}.$$

On the finite-difference mesh the values of $\mathbf{u}^{n+1}$ at the time, t^{n+1}, and positions, $r = r_{j-1}, r_j, r_{j+1}$, may be related by the matrix equation

$$[-\mathbf{A}_j \cdot \mathbf{u}_{j+1} + \mathbf{B}_j \cdot \mathbf{u}_j - \mathbf{C}_j \cdot \mathbf{u}_{j-1}]^{n+1} = \mathbf{D}_j^n, \tag{20}$$

where $\mathbf{A}$, $\mathbf{B}$, and $\mathbf{C}$ are 4×4 matrices and $\mathbf{D}$ is a 4-vector. Equation (20) is solved as a linear system by means of the algorithm given by Richtmyer and Morton (1967)

$$\mathbf{u}_j^{n+1} = \mathbf{G}_j^{n+1} \cdot \mathbf{u}_{j+1}^{n+1} + \mathbf{H}_j^{n+1}, \quad j = N-1, N-2, \ldots, 2, \tag{21}$$

where the matrix, $\mathbf{G}$, and the vector, $\mathbf{H}$, are determined from the recurrence relations

$$G_j^{n+1} = (B_j^{n+1} - C_j^{n+1} \cdot G_{j-1}^{n+1})^{-1} \cdot A_j^{n+1}, \qquad (22)$$

$$H_j^{n+1} = (B_j^{n+1} - C_j^{n+1} \cdot G_{j-1}^{n+1})^{-1} \cdot (D_j^n + C_j^{n+1} \cdot H_{j-1}^{n+1}). \qquad (23)$$

The calculation involves inverting a 4×4 matrix at each mesh point. Given the boundary conditions in finite-difference form this can be performed easily (Potter, 1973).

As a result of the nonlinear nature of Eq. (20) with the coefficients of u^{n+1} being functions of u^{n+1}, it is necessary to solve Eq. (20) iteratively in order to obtain accurate values for the coefficients. Initially, A^n, B^n, and C^n are evaluated at time, t^n, and a trial solution, $u^{(1)}$ is obtained. The iteration is repeated with $A^{(1)}$, $B^{(1)}$, $C^{(1)}$ to obtain $u^{(2)}$ and is continued using $A^{(p-1)}$, $B^{(p-1)}$, $C^{(p-1)}$ to determine $u^{(p)}$ until the solution approximates u^{n+1} to within the desired accuracy. Convergence of the iterative procedure is checked by imposing that the fractional change in u at the pth and $(p-1)$th iteration be less than the specified convergence tolerance, δu:

$$\text{Max}\{(||u_j^{(p)}| - |u_j^{(p-1)}||)/(|u_j^{(p)}| + |u_j^{(p-1)}|)\} \leq \delta u, \qquad \text{for all} \quad 1 < j < N.$$

If the relative deviations are below specified values, convergence is assumed to be established. If convergence is not achieved within a fixed number of iterations the complete time-step is repeated with a reduced value for the time-step. At the completion of a time-step the conservation of particle number and total energy is checked. Since the total energy is not a variable that is involved in Eqs. (5)–(10), this check provides a very sensitive guide to the behavior of the numerical procedure.

The implicit method of solving parabolic equations guarantees a numerically stable solution (Richtmyer and Morton, 1967) and the choice of the time-step only serves to determine the accuracy. To monitor the temporal variation of u, the time-step is restricted according to

$$\Delta t^n \leq a \min\{(|u_j^{n+1}| + |u_j^n|)/(||u_j^{n+1}| - |u_j^n||)\}\Delta t^{n-1}, \qquad \text{for all} \quad 1 < j < N.$$

The variation in the time-step is restricted by

$$1/a_0 \leq \Delta t^n/\Delta t^{n-1} \leq a_0,$$

where a and a_0 are constants $O(1)$.

IV. Programming Techniques

A. THE OLYMPUS SYSTEM

The purpose of the OLYMPUS programming system (Roberts, 1974; Christiansen and Roberts, 1974) is to establish a clear standard structure for FORTRAN programs of a similar kind. The first application is to the generalized initial-value problem [Eq. (2)] which predicts the evolutionary behavior of a physical system. Initial-value problems are familiar in many areas of classical physics and the Tokamak diffusion calculation for which ICARUS has been designed provides a typical example. The concepts on which OLYMPUS is based are adapted from disciplines such as mathematics, theoretical physics, engineering, and architecture and from the field of scientific publishing (Roberts, 1969, 1971). It is intended that the programs together with their write-ups should be published in tested and refereed form in journals such as *Computer Physics Communications* and the associated CPC Program Library (Burke, 1970). This has the advantage of making the programs available for general use and criticism just like other scientific publications. Particular attention is paid therefore to methods of documentation and layout, the eventual aim being to reach a level of program intelligibility similar to that expected of a mathematical textbook (Roberts, 1969).

Standardization is practicable because all programs that solve problems of type (2) have many similar duties to perform, irrespective of the specific equations that are used. Once-and-for-all decisions can be made concerning questions such as overall program structure, control techniques, terminology, notation, and layout, and a powerful library of general-purpose routines can be provided to deal with the frequently encountered housekeeping duties. Automatic techniques for program generation also become possible. All this can relieve the individual programmer of a considerable amount of effort provided that the system has been sufficiently developed and the programmer has understood how to use it. The programmer's work is made more accessible to other people since they know what to expect.

Similar ideas have been accepted in mathematics and theoretical physics for a long time and are known to be very powerful. Typical examples are the abstract formalisms of group theory and of Hamiltonian dynamics. One of the aims of the OLYMPUS system is to translate these ideas, so far as is possible, into programming practice. The Hamiltonian analogy is particularly useful since it specifically applies to equations of type (2).

Each OLYMPUS physics program consists of a set of subprograms written in Standard FORTRAN (National Computing Centre Limited, 1970) and running under the supervision of a universal main program and an associated control routine, COTROL, which form part of a Standard Control and Utility Package (Christiansen and Roberts, 1974). This package is contained in a local system library and a version is provided for each different type of computer system on which OLYMPUS has been implemented (currently, OLYMPUS has been installed on seven types of computer system). The versions are written mostly in FORTRAN and are similar. However, they need to take account of system-dependent features such as word and byte length, channel conventions and supervisor calls. Segregation within the OLYMPUS package of system-dependent features makes it possible to transfer large programs from one type of computer system to another with only minor changes, provided that the recommended conventions have been followed.

The control routine COTROL described by Christiansen and Roberts (1974) expects the programmer to supply a set of primary subprograms with the names and functions defined in Table III: standardization of the nomenclature and overall structure is ensured. These primary subprograms can *call* secondary subprograms whose names and functions are chosen by the programmer to reflect the type of calculation performed (Table IV and Fig. 4). Dummy versions of the primary subprograms are provided within the package as part of the OLYMPUS library and are useful during program development. Taken in their entirety they constitute a dummy program called CRONUS

TABLE III

THE NAMES, IDENTIFICATION NUMBERS, AND TITLES OF THE PRIMARY SUBROUTINES CALLED BY THE MAIN CONTROL SUBROUTINE ⟨0.3⟩ COTROL

Name	Number	Title
LABRUN	1.1	Label the run
CLEAR	1.2	Clear variables and arrays
PRESET	1.3	Set default values
DATA	1.4	Define data specific to run
AUXVAL	1.5	Set auxiliary values
INITAL	1.6	Define physical initial conditions
RESUME	1.7	Resume run from previous record
START	1.8	Start the calculation
STEPON	2.1	Step on the calculation
OUTPUT	3.1	Control the output
TESEND	4.1	Test for completion of run
ENDRUN	4.2	Terminate the run
EXPERT	0.4	Modify standard operation of program

TABLE IV

THE NAMES, IDENTIFICATION NUMBERS, AND TITLES OF THE CLASS 2
SUBROUTINES WHICH PERFORM THE DIFFUSION CALCULATION
IN THE COMPUTER CODE, ICARUS

Name	Number	Title
STEPON	2.1	Step on the calculation
BLKSET	2.2	Set up the matrix form of the difference equations
BNDRY0	2.3	Set up the boundary conditions at radius, $r = 0$
BNDRYN	2.4	Set up the boundary conditions at radius, $r = a$
PHYSIN	2.5	Organize the determination of the transport model
TRANS	2.6	Evaluate the neoclassical transport coefficients
SOURCE	2.7	Evaluate the neoclassical source terms
RIPPLE	2.8	Evaluate the transport resulting from modulations in the toroidal field
TRAP	2.9	Evaluate the transport resulting from trapped particle instabilities
RAD	2.10	Evaluate the radiation losses
ALPHA	2.11	Evaluate the effect of alpha particles
INJECT	2.12	Evaluate the effect of the injection of a beam of neutrals
BLKSLV	2.13	Solve the block tridiagonal system of linear equations
LUF	2.14	Form the lower and upper factors
FANDB	2.15	Forward and backward substitution
FANDBV	2.16	Vector forward and backward substitution
CHECK	2.17	Check the accuracy of the solution
RESET	2.18	Reset variables so that the time-step may be repeated
ADCOMP	2.19	Evaluate the effect of adiabatic compression
STORE	2.20	Determine the variables to be stored periodically in time

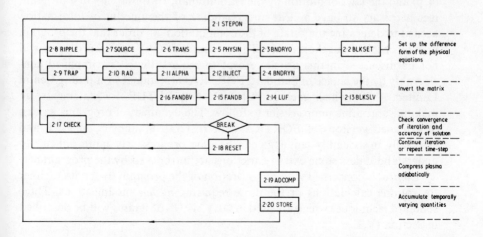

FIG. 4. Flow diagram for the calculation subroutines (Class 2).

that has the same structure as other members of the OLYMPUS family but solves a null set of equations. An OLYMPUS program package also contains a number of utility and diagnostic routines in binary form that may be regarded as an extension of the library of Standard FORTRAN Functions (National Computing Centre Limited, 1970), a set of standard COMMON blocks, a preprocessor program (where this is not already provided by the computer system itself (Hughes *et al.*, 1975)), test programs and data, and documentation files.

B. Program Structure

Four aspects of program structure are taken into account in the design of an OLYMPUS program, namely, architecture, dynamics, control, and diagnostics.

It is often convenient to exploit a mechanical or electrical analogy in which the program (software) is compared to a real physical machine (hardware).

Architecture is a static concept and is concerned with the underlying plan on which the program is based. In FORTRAN this plan is the division of the program into subprograms and COMMON blocks, each of which has a specific duty to perform. This corresponds to the design of a real machine as represented by a set of blueprints or a stationary model. The architecture of OLYMPUS programs is described further in Section IV, C.

Dynamics is concerned with how the program actually works. It is useful to exploit the analogy with Hamiltonian dynamics in which there exists the concept of the state of a dynamical system changing with time according to certain laws; the state is represented by a set of coordinates and momenta $(\mathbf{q}, \mathbf{p})$ and the laws of motion by the Hamiltonian, H. In the case of a program it is necessary to have a clear understanding of the amount of information needed to determine the "state of the computation" in order that the program may be checked out thoroughly. The dynamics of OLYMPUS programs will not be discussed in this article, except to remark that it is simplified considerably by the choice of Standard FORTRAN (National Computing Centre Limited, 1970) which excludes ENTRY and RETURN i statements, so making subroutine jumps easier to follow. The dynamics of programs written in extended versions of FORTRAN, in ALGOL 60 and 68, in PL/I, and especially in assembly language can become progressively more difficult to understand unless some extra restrictions are introduced by the programmer.

Control is concerned with the operation of the program by the user, either for routine calculations or for those requiring *ad hoc* modifications. Three standard techniques which are used in OLYMPUS programs will be described in Section IV, E.

Diagnostics may be compared to the measurements or observations that are made on a physical mechanism or electrical apparatus in order to monitor its working and to check that it is performing correctly. Usually these measurements do not disturb the working of the system, and the same criterion should be applied to the diagnostics in a computer program (although there will result a slight reduction in running speed and a small increase in the storage requirements in order to accommodate the diagnostic routines themselves). The diagnostic facilities of a program are required to monitor (a) the working of the program; (b) the validity of the numerical scheme; and (c) the behavior of the physical model which the calculation is describing. The OLYMPUS system *per se* deals only with (a).

C. Architecture

Every OLYMPUS program has a similar architecture which consists of two main parts

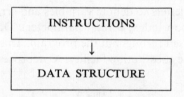

The instructions are represented by the FORTRAN subprograms which are currently organized into seven *classes*

0	Control	4	Epilogue
1	Prologue	5	Diagnostics
2	Calculation	U	Utilities
3	Output		

Each subprogram is assigned a decimal number $\langle m.n \rangle$ as indicated in Table III. The control class 0 is common to all programs as also is the utility class U and most of the diagnostic class 5. The subprograms of classes 1–4, which represent respectively the initialization, calculation, output, and termination operations of the code, will differ from one program to another although the names of the primary subprograms must always be present since these are called by the permanent control subprogram $\langle 0.3 \rangle$ COTROL. As a real program is developed, working subprograms replace the dummy versions otherwise loaded from a library. In this way a program is constructed in a standard and methodical way enabling sections of the program to be tested independently.

The data structure is represented by labeled COMMON blocks which are organized into five *groups*: (1) general OLYMPUS data; (2) physical problem; (3) numerical scheme; (4) housekeeping; and (5) input, output, and diagnostics. These groups are subdivided into decimally numbered blocks [Cr.s] as indicated in Table V. Blank COMMON is not normally used but has the label [C9.0]. It is useful to visualize the COMMON blocks as *operands* and the subprograms as *operators* which act on the data and change the values of the variables and arrays in the same way as the Hamiltonian, H, acts on the coordinates and momenta $(\mathbf{q}, \mathbf{p})$ in dynamics. In order to take advantage of this simple picture all information relevant to the "state of the calculation" is stored in COMMON rather than in local storage, and is transmitted via COMMON rather than through subprogram argument lists. This is easier to understand and has the advantage that the diagnostic routines are able to examine COMMON and provide a "snapshot" of the current state. Alphanumeric indices and subprogram maps are used to make the architecture as clear as possible.

Precautions are taken to avoid the duplication of nonlocal names by the use of different initial letters to denote variables and arrays of different status (COMMON, local, loop index, formal parameter). Only one copy of each COMMON block is used throughout the program. This is inserted automatically by means of a file substitution facility. A preprocessor control statement

TABLE V

THE LABELED COMMON BLOCKS WHICH DEFINE THE "STATE OF THE CALCULATION" IN THE COMPUTER CODE, ICARUS

Name	Number	Title
COMBAS	1.1	Basic system parameters
COMFUN	1.4	Fundamental constants
COMDDP	1.9	Development and diagnostic parameters
COMPHY	2.1	Physical variables
COMCNS	2.2	Physical constants
COMACH	2.3	Machine parameters
COMFLX	2.4	Flux variables
COMSRC	2.5	Source variables
COMESH	3.1	Spatial mesh variables
COMSTO	3.2	Temporal mesh variables
COMXSN	3.3	Averages over the minor cross section
COMFLO	3.4	Flow out of cross section
COMCON	4.1	Calculation control variables
COMOUT	5.1	Output control variables

such as

// SUBSTITUTE COMTOK

for the ICL 4/70 or IBM 360/370 computers results in the named file being inserted in place of the control card. Corresponding facilities are built into the CDC and Univac systems.

The COMMON blocks in Group 1 are intended to be standard library versions which are available to all programs. So far two have been published, namely, [C1.1] COMBAS which contains basic system parameters and [C1.9] COMDDP which contains parameters used during the development and checking of a program. Other blocks will contain fundamental physical constants, character codes, and other general-purpose information.

The physics of the step-by-step calculation is controlled always by SUBROUTINE ⟨2.1⟩ STEPON, which in turn calls on other Class 2 subprograms to perform the actual calculation as illustrated in Fig. 4 and Table IV. There is considerable advantage in separating this part from the rest of the program, which is concerned mainly with practical matters (although ⟨1.6⟩ INITAL sets up the physical initial conditions).

D. INITIALIZATION

Experience indicates that in order to make a program easy to develop, to understand, and to use, it is important to initialize the data structure in a logical way. Although programmers often employ a sequence of data cards which may have quite complex and varied formats and are not necessarily all read at the same position in the program, it is possible to use in most cases, a much simpler scheme which requires only a single READ statement

$$\text{READ (NREAD, NEWRUN)} \qquad (24)$$

Statement (24) is the same for all OLYMPUS programs: NREAD is the input channel number, and NEWRUN is the name of a NAMELIST containing a list of all variables and arrays whose default values may require to be changed.

Data initialization is the main purpose of the Prologue (Class 1) which functions as follows: variables and arrays are cleared to zero in ⟨1.2⟩ CLEAR prior to a sequence of three subroutines being entered

⟨1.3⟩ PRESET

⟨1.4⟩ DATA

⟨1.5⟩ AUXVAL

PRESET assigns suitable default values to all those variables which can be independently set. For clarity this is performed block-by-block with the names arranged in alphanumeric order. DATA is required to change any of these default values, and finally AUXVAL computes the values of auxiliary variables which depend on the input data, such as the determination of

$$NM1 = N - 1,$$

where N is the number of mesh points. This procedure is organized block-by-block where practicable.

NAMELIST data input has the effect of overwriting the values of variables and array elements that are referenced by name and leaving all others unchanged. This facility is flexible, concise, and symbolic so that it is well adapted to on-line working. The input file can be kept quite short if appropriate default values are chosen, as illustrated in Table VI.

⟨1.6⟩ INITAL is intended to provide the physical initial conditions for the calculation. Frequently this is performed best by using function subprograms such as FRHO (density), FTE (electron temperature) which contain generalized functions whose parameters are acceptable in the NAMELIST NEWRUN. ⟨1.8⟩ START performs any housekeeping duties necessary before the run can begin.

E. Control

Three standard methods are used to control an OLYMPUS calculation: (1) NAMELIST data; (2) modification of the initial functions; and (3) the EXPERT facility. Method 1 has been explained in Section IV,D, while Method 2 simply requires that a new load module be generated with a different choice of one or more of the initial functions FRHO, FTE, etc. Usually this means copying the source file and changing a few well-identified statements using an editor facility. Method 3 is more powerful, and its purpose is to enable extensive *ad hoc* modifications to be made to an existing source program without compromising the original version, which may have been carefully tested and which may be in current use by other people.

At appropriate points throughout an OLYMPUS program calls are inserted of the form

$$\text{CALL EXPERT(ICLASS, ISUB, IPOINT)} \qquad (25)$$

The formal parameters, ICLASS and ISUB, are local variables defining the decimal code number ⟨ICLASS.ISUB⟩ of the subprogram from which the call is made, and IPOINT is an integer which specifies the precise point within that subprogram. These calls are not placed of course inside inner loops.

TABLE VI

The Namelist NEWRUN Used as Input Data for the Computer Code, ICARUS, in Order to Model (a) Pseudoclassical Diffusion in a Small Tokamak, (b) Modified Pfirsch–Schlüter Diffusion in a Small Tokamak, and (c) Neoclassical Diffusion in a Large Tokamak

(a) Small Tokamak run 75/01/01 Pseudoclassical diffusion		(b) Small Tokamak run 75/01/01 Modified Pfirsch–Schlüter		(c) Large Tokamak run 75/01/01 Neoclassical diffusion	
& NEWRUN		& NEWRUN		& NEWRUN	
NLOMT2(12)	= T,	NLOMT2(12)	= T,	RMAJOR	= 2.93,
CEOB	= 0.0,	CEOB	= 0.0,	RMINOR	= 1.28,
XN	= 1.0,	XE	= 200.0,	RPLSMA	= 1.28,
CB11	= 0.0,	XN	= 0.0,	CURENT	= 3.0E + 06,
CB12	= 0.0,	CB11	= 0.0,	DENMAX	= 9.1E + 19,
CB13	= 0.0,	CB12	= 0.0,	TEMAX	= 1000.0,
CB22	= 0.0,	CB13	= 0.0,	TIMAX	= 1000.0,
CB23	= 0.0,	CB22	= 0.0,	NLOMT2(12)	= T,
CBI	= 0.0,	CB23	= 0.0,	CB11	= 0.0,
CA11	= 0.0,	CBI	= 0.0,	CB12	= 0.0,
CA12	= 0.0,	CA11	= 0.0,	CB13	= 0.0,
CA22	= 0.0,	CA12	= 0.0,	CB22	= 0.0,
CAY	= 0.0,	CA22	= 0.0,	CB23	= 0.0,
CBY	= 0.0,	CAY	= 0.0,	CBI	= 0.0,
CC11	= 1.0,	CBY	= 0.0,	& END	
CC12	= 1.5,	CC11	= 1.0,	& RESET	
CC22	= 3.25,	CC12	= 1.5,	& END	
CCY	= 1.5,	CC22	= 3.25,		
PIQ(12)	= 2.0,	CCY	= 1.5,		
PIQ(13)	= 1.0,	PIQ(12)	= 2.0,		
PIQ(14)	= 40.0,	PIQ(13)	= 1.0,		
& END		PIQ(14)	= 40.0,		
& RESET		& END			
& END		& RESET			
		& END			

Subroutine ⟨0.4⟩ EXPERT(KCLASS, KSUB, KPOINT) is contained in skeleton form in the OLYMPUS library and is used to control the diagnostics, but the user is free to provide his own version which can be used to insert *ad hoc* sections of coding. If an insertion is to be made at the point (2.13.3) (and possibly also at other points) a convenient technique is to employ the statements

$$ICODE = 10000*KCLASS + 100*KSUB + KPOINT$$

⋮

$$IF(ICODE.EQ.21303)GO\ TO\ 21303$$

⋮

RETURN

21303 CONTINUE

[Additional coding for point (2, 13, 3)]

RETURN

⋮

END

⟨0.4⟩ EXPERT has access to all COMMON variables and arrays (provided that the corresponding // SUBSTITUTE statements are included) and it can call subprograms itself, if necessary. However, it must not call any subprograms that already contain statements of the form (25). This difficulty can be avoided by including a switch of the form

$$IF(NLEXPT)\ CALL\ EXPERT\ (ICLASS, ISUB, IPOINT). \tag{26}$$

This is turned off as soon as EXPERT is entered and restored to its original setting as soon as the return is made.

Provided that EXPERT is arranged in a logical way, a listing of this file together with that of any additional subprograms constitutes a generalized form of data input and sufficiently defines the *ad hoc* modifications that have been made.

A further technique that can be used is to maintain a master version of EXPERT on-line in a file called MASTER, which also contains all the control statements needed to compile EXPERT, to link-edit or compose the new version of the program and to run the job. The line numbers in this file are coordinated with the decimal numbering scheme used by OLYMPUS so that it is easy to incorporate changes without making mistakes. The ICL 4/70 Multijob Editor uses a command of the form

$$EDIT\ B/C,,,,A$$

in order to edit corrections contained in file A into file B to produce a new file C. The files A and B and the line numbers are left unchanged. In order to insert *ad hoc* modifications and to run a new case commands of the following form need to be typed

[Edit file MOD2]

EDIT MASTER/CASE2,,,,MOD2

REMJOB CASE2(S)

The job will be queued in the appropriate remote batch stream and executed. Alternatively, it could be executed on-line.

F. Diagnostics

Facilities are provided by OLYMPUS for tracing the flow of a program, for switching off selected subprograms, for printing messages, and for displaying the contents either of selected COMMON blocks or of individual variables or arrays. Diagnostic output is arranged in convenient form and can largely be controlled by switches set in NAMELIST data input. For example, the input line

NLREPT = T, NPDUMP = 20701, NVDUMP = 100,

would cause all COMMON variables (but not arrays) to be displayed at the point (2, 7, 1), arranged decimally by blocks and alphanumerically within the blocks in the form

⟨name⟩ = ⟨value⟩.

This output forms a properly organized index to the current state of the calculation. Other switch settings can be used to select only certain of the blocks, or to display the arrays.

If the local system supports the facility, certain classes of error cause the supervisor to reenter the program at a point where the FORTRAN statements

CALL CLIST(0, 0)

CALL ARRAYS(0, 0)

are used to generate an index for all COMMON variables and arrays. This is preferable to the octal or hexadecimal dump that is provided by most computer systems.

These facilities are useful during program check-out: the switches can be visualized as part of an "engineer's test panel" and calls of the form (26) as the "cables" which connect the panel to the program under test. The OLYMPUS utility routines are also useful during *ad hoc* modifications, since, for example, the statement

CALL RARRAY(8HBTHETA   , BTHETA, MAXMSH)

in EXPERT will output the array containing the values of the poloidal magnetic field, B_θ, without the need to construct FORMAT statements in the usual way.

G. Documentation

OLYMPUS programs together with their documentation are arranged in a standard format as illustrated in Fig. 5. This format is based on that of a mathematical textbook and is intended to be as easy to follow as possible. The subprograms are divided into decimally numbered sections and subsections, each of which has an appropriate heading. Indentation is used to improve the layout. In Fig. 5 the comments have been printed in lower case on a General Electric Termi Net 300 to make them stand out from the statements. It is planned to publish a number of programs in this form for training purposes.

V. Applications

A. Introduction

Recent publications (Dimock et al., 1973; Hinnov et al., 1973; Berry et al., 1974) on the behavior of Tokamak plasmas in a variety of devices have provided a detailed description of the temporal behavior of the discharge. In particular, detailed measurements of not only the ohmic heating current and voltage, but also the radial profiles of the electron temperature and density have been obtained.

The predictions of neoclassical theory can be examined and are found to be incorrect. It is therefore of paramount importance to be able to selectively omit parts of the model, and to replace complete sections of the model. The development of the physical model to one which simulates well the behavior of an ST discharge (Dimock et al., 1973) is monitored by the method of introduction of additional flux and source terms into the code ICARUS. Physical effects that may be present in the next generation of Tokamaks can be incorporated readily into the code.

Fig. 5. Listing of subroutine ⟨2.10⟩ RAD in which are calculated the source terms that result from bremsstrahlung and cyclotron radiation losses.

```
C
            SUBROUTINE RAD (K, KNIT)
C
C 2.10 Evaluate the radiation losses
C
//  SUBSTITUTE COMFUN
//  SUBSTITUTE COMDDP
//  SUBSTITUTE COMCNS
//  SUBSTITUTE COMSRC
//  SUBSTITUTE COMESH
//  SUBSTITUTE COMXSN
//  SUBSTITUTE COMCON
//  SUBSTITUTE COMOUT
C----------------------------------------------------------------
            DIMENSION ZFIN(5,2)
C
            DATA ICLASS, ISUB/2, 10/
C
            DATA ZFIN/0.38,1.0,8.2,120.0,2500.0,0.13,0.2,1.2,13.0,220.0/
C----------------------------------------------------------------
            IF (NLOMT2(ISUB)) RETURN
C----------------------------------------------------------------
CL          1.          Prologue
C
C     Initialize storage of energy changes at point *1*
            IF (K .EQ. 1) GO TO 500
C
C     Dependent variables at point *k*
            ZN = DENE(K)
            ZPE = PE(K)
C
            ZEFF = EFFZ(K)
C
C     Evaluate the independent variable
            ZR = R(K)
            ZDR = DELTAR(K)
            ZFAC = ZN * ZR * ZDR
C
C     Calculate the electron temperature in electron volts
            ZTE = (ZPE / ZN) / CHARGE
            ZRTTE = SQRT (ZTE)
C
C     Calculate the electron temperature in joules
            ZTE = ZTE * CHARGE
C
C----------------------------------------------------------------
CL          2.          Bremsstrahlung radiation loss
C
            ZBREM = CBREM * ZEFF * ZRTTE * ZN
C
C----------------------------------------------------------------
CL          3.          Cyclotron radiation loss
C
            ZCYC1 = CYC1 * ZN
C
C     Initialize accumulator
            ZSUM=0.0
C     Scan over *i*
            DO 302 J1 = 1, 2
C     Scan over harmonics
            DO 301 J2 = 1, 5
C     Rosenbluth zlambda function
            ZLIN = ZCYC1 * ((0.625E+13 * ZTE) ** (J2-1)) * ZFIN(J2,J1)
C     Contribution of this harmonic
            ZHARM=(J2*J2*J2)*ZLIN/(1.0+ZLIN)
C     Accumulate cyclotron loss
  301       ZSUM = ZSUM + ZHARM
  302       CONTINUE
C
C     Cyclotron power loss
            ZCYC = CYC2 * ZTE *ZSUM * (1.0 - CYCFAC) / ZN
C
C----------------------------------------------------------------
CL          4.          Source terms
C
CL          4.1         Electron pressure equation
            SCPENE = SCPENE - (ZBREM + ZCYC) * CGM1
C
C----------------------------------------------------------------
CL          5.          Store quantities for diagnostic purposes
C
C     Return if not called during the first iteration
  500       IF (KNIT .NE. 0) RETURN
C
C     Initialise for energy changes this timestep
            IF (K .NE. 1) GO TO 501
            XBREM = 0.0
            XCYC = 0.0
            RETURN
  501       CONTINUE
C
C     Determine the energy changes this timestep
C     Bremsstrahlung radiation loss
            XBREM = XBREM + ZBREM * ZFAC
C
C     Cyclotron radiation loss
            XCYC = XCYC + ZCYC * ZFAC
C
C     Determine the energy changes since time, t=0
            IF (K .NE. NM1) RETURN
            XBREMT = XBREMT + XBREM * DT
            XCYCT = XCYCT + XCYC * DT
C
            RETURN
            END
```

B. Present Generation Tokamaks

1. *The Results of Neoclassical Simulations*

The various codes which solve the full set of neoclassical equations give essentially the same results. In Fig. 6 the radial profiles of particle density, n, electron and ion temperatures, T_e and T_i, as predicted by the Düchs code (Hinton *et al.*, 1972) and ICARUS (Watkins *et al.*, 1976) are plotted for conditions applicable to a low density discharge in a device such as ST at a time of 60 msec.

The codes predict narrow, almost linear, density profiles coupled with much broader thermal distributions. The density distribution results from the dominance in the neoclassical equations of the Ware pinch effect (Ware, 1970), which reduces the density in the outer regions of the discharge while the density near the axis increases. Since the temperature gradients exist only near the edge where the density becomes low, the thermal losses are also low.

However, the numerical predictions for existing Tokamaks are not in accordance with experimental results. For conditions applicable to a high

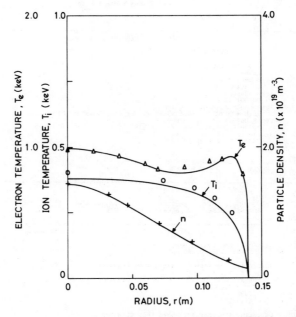

Fig. 6. Comparison of the radial profiles of particle density, n, electron temperature, T_e, and ion temperature, T_i, for conditions applicable to a low density discharge in a small Tokamak ($R = 1.09$ m, $a = 0.14$ m, $\langle n \rangle = 6 \times 10^{18}$ m^{-3}, $I_\varphi = 40$ kA, $B_\varphi = 3$ T) at 60 msec as given by the code, ICARUS (———), and that of Düchs ($+, 0, \Delta$), using the neoclassical diffusion model.

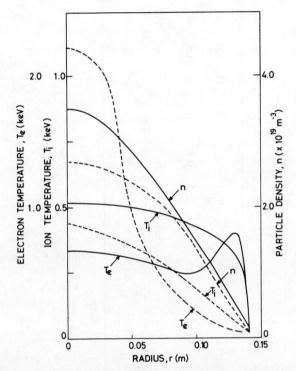

FIG. 7. Comparison of the radial profiles of particle density, n, electron temperature, T_e, ion temperature, T_i, for conditions applicable to a high density discharge in a small Tokamak ($R = 1.09$ m, $a = 0.14$ m, $\langle n \rangle = 1.65 \times 10^{19}$ m^{-3}, $I_\varphi = 60$ kA, $B_\varphi = 3.7$ T) given at 35 msec by the code, ICARUS, using the neoclassical diffusion model (———), and the laboratory experiment (Dimock et al., 1973) (– – – –).

density discharge in a device such as ST (Dimock et al., 1973) the experimental profiles at 35 msec are very different from those predicted numerically (Fig. 7). As a result the energy balance is found experimentally to be dominated by electron thermal conduction and recycling of particles rather than by ion thermal conduction and radiation as is predicted numerically. The experimental replacement times for the particle density, τ_p, electron and ion energies, τ_{Ee}, τ_{Ei}, and total energy, τ_E', are at least an order of magnitude smaller than those obtained numerically (Table VII).

A further deficiency of neoclassical theory is its inability to describe the current rise phase in Tokamaks. In this case, if the transport were purely neoclassical the skin current would persist for extremely long times (Düchs et al., 1971). This is not observed experimentally although the ST discharge (Dimock et al., 1973) appears to exhibit a mild skin effect which is probably not negligible for the purposes of a detailed analysis of the energy balance,

TABLE VII

GENERAL CHARACTERISTICS OF A HIGH DENSITY DISCHARGE IN A SMALL TOKAMAK
($R = 1.09$ m, $a = 0.14$ m, $\langle n \rangle^a \sim 1.7 \times 10^{19}$ m^{-3}, $I_\varphi = 60$ kA, $B_\varphi = 3.7$ T)

Scaling law	Central values					Mean values[a]					Replacement times (ms)[b]			
	n ($\times 10^{19}$ m^{-3})	T_e (keV)	T_i (keV)	E_z (Vm^{-1})	q	$\langle n \rangle$ ($\times 10^{19}$ m^{-3})	$\langle T_e \rangle$ (keV)	$\langle T_i \rangle$ (keV)	$\beta_{\theta e}$	$\beta_{\theta i}$	τ_p	τ_E'	τ_{Ee}	τ_{Ei}
ST experiment (35 ms)	2.7	2.2	0.46	0.36	0.6	1.8	0.76	—	0.75	—	16	13	19	—
Neoclassical theory (34 ms)	3.5	0.67	0.51	0.08	3.2	1.65	0.58	0.45	0.52	0.40	1506	106	604	98
Pseudoclassical theory (33 ms)	2.0	1.03	0.46	0.22	2.0	1.65	0.65	0.35	0.58	0.31	23	20	55	23
Pfirsch–Schlüter × 200 (34 ms)	2.6	2.6	0.50	0.20	0.58	1.65	0.62	0.28	0.56	0.25	6	11	44	22

[a] $\langle n \rangle = \dfrac{2}{a^2} \int_0^a nr\,dr,$

$\langle T \rangle = \int_0^a nkTr\,dr \Big/ \int_0^a nr\,dr,$

$\beta_\theta = 16\pi^2 \int_0^a nkTr\,dr \Big/ \mu_0 I^2.$

[b] $\tau_p = \int_0^a nr\,dr \Big/ [r\Gamma]_{r=a},$

$\tau_E' = \int_0^a [nk(T_e + T_i) r\,dr] \mu_0 \Big/ [(\gamma-1) E_\varphi B_\theta]_{r=a},$

$\tau_{Ee} = \int_0^a nkT_e r\,dr \Big/ [(\gamma-1) r Q_e]_{r=a},$

$\tau_{Ei} = \int_0^a nkT_i r\,dr \Big/ [(\gamma-1) r Q_i]_{r=a}.$

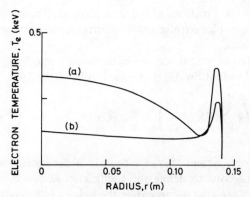

FIG. 8. The radial profiles of the electron temperature, T_e, for conditions applicable to a high density discharge in a small Tokamak ($R = 1.09$ m, $a = 0.14$ m, $\langle n \rangle = 1.65 \times 10^{19}$ m^{-3}, $I_\varphi = 60$ kA, $B_\varphi = 3.7$ T), given by the code, ICARUS, to indicate the neoclassical "skin-effect" when (a) a 5% density pedestal is maintained at $r = a$ for all time, and (b) the bremsstrahlung radiation losses are enhanced by a factor of 10000.0. Times are (a) 2.7 sec, (b) 2.0 sec.

but which reaches a saturation level. It is found (Rosenbluth and Kaufman, 1958) that in a vacuum region the magnetohydrodynamic equations break down as a result of the joule heating term (which is independent of number density) being shared between fewer and fewer particles, with the result that a singularity in the electron temperature can be produced. In the case of the neoclassical equations, Furth et al. (1970) have indicated that the skin effect during the rise of the plasma current could result in the excitation by surface heating of an unstable thermal mode which will then augment and perpetuate the skin effect.

To circumvent this difficulty, the computer codes usually run with the current initially distributed throughout the plasma volume and with the total current maintained constant for all time. Nevertheless, skin currents can result if too low a value for the density pedestal is maintained (Fig. 8).

2. The Development of the Physical Model

The model of a Tokamak plasma as formulated by the neoclassical theory gives the discharge a temporal behavior which relates very little to an experimental discharge. It is necessary to be able to omit or replace easily sections of the model. This implies that the solution be determined in such a way that the physical model is separated from the numerical model in the sense that changes in the physical model can be undertaken readily and will be transmitted automatically in the correct form to the numerical model. An interface may be constructed which will transform the flux, **F**, and source, **S**, terms supplied by a user as functions of the dependent variables and of their spatial gradients into the block-tridiagonal, matrix representation of the finite-difference equations.

The fluxes and source terms are determined respectively as coefficients, FX$\alpha\beta$ and SC$\alpha\beta$ so that Eq. (3) is assumed to be of the form

$$\frac{\partial \alpha}{\partial t} + \frac{1}{r}\frac{\partial}{\partial r}\left(r \sum_\beta \text{FX}\alpha\beta \frac{\partial \beta}{\partial r}\right) = \sum_\beta \beta \text{SC}\alpha\beta, \quad \alpha, \beta \equiv \{n, p_e, p_i, B_\theta\}. \quad (27)$$

In general, for a system of N equations in N dependent variables it would be most advantageous to define the coefficients as two-dimensional arrays FX(α, β), SC(α, β) so that looping over the indices 1-N may be used to introduce even further independence.

3. Additional Flux Terms

a. Pseudoclassical Diffusion. The pseudoclassical scaling laws (Yoshikawa, 1970; Artsimovich, 1971) may be incorporated into the code ICARUS in a number of different ways which highlight the versatility of the OLYMPUS system. The definitions for the fluxes of particle density, electron and ion thermal energy densities, and current density are

$$\Gamma = nV_r = -KD_\theta^c(1 + T_i/T_e)\,\partial n/\partial r,$$

$$\tilde{Q}_e = -KD_\theta^c(1 + T_i/T_e)(\partial p_e/\partial r - kT_e\,\partial n/\partial r),$$

$$\tilde{Q}_i = -[(m_i/m_e)(T_e/T_i)]^{1/2}D_\theta^c(\partial p_i/\partial r - kT_i\,\partial n/\partial r)(\text{CAI}\varepsilon^{1/2} + \text{CBI}\varepsilon + \text{CCI}),$$

$$Q_e = \tilde{Q}_e + \tfrac{3}{2}\Gamma kT_e, \quad (28)$$

$$Q_i = \tilde{Q}_i + \tfrac{3}{2}\Gamma kT_i,$$

$$J_\varphi = K^{-1}\sigma_s(\text{CA33}\varepsilon^{1/2} + \text{CB33}\varepsilon + \text{CC33})E_\varphi,$$

where (Spitzer, 1962)

$$K \sim Z_{\text{eff}}[0.457/(1.077 + Z_{\text{eff}}) + 0.29]. \quad (29)$$

The expressions for the fluxes incorporate no cross terms so that Γ is dependent on $\partial n/\partial r$ only and $\tilde{Q}$ is dependent on $\partial T/\partial r$ only. In particular the trapped particle pinch terms in the neoclassical expressions are not present. Since the ion transport is found experimentally to be predominantly neoclassical, these terms are retained. The effect of a relatively small population of heavy impurity ions (such as iron or tungsten) is to dominate the effective ionic charge, Z_{eff}, of a hydrogen plasma and so modify the resistivity and the other transport coefficients in accordance with Eqs. (28) and (29) (Spitzer, 1962).

For a set of fluxes as given by these equations, the coding for the neoclassical fluxes contained in the subroutine $\langle 2.6 \rangle$ FLUXES (Fig. 4) may be completely replaced by a new version. This would be the procedure if the neoclassical version were now to be disregarded and the code were to be based completely on the pseudoclassical model. The *ad hoc* inclusion of the pseudoclassical model could be achieved by a suitable call to the EXPERT subroutine after the determination of the neoclassical fluxes. The appropriate section of the user-supplied EXPERT would overwrite or append the fluxes with those appropriate to the pseudoclassical model.

However, for the particular pseudoclassical model given by Eqs. (28) and (29) examination of the equations indicates that they form a subset of the basic model, provided that the numerical coefficients are chosen correctly. In Table VI(a) the changes necessary to run the code in the pseudoclassical mode are listed in the NAMELIST NEWRUN that may be read in as input data.

The NAMELIST NEWRUN indicates additionally how this facility enables certain parts of the physical model to be temporarily removed in order to estimate the effect of these terms. For example, the setting of CEOB equal to zero enables all trapped particle pinch terms to be omitted from the calculation. The selective omission of terms in an equation facilitates both the debugging of the code during development work and the understanding of the model equations during production work.

b. Pfirsch–Schlüter Diffusion. The model for plasma diffusion in the highly collisional regime ($v_{ei} \gg v_t$) (Rosenbluth and Kaufman, 1958; Taylor, 1961) and modified by Pfirsch and Schlüter (1962) to take into account the effects of toroidal geometry is known to give diffusion losses which are too small in comparison with those observed in present-day Tokamaks. Nevertheless, C. Mercier and Soubbaramayer (private communication, 1974) have found that the Pfirsch–Schlüter form of the diffusion coefficient modified by a large multiplicative factor, $\lambda \gg 1$, is sufficient to simulate Tokamak behavior in TFR (Rebut *et al.*, 1973) and ST (Dimock *et al.*, 1971) experiments. The modified Pfirsch–Schlüter diffusion may be represented by the following definition of the diffusion coefficient

$$D_{\text{MPS}} = \lambda(1 + 1.6q^2) v_{ei} \rho_{e\varphi}^2, \tag{30}$$

where q and $\rho_{e\varphi}$ are, respectively, the Tokamak safety factor and the electron gyroradius in the toroidal magnetic field

$$q = rB_\varphi/RB_\theta, \qquad \rho_{e\varphi}^2 = 2m_e kT_e/e^2 B_\varphi^2.$$

The code ICARUS may be used to investigate the behavior of a collision-free plasma in which the electron diffusion is determined by the modified Pfirsch–Schlüter diffusion coefficient, and the ion thermal transport is determined by that given by the neoclassical model. The changes necessary to run the code in this mode are given in the NAMELIST reproduced in Table VI(b).

However, the use of neither a pseudoclassical nor a modified Pfirsch–Schlüter model is sufficient to be able to describe the "bell-shaped" electron temperature profile observed in the ST experimental discharge (Fig. 7). Although the replacement times are reduced as a result of the enhanced diffusion processes (Table VII), the diffusion alone is unable to induce central peaking of the electron temperature, even though the Pfirsch–Schlüter diffusion coefficient peaks outwardly. It is necessary to invoke additional mechanisms to reproduce the experimental profiles.

4. Additional Source Terms

The addition of source terms to those already incorporated in the basic model may be accomplished in the first instance by the use of the EXPERT facility which will contain coding that modifies the arrays that contain the relevant source terms. Subsequent to the successful inclusion of these terms, the coding in EXPERT may be transferred to a single, self-contained subroutine which is concerned exclusively with a particular aspect of the physical model. For example, it is reasonable to define in a single subroutine the radiation losses associated with bremsstrahlung and cyclotron processes (Fig. 5). These are purely source terms with power losses given respectively by Rose and Clark (1961) and Rosenbluth (1970)

$$P_{\text{bremsstrahlung}} = (3/2)^{1/2} (Z_{\text{eff}}^2 e^6/24\pi)(\mu_0^3 c^3/m_e h) n_e n_i (8kT_e/\pi m_e)^{1/2}$$

$$= \text{CBREM}\, Z_{\text{eff}}\, n_e^2 T_{e(\text{ev})}^{1/2}, \qquad (31)$$

$$P_{\text{cyclotron}} = 0.16 \times 10^{16} \times (B_\varphi^3/R) T_e \sum_{i=1}^{2} \sum_{k=1}^{\infty} k^3 \Lambda_k^i/(1+\Lambda_k^i), \qquad (32)$$

where

$$\Lambda_k^i = 5.0 \times 10^{-16} (nR/B_\varphi)(10^{-6} T_{e(\text{ev})})^{k-1} f_k^{(i)},$$

$$f_k^{(1)} = (ke)^{k-2}, \qquad f_k^{(2)} = f_k^{(1)}/(2k+1),$$

where h is Planck's constant and c is the velocity of light. In the last expression above, e is the base for natural logarithms, and *not* the charge on the electron.

The coefficient CBREM is a function only of the fundamental constants. During the initialization procedure it is cleared to zero in subroutine ⟨1.2⟩ CLEAR, defined equal to unity in subroutine ⟨1.3⟩ PRESET, and calculated as the product of CBREM and the fundamental constants in subroutine ⟨1.5⟩ AUXVAL. Accordingly, the magnitude of the bremsstrahlung losses may be modified in subroutine ⟨1.4⟩ DATA by defining the value of CBREM different from unity. In particular, setting CBREM equal to zero allows the removal of bremsstrahlung radiation effects.

The complete omission of all radiation effects contained in subroutine ⟨2.10⟩ RAD may be achieved by setting in the NAMELIST NEWRUN the logical array element

$$\text{NLOMT2(10)} = \text{T}.$$

A logical IF statement at the beginning of each CLASS 1, CLASS 2, and CLASS 3 subroutine can allow the return to the calling subroutine before any calculation is performed. Any physical effect that is incorporated entirely within one subroutine can be suppressed by means of this facility.

Although radiation cooling is a process that will tend to make the Tokamak discharge thermally unstable (Furth *et al.*, 1970) its effect is to reduce the central temperature of the discharge and encourage outward peaking of the electron temperature (Fig. 8). If a cooling mechanism is required to induce central peaking of the electron temperature then it is necessary that this mechanism be effective in the outer, low density regions of the discharge. Experimentally, it is observed that although there is a large outflow of particles to the limiter of the Tokamak, recycling of particles occurs to such an extent that the total number of particles contained in the torus remains approximately constant. Penetration into the plasma of these recycled particles occurs, with the result that they are ionized and afford a source to the particle distribution and a sink to the electron energy distribution.

Detailed calculations of the effect of a neutral gas component have been performed by Düchs *et al.* (1972), Girard *et al.* (1972), and Hogan and Dory (1972), and take into account both the ionization of the neutral population and the production of hot neutrals as a result of charge exchange with the background plasma ion population. The source terms which enter the particle density equation (5) and electron energy equation (6) as a result of the ionization of the neutral species may be written

$$\delta n/\delta t = n n_\text{n} \langle \sigma v \rangle_\text{I}$$
$$\delta p_\text{e}/\delta t = -p_\text{e} n_\text{n} \langle \sigma v \rangle_\text{I} (E_\text{I}/kT_\text{e}), \tag{33}$$

where the interaction of the background plasma population, n, with the neutral species, n_n, is determined by the cross section, $\langle \sigma v \rangle_I$, for ionization, together with an average electron energy loss of E_I per ionizing collision. To indicate the predominant effect of the neutral gas, namely, the depression of the electron temperature on the outside of the discharge and the elevation of the electron temperature in the central region of the discharge, a very simple model for the neutrals will be used for the purposes of this article. The source terms in Eq. (33) are replaced by an heuristic model which maintains constant the total particle content of the torus

$$\delta n/\delta t = [\Gamma(r=a)(m+2)/a](r/a)^m$$
$$\delta p_e/\delta t = -[\Gamma(r=a)(m+2)/a]E_I(r/a)^m. \quad (34)$$

The penetration of the neutral species may be achieved by setting m to an appropriate value. In ST the neutrals penetrate well into the plasma (Dimock et al., 1973) so it is reasonable to choose $m = 2$, which corresponds to a distribution that is parabolic in r. Although both the pseudoclassical and modified Pfirsch–Schlüter diffusion models give electron temperature profiles that have an inflexion point (Figs. 9 and 10), the experimental values for the

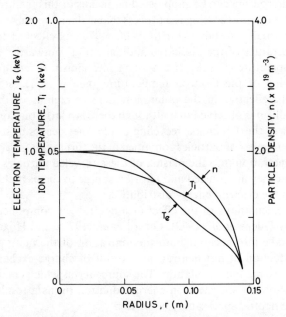

FIG. 9. The radial profiles of particle density, n, electron temperature, T_e, ion temperature, T_i, for conditions applicable to a high density discharge in a small Tokamak ($R = 1.09$ m, $a = 0.14$ m, $\langle n \rangle = 1.65 \times 10^{19}$ m^{-3}, $I_\varphi = 60$ kA, $B_\varphi = 3.7$ T) given at 33 msec by the code, ICARUS, using the pseudoclassical diffusion model and a "neutral species" represented by $m = 2$ and $E_I = 40$ eV.

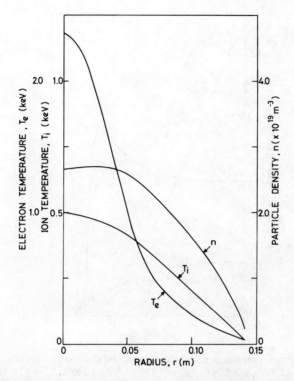

FIG. 10. The radial profiles of particle density, n, electron temperature, T_e, ion temperature, T_i, for conditions applicable to a high density discharge in a small Tokamak ($R = 1.09$ m, $a = 0.14$ m, $\langle n \rangle = 1.65 \times 10^{19}$ m^{-3}, $I_\varphi = 60$ kA, $B_\varphi = 3.7$ T) given at 34 msec by the code, ICARUS, using the Pfirsch–Schlüter diffusion model with $\lambda = 200.0$ and a "neutral species" represented by $m = 2$ and $E_I = 40$ eV.

particle density and temperatures are approximated more accurately by the modified Pfirsch–Schlüter scaling law (Table VII) as a result of the greater influx of particles which induces more peaking of the electron temperature on axis.

C. Next Generation Tokamaks

Extrapolations of the results of both neoclassical and pseudoclassical theory to the regime applicable to the next generation of larger Tokamaks give optimistic predictions. In both cases conditions are favorable for thermonuclear burning to take place without the aid of additional heating mechanisms and provided that the mean particle density is less than 5×10^{19} m^{-3} and the toroidal current is greater than 3 MA (Fig. 11). On the other hand, the Pfirsch–Schlüter model, which gave the best simulation of the ST device, produces less

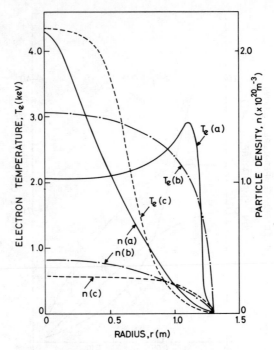

FIG. 11. The radial profiles of number density, n, and electron temperature, T_e, for a large Tokamak ($R = 2.82$ m, $a = 1.28$ m, $\langle n \rangle = 5.0 \times 10^{19}$, $I_\varphi = 3$ MA, $B_\varphi = 3$T). The results are obtained by the code, ICARUS, using (a) the neoclassical model (———) at 5.4 sec, (b) the pseudoclassical model (—·—·—) at 4.7 sec, (c) the Pfirsch–Schülter model, with $\lambda = 200.0$ (— — —) at 4.8 sec. In (b) and (c) the particle content is maintained constant by a "neutral species" represented by $m = 6$ and $E_1 = 40$ eV.

favorable results when applied to a large Tokamak. Shallow penetration ($m = 6$) of a large influx of "neutrals" cools the plasma considerably near radius, $r = a$, with the result that the mean temperature is low (Table VIII).

The neoclassical model has the advantage that an isolated pinch is formed as a result of the Ware pinch effect. The input energy is dissipated almost entirely by bremsstrahlung radiation losses so that the predicted energy replacement time is comparatively long (Table VIII). However, the neoclassical solution is thermally unstable (Düchs et al., 1971) with off-axis peaking of the temperatures occurring to different extents depending on the initial conditions. Although the replacement times predicted by the pseudoclassical diffusion model are less than those corresponding to the neoclassical formulation, the solution is independent of the initial conditions and greater central and mean temperatures are obtained as a result of channeling of the toroidal current along the minor axis.

TABLE VIII

GENERAL CHARACTERISTICS OF A DISCHARGE IN A LARGE TOKAMAK
($R = 2.93$ m, $a = 1.28$ m, $\langle n \rangle^a = 5.0 \times 10^{19}$ m^{-3}, $I_\varphi = 3$ MA, $B_\varphi = 3T$)

Scaling law	Central values					Mean values^a					Replacement times^b (s)			
	n ($\times 10^{19}$ m^{-3})	T_e (keV)	T_i (keV)	E_z (Vm^{-1})	q	$\langle n \rangle$ ($\times 10^{19}$ m^{-3})	$\langle T_e \rangle$ (keV)	$\langle T_i \rangle$ (keV)	$\beta_{\theta e}$	$\beta_{\theta i}$	τ_p	τ_E'	τ_{Ee}	τ_{Ei}
Neoclassical theory with $T_{e\,max}$ set initially	100 eV (5.5 s)	1.9	1.9	0.014	1.28	5.1	2.0	2.0	0.19	0.19	−942	16	7349	500
	4000 eV (5.4 s)	2.1	2.1	0.013	1.17	5.2	2.2	2.2	0.21	0.21	−325	11	−5479	447
Pseudoclassical theory (4.6 s)	8.1	3.1	3.1	0.012	0.73	5.0	2.6	2.6	0.24	0.24	147	6	399	31
Pfirsch–Schlüter ×200 (4.0 s)	6.0	3.6	3.6	0.011	0.65	5.0	1.3	1.3	0.12	0.12	0.2	1	5	7

^{a} $\langle n \rangle = \dfrac{2}{a^2} \int_0^a nr\,dr,$

$\langle T \rangle = \int_0^a nkTr\,dr \Big/ \int_0^a nr\,dr,$

$\beta_\theta = 16\pi^2 \int_0^a nkTr\,dr \Big/ \mu_0 I^2.$

^{b} $\tau_p = \int_0^a nr\,dr \Big/ [r\Gamma]_{r=a},$

$\tau_E' = \int_0^a nk(T_e + T_i)r\,dr \mu_0 \Big/ [(\gamma-1)E_\varphi B_\theta]_{r=a},$

$\tau_{Ee} = \int_0^a nkT_e r\,dr \Big/ [(\gamma-1)rQ_e]_{r=a},$

$\tau_{Ei} = \int_0^a nkT_i r\,dr \Big/ [(\gamma-1)rQ_i]_{r=a}.$

Within the context of the simple model for the neutral recycling, the pseudoclassical model produces the most optimistic results. Nevertheless it is anticipated that additional heating mechanisms will be necessary in order to ensure thermonuclear burning in future Tokamaks. The injection of a neutral beam of particles into a Tokamak (Kelley et al., 1972; Stewart et al., 1973; Aldcroft et al., 1973; Dei-Cas et al., 1973; McNally, 1973; Cordey et al., 1974; Rome et al., 1974) in order to elevate the temperature above that set by ohmic heating alone is an illustration of an experimental technique for which theoretical progress is very rapid (Sweetman, 1973; Callen et al., 1974; Connor and Cordey, 1974; Cordey, 1974; Cordey and Core, 1974; Rome et al., 1974).

The incorporation into a diffusion code of the effects of a beam of fast neutral particles may be performed in a sequence of separate operations which illustrate the versatility of the OLYMPUS system.

Provided that the energies of the fast neutrals are sufficiently high

$$E_f > 2 \times 10^4 A_f \quad [eV],$$

where A_f is the atomic mass of the fast neutrals, the number $n_f(r, t)\, dr\, dt$ of fast neutrals that are deposited in the radial interval $r \to r + dr$ and the time interval $t \to t + dt$ is given by Riviere (1971)

$$n_f(r, t) = \left[\frac{I_f}{q_f L} n_e(r)\right] \left\{\exp\left[\int_0^{x=r} \frac{n_e}{L}(x)\, dx\right] + \exp\left[-\int_0^{x=r} \frac{n_e}{L}(x)\, dx\right]\right\}$$
$$\cdot \exp\left[-\int_0^{x=a} \frac{n_e}{L}(x)\, dx\right] \tag{35}$$

where I_f and q_f are, respectively, the neutral beam equivalent current and charge, and n_e is the electron number density.

The function, L, defined according to

$$L = L_f \sin \delta / Z_{\text{eff}}(r)$$

is a measure of the penetration of the beam and is affected by the effective charge, Z_{eff}, the angle, δ, at which the beam is inclined to the magnetic field, and the penetration "length," L_f, which for sufficiently high energies may be approximated (Sweetman, 1973) by

$$L_f = 5.5 \times 10^{14} E_f / A_f \quad [m^{-2}].$$

The determination of $n_f(r, t)$ forms a well-defined unit that is not affected by the calculations that follow. It is performed therefore in a single subroutine

⟨2.12⟩ INJECT which, during the initial stages of development, called ⟨0.4⟩ EXPERT in order to calculate the birth profile, $S(r, t)$, of electrons resulting from the ionization of the neutral beam. This involves distributing the beam around the toroidal and poloidal azimuths. A simplified model, applicable to a finite pencil beam of neutrals, has been incorporated into the code ICARUS. This model defines

$$S(r, t) = n_f(r, t)/4\pi rR, \tag{36}$$

where R is the major radius of the torus. The use of the EXPERT facility allows the model for the calculation of the birth profile to be subsequently developed—even by physicists who have no other connection with the rest of the code.

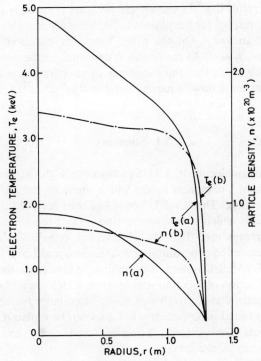

FIG. 12. The radial profiles of number density, n, and electron temperature, T_e, for a large Tokamak ($R = 2.82$ m, $a = 1.28$ m, $\langle n \rangle = 5.0 \times 10^{19}$ m^{-3}, $I_\varphi = 3$ MA, $B_\varphi = 3$ T). The results are obtained by the code, ICARUS, using the pseudoclassical diffusion model in which the particle content is maintained constant by a "neutral species" represented by $m = 6$ and $E_1 = 40$ eV, and increased by the injection of a beam of fast neutrals at (a) high energies ($E_f = 80$ keV, $I_f = 10$ A, ———), at 5.7 sec, and (b) low energies ($E_f = 20$ keV, $I_f = 40$ A, ——·——), at 6.4 sec.

The distribution of injected power to the electrons and ions can be determined analytically (Stix, 1972; Sweetman, 1973). The model of Sweetman (1973) is incorporated at present into ICARUS with the electrons and ions receiving fractions f_e and f_i of the total injected power

$$f_i = [1 + 0.34 X^{1.2}]^{-1}, \qquad f_e = 1 - f_i,$$
$$X = [A_i^{2/3}/14.6 A_f] E_f/T_{e(\text{ev})}, \tag{37}$$

where A_i and A_f are, respectively, the atomic masses of the plasma ions and fast neutrals.

The effect on the temperature distributions of a beam of injected hot neutrals is dependent on the penetration of the beam (Fig. 12). For injection at high energy ($E_f = 8 \times 10^4$ eV, $I_f = 10$ A) there is deep penetration of the beam with the result that the central temperature is strongly increased. This can result in a tendency for the plasma to be thermally unstable when $q(r = 0)$ becomes less than unity. On the other hand, for injection at low energy ($E_f = 2 \times 10^4$ eV, $I_f = 40$ A) the energy is deposited in the outer, low density regions, with the result that the elevations of temperature and particle concentration increase the flow of particles and energy out of the discharge.

VI. Summary

Using the Tokamak code ICARUS as an example, this article has described the construction of a computer model which simulates the physical behavior of a plasma discharge. The ICARUS code has been designed to be a versatile research apparatus which is able to examine various theoretical assumptions that may be represented by sets of coupled one-dimensional diffusion equations. By using the programming conventions and library facilities provided by the OLYMPUS system, it is possible to insert into the basic version of the code a selection of flux and source terms which are additional to those contained in the neoclassical diffusion model. Particular features of the code not only can be tested independently but also can be replaced easily, so that different physical assumptions, numerical techniques, and input/output facilities can readily be used.

The predictions of neoclassical diffusion in an axisymmetric, toroidal containment device have been compared with the results of experiment, with the predictions of a pseudoclassical model, and with those of the Pfirsch–Schlüter model modified by a numerical factor chosen to make the associated diffusion coefficient comparable with that for pseudoclassical diffusion at radius $r \approx 0.5a$. Within the context of an heuristic model for neutral recycling, which serves to maintain constant the total particle content and to provide an energy sink for the electrons, the modified Pfirsch-Schlüter diffusion provides the best description of a small Tokamak discharge. On the other hand the pseudoclassical model provides the most optimistic predictions for the behavior of a future generation large Tokamak.

In extrapolating from one regime to another it is of paramount importance both to have confidence in the behavior of the computer code and to have the theoretical assumptions clearly stated. To this end the OLYMPUS system imposes rigorous standards of structure and documentation which enable the physics, numerical techniques, and programming methods to be understood and the code to be run by other workers on their own computers.

Acknowledgments

The authors are happy to acknowledge discussions with Dr J. A. Reynolds and Dr T. E. Stringer, and would like to thank Dr D. F. Düchs for making versions of his code available to them for Tokamak calculations at Culham.

References

Aldcroft, D., Burcham, J., Cole, H. C., Cowlin, M., and Sheffield, J. (1973). *Nucl. Fusion* **13**, 393.
Artsimovich, L. A. (1971). *Sov. Phys.—JETP Lett.* **13**, 70.
Artsimovich, L. A., Anashin, A. M., Gorbunov, E. P., Ivanov, D. P., Petrov, M. P., and Strel'kov, V. S. (1969). *Sov. Phys.—JETP Lett.* **10**, 82.
Bateman, G., Schneider, W., and Grossmann, W. (1974). *Nucl. Fusion* **14**, 669.
Berry, L. A., Clarke, J. F., and Hogan, J. T. (1974). *Phys. Rev. Lett.* **32**, 362.
Bol, K., Ellis, R. A., Eubank, H., Furth, H. P., Jacobsen, R. A., Johnson, L. C., Mazzucato, E., Stodiek, W., and Tolnas, E. L. (1972). *Phys. Rev. Lett.* **29**, 1495.
Braginskii, S. I. (1965). *Rev. Plasma Phys.* **1**, 205-311.
Burke, V. M. (1970). *Comput. Phys. Commun.* **1**, 473.
Callen, J. D., Colchin, R. J., Fowler, R. H., McAlees, D. G., and Rome, J. A. (1974). *Proc. Int. Conf. Plasma Phys. Control. Nucl. Fusion Res., 5th, 1975* Vol. 1, p. 645.
Chapman, S., and Cowling, T. G. (1953). "The Mathematical Theory of Non-Uniform Gases." Cambridge Univ. Press, London and New York.
Christiansen, J. P., and Roberts, K. V. (1974). *Comput. Phys. Commun.* **7**, 245.
Christiansen, J. P., Ashby, D. E. T. F., and Roberts, K. V. (1974). *Comput. Phys. Commun.* **7**, 271.
Christiansen, J. P., Ashby, D. E. T. F., and Roberts, K. V. (1975). *Comput. Phys. Commun.* **10**, 251.
Connor, J. W., and Cordey, J. G. (1974). *Nucl. Fusion* **14**, 185.
Cordey, J. G. (1974). *Proc. Int. Conf. Plasma Phys. Control. Nucl. Fusion Res., 5th, 1975* Vol. 1, p. 623.
Cordey, J. G., and Core, W. G. F. (1974). *Phys. Fluids* **17**, 1626.
Cordey, J. G., Hugill, J., Paul, J. W. M., Sheffield, J., Speth, E., Stott, P. E., and Tereshin, V. I. (1974). *Nucl. Fusion* **14**, 441.

Dei-Cas, R., De Sacy, S., Druaux, J., Marty, D., and Rebut, P. H. (1973). *Proc. Int. Symp. Toroidal Plasma Confinement, Garching, 3rd, 1973* E9.
Dimock, D., Eckhartt, D., Eubank, H., Hinnov, E., Johnson, L. C., Meservey, E., Tolnas, E., and Grove, D. J. (1971). *Proc. Int. Conf. Plasma Phys. Control. Nucl. Fusion Res., 4th, 1971* Vol. 1, p. 451.
Dimock, D. L., Eubank, H. P., Hinnov, E., Johnson, L. C., and Meservey, E. B. (1973). *Nucl. Fusion* **13**, 271.
Dnestrovskii, Y. N., Kostomarov, D. P., and Pavlova, N. L. (1970). *Proc. Eur. Conf. Control. Fusion Plasma Phys., 7th, 1970* Vol. 1, p. 17.
Dory, R. A., and Widner, M. M. (1970). *Bull. Amer. Phys. Soc.* [2] **11**, 1418.
Drummond, W. E., Nielsen, P., Phillips, P., Medley, S., Jancarik, J., and Bengston, R. (1973). *Proc. Int. Symp. Toroidal Plasma Confinement, Garching, 3rd, 1973* B6.
Düchs, D. F. (1970). *Bull. Amer. Phys. Soc.* [2] **11**, 1488.
Düchs, D. F., Furth, H. P., and Rutherford, P. H. (1971). *Proc. Int. Conf. Plasma Phys. Control. Nucl. Fusion Res., 4th, 1971* Vol. 1, p. 369.
Düchs, D. F., Furth, H. P., and Rutherford, P. H. (1972). *Proc. Eur. Conf. Control. Fusion Plasma Phys., 5th, 1972* Vol. 1, p. 14.
Feneberg, W., and Lackner, K. (1973). *Nucl. Fusion* **13**, 549.
Friedberg, J. P. (1970). *Phys. Fluids* **13**, 1812.
Furth, H. P., Rosenbluth, M. N., Rutherford, P. H., and Stodiek, W. (1970). *Phys. Fluids* **13**, 3020.
Galeev, A. A., and Sagdeev, R. Z. (1968). *Sov. Phys.—JETP* **26**, 233.
Gibson, A., Bickerton, R. J., Cole, H. C., Haegi, M., Hugill, J., Paul, J. W. M., Reynolds, P., Sheffield, J., Speth, E., and Stott, P. E. (1973). *Proc. Int. Symp. Toroidal Plasma Confinement, Garching, 3rd, 1973* B16-I.
Girard, J. P., Khelladi, M., and Marty, D. (1972). *Proc. Eur. Conf. Control. Fusion Plasma Phys., 5th, 1972* Vol. 1, p. 105.
Goedbloed, J. P., and Hagebeuk, H. J. L. (1972). *Phys. Fluids* **15**, 1090.
Hazeltine, R. D., Hinton, F. L., and Rosenbluth, M. N. (1973). *Phys. Fluids* **16**, 1645.
Hinnov, E., Dimock, D. L., Johnson, L. C., and Meservey, E. B. (1973). *Proc. Int. Symp. Toroidal Plasma Confinement, Garching, 3rd, 1973* B13.
Hinton, F. L., Wiley, J. C., Düchs, D. F., Furth, H. P., and Rutherford, P H. (1972). *Phys. Rev. Lett.* **29**, 698.
Hogan, J. T., and Dory, R. A. (1972). *Proc. Eur. Conf. Control. Fusion Plasma Phys., 5th, 1972* Vol. 1, p. 40.
Hogan, J. T., Widner, M. M., and Dory, R. A. (1971). *Phys. Rev. Lett. A* **36**, 217.
Hughes, M. H., Roberts, K. V., and Roberts, P. D. (1975). *Comput. Phys. Commun.* **9**, 51.
Itoh, S., Fujisawa, N., Funahashi, A., Kunieda, S., Takeda, T., Matoba, T., Kasai, S., Sugawara, T., Toi, K., Suzuki, N., Maeno, M., Inoue, K., Ohta, M., Matsuda, S., Ohga, T., Arai, T., Yokokura, K., and Mori, S. (1973). *Proc. Int. Symp. Toroidal Plasma Confinement, Garching, 3rd, 1973* B4.
Kadomtsev, B. B., and Pogutse, O. P. (1971). *Nucl. Fusion* **11**, 67.
Keeping, P. M., Grimm, R. C., and Killeen, J. (1972). *Proc. Eur. Conf. Control. Fusion Plasma Phys., 5th, 1972* Vol. 1, p. 38.
Kelley, G. G., Morgan, O. B., Stewart, L. D., and Stirling, W. L. (1972). *Nucl. Fusion* **12**, 169.
Kelley, G. G., Barnett, C. F., Berry, L. A., Bush, C. E., Callen, J. D., Clarke, J. F., Colchin, R. J., Dory, R. A., England, A. C., Hogan, J. T., McNally, J. R., Murakami, M., Neidigh, R. V., Roberts, M., Rome, J. A., and Wing, W. R. (1973). *Proc. Int. Symp. Toroidal Plasma Confinement, Garching, 3rd, 1973* B3-I.
Knorr, G. (1965). *Phys. Fluids* **8**, 1334.

McNally, J. R. (1973) "Neutral Injection Heating of Tokamaks," Report ORNL TM 4363. Oak Ridge Nat. Lab., Oak Ridge, Tennessee.
Mercier, C., and Soubbaramayer. (1970). *Proc. Eur. Conf. Control. Fusion Plasma Phys.*, *4th, 1970* Vol. 1, p. 16.
National Computing Centre Limited. (1970). "Standard Fortran Programming Manual." Nat. Comput. Cent. Ltd., Manchester, England.
Newcomb, W. A. (1960). *Ann. Phys. (New York)* **10**, 232.
Pfirsch, D., and Schlüter, A. (1962). "The Effect of Electrical Conductivity on the Equilibrium of Low Pressure Plasmas in Stellarators," Report MPI/PA/7/62. Max-Planck Inst. Phys. Astrophys., Munich.
Potter, D. E. (1973). "Computational Physics." Wiley, New York.
Potter, D. E., and Tuttle, G. H. (1973). *Proc. Eur. Conf. Control. Fusion Plasma Phys.*, *6th, 1973* Vol. 1, p. 217.
Rebut, P. H., Bariaud, A., Breton, C., Bussac, J. P., Crenn, J. P., Dei-Cas, R., Michelis, C. De., Delmas, M., Ginot, P., Girard, J. P., Gourdon, C., Hennion, F., Huguet, M., Launois, D., Lecoustey, P., Marty, D., Mattioli, M., Mercier, C., Morriette, P., Platz, P., Plinate, P., Sledziewski, Z., Smeulders, P., Soubbaramayer, Tachon, J., Torossian, A., and Ya'akobi, B. (1973). *Proc. Eur. Conf. Control. Fusion Plasma Phys.*, *6th, 1973* Vol. 2, p. 20.
Richtmyer, R. D., and Morton, K. W. (1967). "Difference Methods for Initial-Value Problems." Wiley, New York.
Riviere, A. C. (1971). *Nucl. Fusion* **11**, 363.
Roberts, K. V. (1969). *Comput. Phys. Commun.* **1**, 1.
Roberts, K. V. (1971). *Comput. Phys. Commun.* **2**, 385.
Roberts, K. V. (1974). *Comput. Phys Commun.* **7**, 237.
Rome, J. A., Callen, J. D., and Clarke, J. F. (1974). *Nucl. Fusion* **14**, 141.
Rose, D. J., and Clark, M. (1961). "Plasmas and Controlled Fusion." Wiley, New York.
Rosenbluth, M. N. (1970). *Nucl. Fusion*, **10**, 340.
Rosenbluth, M. N., and Kaufman, A. N. (1958). *Phys. Rev.* **109**, 1.
Rosenbluth, M. N., Hazeltine, R. D., and Hinton, F. L. (1972). *Phys. Fluids* **15**, 116.
Rutherford, P. H. (1970). *Phys. Fluids* **13**, 482.
Shafranov, V. D. (1970). *Sov. Phys.—Tech. Phys.* **15**, 175.
Sharp, W., and Taylor, J. C. (1973). *Proc. Eur. Conf. Control. Fusion Plasma Phys.*, *6th, 1973* Vol. 1, p. 45.
Spitzer, L. (1962). "Physics of Fully Ionized Gases." Wiley (Interscience), New York.
Stewart, L. D., Callen, J. D., Clarke, J. F., Davies, R. C., Dory, R. A., Hogan, J. T., Jernigan, T. C., Morgan, O. B., Rome, J. A., and Stirling, W. L. (1973). *Proc. Int. Symp. Toroidal Plasma Confinement, Garching, 3rd, 1973* E8.
Stix, T. H. (1972). *Plasma Phys.* **14**, 367.
Sweetman, D. R. (1973). *Nucl. Fusion* **13**, 157.
Sykes, A., and Wesson, J. A. (1974). *Nucl. Fusion* **14**, 645.
Taylor, J. B. (1961). *Phys. Fluids* **4**, 1142.
von Hagenow, K., and Lackner, K. (1973). *Proc. Int. Symp. Toroidal Plasma Confinement, Garching, 3rd, 1973* F7.
Ware, A. A. (1970). *Phys. Rev. Lett.* **25**, 916.
Watkins, M. L., Hughes, M. H., Keeping, P. M., and Killeen, J. (1976). *Comput. Phys. Commun.* (to be submitted for publication).
Winsor, N. K., Johnson, J. L., and Dawson, J. M. (1970). *J. Comput. Phys.* **6**, 340.
Wooten, J., Hicks, H. R., Bateman, G., and Dory, R. A. (1974). "Preliminary Results of the 3-D Nonlinear Ideal MHD Code," Report ORNL TM4784. Oak Ridge Nat. Lab., Oak Ridge, Tennessee.
Yoshikawa, S. (1970). *Phys. Rev. Lett.* **25**, 353.

Equilibria of Magnetically Confined Plasmas

BRENDAN MCNAMARA

LAWRENCE LIVERMORE LABORATORY
UNIVERSITY OF CALIFORNIA
LIVERMORE, CALIFORNIA

I. Introduction	211
II. Toroidal Equilibrium	215
A. General Theory	215
B. Toroidal Equilibrium Computations	220
III. Anisotropic Pressure Equilibria	231
A. Mirror Machine Scaling Laws from Kinetic Theory	231
B. The Longitudinal Invariant	232
C. Mirror Energetics and the MIRICLE Machine	235
D. Guiding Center Equilibria	235
E. Numerical Methods For Guiding Center Equilibria	239
References	249

I. Introduction

THE TOKAMAKS, MIRROR MACHINE, high-beta pinches, Stellarators, and radio frequency confinement experiments of today are the survivors of twenty years of evolution in the development of a fusion reactor. This review describes many of the numerical methods developed for computing plasma equilibria in these devices, but, in order to understand the computations, it is essential to at least indicate the physical reasoning behind the choice of geometries, to collect the basic limitations on plasma parameters and choice of computational problem, and to show the connection between the simple numerical models and the best available theory of equilibrium. These ideas are presented in highly condensed form, as a guide to the theoretical literature. The article deals primarily with numerical and theoretical methods for solving, in many configurations, the guiding center fluid equations relating magnetic field, **B**, current, **J**, and the plasma pressure tensor, P,

$$(1/c)\mathbf{J} \times \mathbf{B} = \nabla \cdot \mathrm{P}, \tag{1}$$

$$\nabla \times \mathbf{B} = (4\pi/c)\mathbf{J}, \tag{2}$$

$$\nabla \cdot \mathbf{B} = 0. \tag{3}$$

These equations are only valid on the short time-scales of thermal motion and Alfven wave propagation, and the basic theoretical problem is to connect this fluid theory with the particle description on all relevant time-scales. This has been done for mirror machines and some toroidal configurations by Hall and McNamara (1975). Mirror equilibria tend to be relatively far from thermodynamic equilibrium and so careful control of the ion distribution function is essential to plasma stability on the very short time-scales of the plasma and ion cyclotron periods. Fortunately, the theory of minimum-B wells (Taylor, 1963, 1964; Grad, 1967c) shows that the long-term stability of mirrors is assured by relatively simple conditions. The earliest computations of plasma equilibria were carried out by Killeen and Whiteman (1966) for axisymmetric systems with $P = P(B)$. There is renewed interest in such devices for their relative simplicity and for the superior energetics of the MIRICLE mirror configuration (Hall and McNamara, 1974). Equilibria in linear magnetic wells of the Baseball type are the most complicated of all in this article, and so the whole subject of mirror equilibria is developed in Section III. Toroidal equilibria are discussed in Section II and a brief account of the theory of the computations is given in Section II,A.

There are many choices of coil configuration for creating and maintaining toroidal equilibria, as indicated in the schematic toroidal device in Fig. 1. Low-beta Tokamaks and Stellarators and high-beta pinches require an ion confinement time that is long compared with ion collision time to produce net fusion power, and so a scalar pressure fluid model seems appropriate. How-

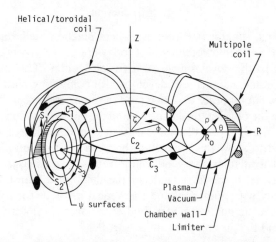

FIG. 1. Schematic view of a toroidal fusion machine showing coils and the coordinate systems: (1) polar: R, ϕ, z; (2) toroidal polars: ρ, θ, z; (3) spherical polars: r, ζ, ϕ; and areas for surface integral, S_1, S_2, S_3, and contours, C_1, C_2, C_3, for line integrals arising in toroidal theory.

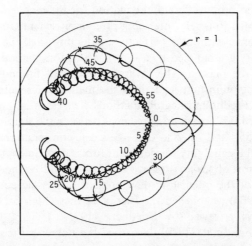

FIG. 2. A confined, blocked particle orbit, projected into the plane $\phi = 0$, in a high shear $l = 3$ Stellarator with windings of average pitch about 45°. The circle $r = 1$ is at the maximum radius of the separatrix. The position of the guiding center as the particle passes $\phi = 0$ is marked by x and the numbers indicate the order of the transits. From Gibson and Taylor (1967).

ever, single particle orbits are generally complicated, as illustrated in Fig. 2 by the orbit of a locally trapped ion in a high shear Stellarator (Gibson and Taylor, 1967), and there seems little hope of developing a fluid treatment of toroidal equilibria which is consistent with the particle model. This subject is treated in depth by Grad (1967a). The pressure profiles are therefore determined on the short time-scales by minimizing the equilibrium energy (Section II, A, 1), and on the long time-scales possibly by diffusion models as described by Hogan in this volume. Several theoretical descriptions of the long-term behavior of toroidal equilibria are sketched in Section II, A, 2. The background to the computational methods is completed in Section II, A, 3 with a brief discussion of the beta and stability limits in axisymmetric systems. The formulas given there delineate the parameter space in which the calculations are performed and provide analytic checks on the computations. A very useful collection of design formulas for circular cross section Tokamaks is given by Mukhovatov and Shafranov (1971).

The description of numerical schemes is limited to unusual or difficult points peculiar to this topic. Standard texts, such as Richtmeyer and Morton (1967), describe the commonly used "alternating direction implicit" method (ADI), and the method of successive overrelaxation is treated with many useful examples by Hockney (1967). Scalar pressure equilibria with axial symmetry are discussed in Section II, B where the equations reduce to a single nonlinear second-order differential equation for the magnetic flux. The nonlinearities have led to a very interesing three-level scheme for treating bifurcations. In systems without a perfectly conducting shell the boundary conditions at infinity are handled by a Green's function method combined with constraints on the plasma. The inverse problem of finding the coils for a desired plasma equilibrium leads into the method of regularization of improperly posed Fredholm integral equations. The essential physical results of this array of methods are: good agreement between the simple formulas given and computations and, in one case, between computations and the ATC experiment, and a capability for design of future experiments.

The discussion of scalar pressure eqilibria would be incomplete without some mention of the interesting development of Scyllac equilibrium codes. Blank et al. (1969) established the theory of these high-beta equilibria and their stability is reviewed by Friedberg (1971). Elegant work by Bloch (1969a,b) and Freidman (1970) developed the numerical techniques of conformal mapping for two-dimensional free boundary problems and these ideas are applied by Betancourt (1974) to three-dimensional Scyllac equilibria. Betancourt uses a variational method to derive difference equations in auxiliary computational domains related to the plasma configuration by a nonconformal mapping. The equilibrium position of the plasma–vacuum interface is found by a steepest descents approach to the minimum energy state. The interesting physical result of this approach is that even in unstable situations the iteration process, starting from a circular cross section plasma as the initial condition, steadily reduces the free energy to a minimum before instability arises and the process diverges. This seems to model the actual behavior of the Scyllac discharges. The details of this numerical approach are given in the article on waterbag methods by Potter in this volume.

Mirror machine equilibria are limited by high frequency stability requirements, adiabaticity of particle orbits, and the energetics of the system, the theory of which is sketched in Sections III, A–C. The fluid theory of anisotropic pressure equilibrium is outlined in Section III, D where the choice of pressure profile and magnetic geometry determine which of four sets of equations is appropriate. The connection with kinetic theory is illustrated by the Baseball drift surfaces which contrast with the Stellarator drifts in Fig. 2. This comprehension of the role of particle drifts in mirror equilibria has led Hall to suggest a simple practical means of constructing the very stable $P(B)$ equilibria.

The numerical methods required for mirror equilibria are described in Section III, E. The Douglas–Gunn version of the ADI method is used for three-dimensional $P(B)$ Baseball equilibria. The more general flux tube equilibria require the solution of magnetic differential equations, which have rather curious properties. A simple approach to the study of the global properties of ADI schemes is illustrated with these equations. These equilibria are described by two basic scalar functions but several vector fields have to be computed in the course of the calculations. Some brief observations are made on experience with vector coding methods for multidimensional equations.

II. Toroidal Equilibrium

A. General Theory

1. *The Kruskal–Kulsrud Variational Principle*

The simplest description of the Tokamak and Stellarator equilibria is given by the fluid equations (1)–(3), with scalar pressure, p, so that

$$\mathbf{J} \times \mathbf{B} = \nabla p. \tag{4}$$

The plasma will move through a succession of such equilibria on the diffusion time-scale and at each moment will adjust to minimize its total energy, W, given by the volume integral in the torus:

$$W = \int dv \left(\frac{B^2}{8\pi} + \frac{p}{\gamma - 1} \right), \tag{5}$$

where γ is the ratio of specific heats of the plasma. The plasma is constrained by boundary conditions, by Eq. (3), and by any other constraints which are assumed to be appropriate. Without these constraints W would be minimized by $B = 0, p = 0$ which would be neither interesting nor true. Most calculations are based on the variational principle of Kruskal and Kulsrud (1958) which assumes smooth, single-valued magnetic surfaces ψ to exist. The complete set of constraints, in the coordinates (ρ, θ, ϕ) shown in Fig. 1 is:

1. ψ has toroidal level surfaces with $\psi = \psi_0$ at the walls, $\min \psi = 0$.
2. ψ represents magnetic surfaces by satisfying the magnetic differential equation

$$\nabla \cdot (\psi \mathbf{B}) = 0. \tag{6}$$

3. $\nabla \cdot \mathbf{B} = 0$ and, with the above condition, is assured by writing

$$\mathbf{B} = \nabla \psi \times \nabla v. \tag{7}$$

4. The magnetic field is frozen to the plasma on time-scales short

compared with the diffusion time and so the magnetic fluxes are constant:

$$\int_{\psi \leq c} dV \nabla \cdot (\phi \mathbf{B}) = c, \tag{8}$$

$$\int_{\psi \leq c} dV \nabla \cdot (\theta \mathbf{B}) = \chi(c). \tag{9}$$

5. The total mass of plasma in a flux tube is constant.

$$\int_{\psi \leq c} dV p^{1/\gamma} = M(c). \tag{10}$$

The minimization of the energy W subject to these constraints leads to the equation (4) and the condition

$$p = p(\psi). \tag{11}$$

In an azimuthally symmetric torus it is convenient to separate the toroidal and poloidal fluxes by writing $\mathbf{B}$ as

$$\mathbf{B} = F(\psi)\nabla\phi + \nabla\phi \times \nabla\psi. \tag{12}$$

The toroidal magnetic field is given by $\mathbf{B}_t = (F/R)\hat{\psi}$, where the caret denotes a unit vector. The $\hat{\psi}$ component of Eq. (4) now reduces to

$$\Delta^*\psi = \nabla \cdot (R^{-2}\nabla\psi) = -4\pi p' - R^{-2} FF', \tag{13}$$

where the prime is a derivative, $p' = \partial p/\partial \psi$, and Δ^* is just the operator $\nabla \cdot (R^{-2} \nabla)$. The right-hand side of Eq. (13) is just $4\pi R \mathbf{J} \cdot \hat{\phi}$, the toroidal current density. Axially symmetric toroidal equilibria are therefore completely determined by solving Eq. (13), once $p(\psi)$ and $F(\psi)$ are given. These two functions are usually guessed to be a plausible shape or could be inferred from radial diffusion calculations. We observe that by restricting the equilibria to being axisymmetric we can no longer guarantee that equilibria found from Eq. (13) are stable in three dimensions. Stability to kink and interchange modes has to be examined separately.

2. *Toroidal Equilibrium on Longer Time-Scales*

In the Adiabatic Toroidal Compressor (ATC) experiment, a vertical field is applied to compress a Tokamak plasma radially inward on a time-scale short compared with the resistive diffusion time-scale but long compared with the thermal and Alfven wave time-scales. The plasma passes adiabatically through a sequence of equilibria which preserve the fluxes, Eqs. (8), (9), and the plasma mass in a flux tube, Eq. (10). Grad *et al.* (1975) replace the pressure and flux profiles in Eq. (13) by

$$4\pi p = \mu(\psi)\dot{\psi}^\gamma, \qquad F/R = \nu(\psi)\dot{\psi}^2, \tag{14}$$

where V is the volume of a flux tube, $\dot{\psi} \equiv \partial\psi/\partial V$, and (μ, ν) are profiles given as initial conditions. Equation (13) becomes

$$\Delta^*\psi = -\mu'\dot{\psi}^\gamma - \nu\nu'\dot{\psi}^2 - (\gamma\mu\dot{\psi}^{\gamma-2} + \nu^2)\ddot{\psi}, \tag{15}$$

which is a new type of generalized differential equation, being second order in space and also in the surface variable V. The surface average of (15) gives an equation for the average pressure balance

$$(U + \gamma_\mu \dot{\psi}^{\gamma-2} + \nu^2)\ddot{\psi} + (\dot{U} + \mu^1 \dot{\psi}^{\gamma-1} + \nu\nu^1\dot{\psi})\dot{\psi} = 0, \tag{16}$$

where $U = \int |\nabla V| dS$. Boundary conditions at the magnetic axis, $V = 0$, and the plasma surface conserve flux in the plasma. The boundary conditions for Eq. (15) at the compression coils determine the plasma position. Numerical methods for solving this unusual pair of equations are still under development, but in a very much more simplified waterbag model Stevens (1974) was able to simulate an ATC experiment with some success. This is the only reported comparison between equilibrium computation and experiment and the development of these techniques is an important new field.

The effects of resistivity in the plasma are to allow the plasma to move across the field, to permit breaking and rejoining of field lines, and, in unstable situations, to do this quite rapidly on the diffusion time-scale. Taylor (1974) has shown how to incorporate these effects in the choice of equilibrium states by using Woltjer's invariant K, given by (Woltjer, 1958)

$$K = \int \mathbf{A} \cdot \mathbf{B} \, dV. \tag{17}$$

The integral is taken over the volume of a flux tube and $\mathbf{A}$ is the magnetic vector potential, $\mathbf{B} = \nabla \times \mathbf{A}$. Maxwell's equation relates $\mathbf{A}$ to the electric field $\mathbf{E}$ and electrostatic potential ϕ by

$$\partial \mathbf{A}/\partial t = \mathbf{E} + \nabla\phi. \tag{18}$$

The time rate of change of K is (Hall and McNamara, 1975)

$$\frac{\partial K}{\partial t} = \int dS[(\mathbf{A} \cdot \mathbf{V} - c\phi)\hat{\mathbf{n}} \cdot \mathbf{B} + (c\mathbf{E} + \mathbf{V} \times \mathbf{B}) \cdot (\mathbf{n} \times \mathbf{A})]$$
$$- 2c \int dV \mathbf{E} \cdot \mathbf{B}, \tag{19}$$

where **V** is the velocity of a surface element of V, and $\hat{\mathbf{n}}$ the unit normal to the surface. If the bounding surface is at infinity, or if it is a fixed tube of force at a conducting wall, the surface integrals vanish. The volume integral can be expressed in terms of the parallel resistivity, $\eta_\parallel$, of the plasma to give

$$\partial K/\partial t = -2c \int dV \eta_\parallel \mathbf{J} \cdot \mathbf{B}. \tag{20}$$

Since $\int |\mathbf{J} \cdot \mathbf{B}| \, dV$ is bounded, even in the presence of resistive instabilities, K decays on the diffusion time-scale, $O(\eta_\parallel^{-1})$, and is therefore an additional constraint on the possible equilibrium states. On including this constraint in the variational principle, via a Lagrange multiplier, μ, Taylor has shown that the minimum energy state is given by

$$c^{-1}\mathbf{J} = (\mu/4\pi)\mathbf{B} + (\mathbf{B} \times \nabla p)/B^2. \tag{21}$$

The equilibrium equation (4) is still valid but now, at low beta when the pressure gradient can be neglected, the final equilibrium state is force free with

$$\nabla \times \mathbf{B} = \mu \mathbf{B}. \tag{22}$$

This equation is easily solved in a straight cylinder—the infinite aspect ratio torus—and K can be directly evaluated from the definition, (14), as a function of μ. In a toroidal discharge, such as ZETA (Robinson, 1969), an azimuthal slit must be made in the torus at $\phi = 0$, across which the voltage is applied. The surface integral in Eq. (19) no longer vanishes and, integrating (19), we see that

$$K = \psi_0 V_s, \tag{23}$$

where V_s is the number of volt-seconds applied. Taylor was able to show that the final equilibrium state of these discharges could have a reversed axial field at the wall or, at larger K, the equilibria bifurcate and the lowest energy state would be helically distorted around the torus. The analytic results agree well with observations in ZETA and HBTX (Bodin et al., 1971) and make the computations of force-free equilibria reported in Section II, B more relevant to experiments.

3. Beta and Stability Limitations on Axisymmetric Toroidal Equilibria

An axisymmetric scalar pressure plasma must be confined by a large toroidal current which generates a poloidal magnetic field which is strongest on the inside of the ring. A vertical or poloidal field, B_p, is therefore required to maintain a toroidal plasma in equilibrium. This field should vary so that the stability index, $\eta_1 = \rho \partial (\ln B_p)/\partial \rho$, lies between $-3/2$ and 0 to ensure stability to radial and vertical displacements. The composite toroidal fusion reactor shown in Fig. 1 therefore has the following elements: a plasma chamber which may or may not be circular in cross section; a limiter which prevents the plasma burning the walls during the discharge; a chamber wall which may be conducting to stabilize the plasma by image currents; multipole windings inside or outside the chamber to stabilize, shape, and position the plasma. Each of these elements appears in the computations to be described through the boundary conditions on ψ.

The Doublet (Ohkawa, 1968) and Belt pinch (Zwicker et al., 1971) configurations are versions of the Tokamak which seek to achieve a higher beta, and therefore higher fusion power density, than the circular cross section Tokamak by raising the total plasma current. This is limited by the well-known Kruskal–Shafranov condition against the kink instability which is given in terms of the safety factor, q, by

$$q = \partial \chi / \partial \psi = L_1 B_T / L_3 B_p > 1. \qquad (24)$$

L_1, L_3 are the lengths of the circuits C_1, C_3, the short and long way round the terms in Fig. 1, and B_T is the toroidal field strength. Artsimovich and Shafranov (1972) pointed out that by making the plasma cross section noncircular the length L_1 could be substantially increased. The noncircular cross sections are more sensitive to ballooning modes (Laval et al., 1970) and a simple form of the limitation on beta is given by Ohkawa and Jensen (1970)

$$\beta < (R/a)(L_1/L_3)^2 \, 1/q^2. \qquad (25)$$

It appears desirable to have L_1/L_3 as large as possible, which is the basis of the Belt pinch where the plasma cross section has a height to width ratio of about 10. The Belt pinch has little shear and is therefore most sensitive to highly localized modes. The Suydam (1958) condition limits beta to

$$\beta \lesssim \tfrac{1}{4}(R/a)(L_1/L_3 \cdot a/q \cdot \partial q/\partial \rho)^2. \qquad (26)$$

These instabilities are essentially three-dimensional and are not seen in the equilibrium calculations. In this volume, R. Grimm reviews methods of finding the stability limits numerically. The Belt pinch is formed by compression and the greatest plasma heating would be achieved for the greatest compression ratio, k = chamber width/plasma width. The limit was sought by Becker (1974) and Okabayashi and Sheffield (1974) in numerical studies of the vertical stability of the pinch and was $k < 2$ for an infinite slab and

$k < 1.3$ for an elliptic cylinder with parabolic current distribution. This is similar to results from analytic solutions of the equilibrium equation (13) by Hernegger and Mashke (1974) but had already been given by Morse and Feshbach (1953) in an equivalent capacitance problem!

B. Toroidal Equilibrium Computations

1. *Axisymmetric Systems*

Axisymmetric, scalar pressure equilibria are described by Eq. (13) with appropriate boundary conditions. This equation can be written in the form

$$\Delta_R \psi + \Delta_Z \psi \equiv \frac{\partial^2 \psi}{\partial R^2} - \frac{1}{R}\frac{\partial \psi}{\partial R} + \frac{\partial^2 \psi}{\partial Z^2}$$
$$= -S(\psi)[4\pi R^2 p'(\psi) + FF']$$
$$= -4\pi R J_\phi S, \qquad (27)$$

where Δ_R and Δ_Z are the R, Z differential operators and S describes the plasma boundary, $\psi = \psi_p$, by

$$S(\psi) = 1, \quad \psi \geq \psi_p,$$
$$= 0, \quad \psi < \psi_p. \qquad (28)$$

Simple pressure profiles could by specified to linearize the right-hand side of Eq. (27) if it were not for the essential nonlinearity introduced by the free boundary as given by S. Before describing solution methods and results it is useful to give some simple relations. The functions p and F are related to the poloidal and toroidal currents by:

$$J_p = -c(B_p F'/4\pi), \qquad (29)$$

$$J_T = -c[Rp' + (FF'/4\pi R)]. \qquad (30)$$

The average pressure balance of a plasma inside a perfectly conducting shell can be examined by integrating Eq. (27) over the total volume V_0 of the torus. The poloidal beta is conveniently defined as

$$\beta_p = 8\pi \bar{p} / \langle B_p^2 \rangle_{\psi_w}, \qquad (31)$$

where $\bar{p}$ is the volume average of the pressure and $\langle B_p^2 \rangle_{\psi_w}$ is the average poloidal field pressure over the wall at $\psi = \psi_w$. The average of Eq. (27) gives

$$\beta_p = 1 + \int_0^{\psi_w} \frac{d\psi V(\psi)}{\langle B_p^2 \rangle_{\psi_w} V_0} \left[2FF'\langle 1/R^2 \rangle_\psi + \langle B_p^2 \rangle_\psi \frac{\partial}{\partial \psi} \left(\ln \frac{V'^2 \langle B_p^2 \rangle}{V} \right)_\psi \right], \qquad (32)$$

where $V(\psi)$ is the volume of the flux tube ψ. The second term in the integral is small at large aspect ratio, A, and so $B_p > 1$ only if FF' is positive. As the sign of F' changes, so the poloidal current, J_p, Eq. (29), reverses. As FF' increases it is apparent from Eq. (30) that the toroidal current may also reverse on the inside of the torus at smaller R, weakening the confining field.

2. Circular Tokamaks

These observations are confirmed by Callen and Dory (1972), who successfully applied the simplest numerical approach of successive overrelaxation to Eq. (27). They used nonlinear profiles for $p(\psi)$, $F(\psi)$, taken as a piece of a cosine function, $\cos(\psi_0 - \psi)$. Some of their flux surfaces are shown in Fig. 3 for various values of β_p and $A = 30$. These sample equilibria give a

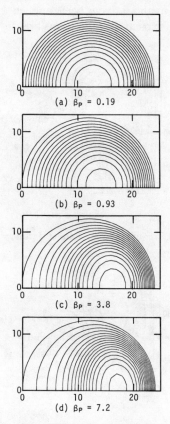

FIG. 3. Numerically computed flux surfaces, separated by equal amounts of flux, in a circular cross section Tokamak with $R_0 = 36, b = 12$. The high-beta cases, (c) $\beta_p \sim R_0/b$ and (d) $\beta_p \sim (R_0/b)^2$, are unstable by Eq. (24) with $q = 0.7$ and 0.5, respectively. From Callen and Dory (1972).

wide range of β_p from A^{-1} to A^2, the upper limit resulting from the fact that the iteration scheme failed to converge. However, the last two cases are unstable to the three-dimensional kink instability since q on the magnetic axes is 0.7 and 0.5, respectively. The situation is only worsened by applying an external vertical magnetic field, B_V, to center the plasma at high $\beta_p \gtrsim A$. This is achieved numerically by applying the boundary condition $\psi_w = R^2 B_V/2 + \text{const}$. The current profile becomes more peaked and q on the axis decreases. The optimum configurations found by Callen and Dory (1972) have $\beta_p \lesssim 2$, a maximum $\beta = 8\pi\bar{p}/B_T^2 \approx 0.1$, with the total current adjusted to keep $q_{\min} = 1$. Other stability criteria, such as those in Section II,A,3, were not examined.

3. Doublet Equilibria

Similar numerical methods have been applied by Chu et al. (1974) to Doublet equilibria and typical configurations are shown in Figs. 4 and 5. An

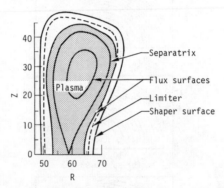

FIG. 4. A moderate $\beta_p = 0.25$ equilibrium in Doublet II with $\gamma = 1$. From Chu et al. (1974).

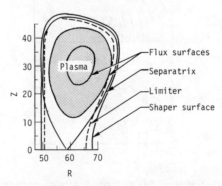

FIG. 5. A Doublet II equilibrium with $\beta_p = 0.25$, $\gamma = 2$. Current peaking in the channel pushes the plasma into the Droplet configuration. From Chu et al. (1974).

additional feature required in these calculations is that plasma on flux surfaces ($\psi < \psi_L$) intersecting the limiter has to be scooped off during the iteration procedure. The azimuthal current used was a linear function of ψ:

$$J_{\theta 1} = j_0 [(R/R_0)\beta_p + (R_0/R)(1 - \beta_p)][1 + (\gamma - 1)(\psi - \psi_L)/(\psi_{\max} - \psi_L)], \tag{33}$$

where j_0, β_p, and γ were constants. The parameter j_0 was adjusted to keep the total current, $I_p = \int J_\phi \, dS_3$, constant during the iteration. The Droplet configurations, Fig. 5, are sometimes seen in experiments when breakdown of the initial plasma occurs mainly in one half of the machine. No results on maximum beta or stability were given.

4. Force-Free Multipole Equilibria

When the azimuthal current, J_ϕ is given as a linear function of ψ, Eq. (27) is separable and can be solved analytically. Boundary conditions on coils (Matsui et al., 1973) or flux shapes such as in Doublet (Dobrott et al., 1973) force approximations to be made. Exact solutions have been studied in straight cylinders (Kerner et al., 1972), elliptic cylinders (Gajewski, 1972), and rectangular cross section tori (Mashke, 1973; Suzuki, 1973).

This approach has been implemented numerically by Feneberg and Lackner (1971) for force-free ($p' \equiv 0$), sharp boundary (at $\psi = \psi_p$) equilibria with a constant current profile ($FF' = j_0$), contained only by multipole coils. The flux function is split into the part ψ_1 due to the plasma and the part ψ_M due to multipole coils. The iterative solution is given, in spherical polar coordinates ($r, x = \sin J$) (cf. Fig. 1) by

$$\psi_1^{n+1} = \sum_{k=0}^{K} \frac{1}{2k+3} \left[\frac{1}{r^{k+1}} \int_0^r r'^{k+2} g_k \, dr' + r^{k+2} \int_r^\infty \frac{g_k}{r'^{k+1}} \, dr' \right] (1 - x^2) P_k^{(1,1)}(x), \tag{34}$$

where K is the minimum number of terms required for accuracy ($K \leq 50$) and

$$g_k = -(j_0/h_k) \int_{-1}^{1} S(\psi_1 + \psi_M) P_k^{(1,1)}(x) \, dx, \tag{35}$$

$P_k^{(\alpha, \beta)}(x)$ are the Jacobi polynomials, and h_k is the normalization factor

$$h_k = \int_{-1}^{1} (1 - x^2)(P_k^{(1,1)}(x))^2 \, dx. \tag{36}$$

This formulation automatically satisfies the boundary condition $\psi_1 = 0$

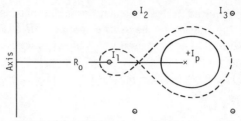

FIG. 6. Multipole equilibrium with circular cross section and stagnation point on the inside. $I_p = 0.926, I_1 = 0.982, I_2 = -0.341, I_3 = -0.15, 4\pi^2 I_0/cR_0 = 35$. From Feneberg and Lackner (1971).

at $r = \infty$ but allows the trivial solution $\psi_1 = 0$. This was avoided by choosing $\psi_p{}^n$ during the iterations as follows:

1. ψ_p was adjusted so that the plasma boundary always intersected the symmetry plane, $z = 0$, at the same point at the outer side of the torus, and j_0 was kept constant. This worked well for approximately circular cross section plasmas.

2. ψ_p was adjusted so that the plasma boundary touched the cone $\zeta = \zeta_1$, $\zeta_1 + \pi/2$. At the same time the total plasma current I_p was kept constant. This worked well for elongated plasmas. Convergence was not obtained by fixing I_p and an outer boundary point as in 1.

A typical result is shown in Fig. 6. The net current in the five multipole windings is zero, minimizing the mutual inductance of the plasma and the coils. The equilibrium exhibits a separatrix of the flux surfaces which will ultimately be required in a reactor as a means of removing impurities from the discharge. The toroidal current density j_T is scaled to a characteristic distance R_0 and current I_0 by $j_0 = (4\pi^2 I_0/R_0 c) j_T$. The shift of the plasma compares well with Shafranov's formula, as can be seen from Fig. 7. Elliptic cross section plasmas have also been examined by Feneberg and Lackner (1973) and equally good agreement with design formulas due to Mukhovatov and Shafranov (1971) is found. This is not surprising since the calculations correspond to zero beta and are at quite large enough aspect ratio to agree with the asymptotic expansions. Similar results are obtained at low beta by Suzuki (1974).

5. Bifurcation

The nonlinearity of the equilibrium, Eq. (27), gives rise to the possibility of multiple solutions: trivial solutions like $\psi_1 = 0$, grossly different ones like the Doublet and Droplet, and failure to converge as a bifurcation point is approached. This was first experienced by Fisher (1971) when studying Astron equilibria for which the flux equation became $\Delta^*\psi = \exp(-\psi)$. In one dimension it was easy to show that there are at least two solutions to this

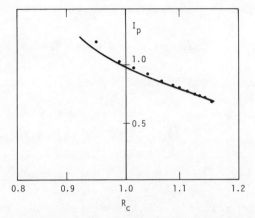

FIG. 7. Equilibrium positions of the plasma column in Fig. 6 as a function of I_p. Results of Shafranov's (1966) formula (solid line), $B_\perp = I_p [\ln(8R_0/a) - 1.25]/4R_0$ and from numerical calculations compare well. From Feneberg and Lackner (1971).

equation, but the simple iteration process without constraints, $\Delta^* \psi^{n+1} = \exp(-\psi^n)$, fails to find the deeper solution. Marder and Weitzner (1970) showed that an eigenvalue of the linearized error operator, $(\Delta^*)^{-1} \exp(-\psi_B)$, passes through unity at the bifurcation point $\psi = \psi_B$, and so the error grows and the iteration is unstable. They showed how a three-step iteration could map the eigenvalues of the error operator into the unit circle, and obtained both classes of Astron equilibria by appropriate choices of initial conditions. This method has been used successfully by Suzuki (1974) (see Fig. 8) for

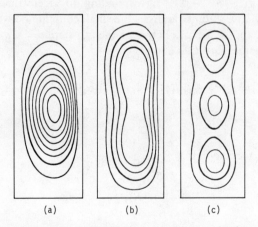

FIG. 8. Belt pinch magnetic surfaces in a rectangular conducting shell with current profiles from Eq. (37) with $\mu = 0.001$ and (a) $\lambda^2 = 1.0 \times 10^{-5}$, (b) $\lambda^2 = 1.2 \times 10^{-5}$, (c) $\lambda = 3.6 \times 10^{-5}$. From Suzuki (1974).

many different cases in which J_ϕ is linear in ψ, of the form

$$4\pi R J_\phi = (\lambda^2 R^2 + \mu)\psi \equiv f(R)\psi. \tag{37}$$

The first step of Suzuki's method is given by the alternating direction implicit (ADI) scheme:

$$(\Delta_R + f(R)\,S(\psi_n) - \alpha_1)\psi^* = -(\Delta_Z + \alpha_1)\psi_n, \tag{38}$$

$$(\Delta_Z + fS(\psi^*) - \alpha_1)\psi_{n+1/3} = -(\Delta_R + \alpha_1)\psi^*. \tag{39}$$

The parameter α_1 introduces a shift of the eigenvalues of the operators, and will be determined by the stability analysis. Each of these equations is a linear, homogeneous, second-order, ordinary differential equation in one independent variable, R or Z, the other coordinate, Z or R, being merely a parameter. The standard finite-difference representations give a pair of tridiagonal matrices which are easily inverted by the usual Gaussian elimination scheme. This step is repeated to generate $\psi_{n+2/3}$ from $\psi_{n+1/3}$ and the third step is to construct the linear sum

$$\psi_{n+1} = (1 - \alpha_2)\psi_n + 2\alpha_2 \psi_{n+1/3} - \alpha_2 \psi_{n+2/3}. \tag{40}$$

The errors in the calculation can be expanded in the eigenfunctions $\rho_i(R), \zeta_j(Z)$ of the operators $-\Delta_R, -\Delta_Z$, corresponding to eigenvalues λ_i, μ_j, where $0 < \lambda_i, \mu_j < \infty$. In a stability analysis the small coupling between eigenmodes can be ignored and so a typical error at the nth step, $\varepsilon = \varepsilon_n \rho_i \zeta_j$, is amplified by the ADI scheme, Eqs. (38) and (39), by

$$\varepsilon_{n+1/3} = \varepsilon_n(\alpha_1 - \lambda_i)/(\alpha_1 + \lambda_i - F_{ij})(\alpha_1 - \mu_j)/(\alpha_1 + \mu_j - F_{ij})$$

$$\equiv E_{ij}\varepsilon_n, \tag{41}$$

where F_{ij} is the average of f:

$$F_{ij} = \int w f(R)\rho_i^2 \zeta_j^2\, dR\, dZ. \tag{42}$$

The weight function for the orthogonal modes (ρ, ζ) is denoted by w. The shift α_1 must be greater than $F = \text{Max}(F_{ij})$ to avoid the poles of the amplification factors E_{ij}. The range of E_{ij} is then

$$-\alpha_1/(\alpha_1 - F) < E_{ij} < [\alpha_1/(\alpha_1 - F)]^2. \tag{43}$$

Inserting formula (41) into the linear combination, Eq. (40), gives

$$\varepsilon_{n+1} = (1 - \alpha_2(1 - E_{ij}))^2 \varepsilon_n \equiv E\varepsilon_n. \tag{44}$$

The scheme will converge if $|E| < 1$ which, using the range in Eq. (43), requires $0 < \alpha_2 < 0.5$. Suzuki (1974) used $\alpha_1 = (3 \sim 4) f(R_{\max})$ and $\alpha_2 = 0.45$. A two-step process would not have worked since E would be given by $E = (1 + \alpha_2) - \alpha_2 E_{ij}$. The amplification factors $E_{ij} > 1$ could be handled but then the small factors $E_{mn} < 1$ would be troublesome. These techniques could be generalized to nonlinear current profiles, $J_\phi(\psi)$, and the different classes of solution found by choice of initial condition $\psi_0(R, Z)$.

6. *The Virtual Casing Principle*

Thus far we have described methods for finding plasma equilibria in a given coil and casing. The inverse problem is equally interesting to solve as part of the overall design problem, namely, to specify the plasma boundary, Γ, and find corresponding coils. If one were to enclose the plasma in a perfectly conducting shell on Γ carrying a surface current

$$\mathbf{J}_s = (c/4\pi)(\mathbf{B}_\tau \times \hat{n}), \tag{45}$$

(where $\mathbf{B}_\tau$ is the equilibrium field at the boundary and $\hat{n}$ an external normal) then the virtual casing principle of Shafranov and Zakharov (1972) observes the following: the magnetic field of this current inside the plasma is just the vacuum magnetic field, $\mathbf{B}_m$, required to maintain the plasma equilibrium, while outside it is equal in magnitude and opposite in sign to the self-field of the plasma currents. If we have solved the equilibrium equation (27) in a casing of the desired shape then the surface current is

$$J_s = (c(\partial\psi/\partial n)/8\pi^2)_\Gamma$$

and the magnetic field of this current is then

$$\mathbf{B}^m(R, Z) = \oint_\Gamma J_s(s) \mathbf{b}(R, Z/s) \, ds. \tag{46}$$

The kernel, $\mathbf{b}$, is the field at the point (R, Z) due to a unit current at a point s

on Γ, and is given by

$$b_r(R,Z|a,\zeta)$$
$$= \frac{2}{c}\frac{Z-\zeta}{r[(a+R)^2+(Z-\zeta)^2]^{1/2}}\left[-K(k)+\frac{a^2+R^2+(Z-\zeta)^2}{(a-R)^2+(Z-\zeta)^2}E(k)\right], \quad (47)$$

$$b_z(R,Z|a,\zeta)$$
$$= \frac{2}{c}[(a+R)^2+(z-\zeta)^2]^{-1/2}\left[K(k)+\frac{a^2-R^2-(z-\zeta)^2}{(a-R)^2+(z-\zeta)^2}E(k)\right], \quad (48)$$

where $k^2 = 4aR/(a+R)^2+(z-\zeta)^2$ and $K(k), E(k)$ are complete elliptic integrals. The flux function for a unit current is

$$\psi_0(R,z|a,\zeta) = (8\pi/c)\{(aR/k^2)[(1-\tfrac{1}{2}k^2)K(k) - E(k)]\}^{1/2}. \quad (49)$$

As an example, Zakharov (1973) starts from a simple polynomial solution of Eq. (27) with p', FF' constant:

$$\psi = -(4\pi^2/c)(p'R^2 + FF')z^2$$
$$+ [R^6 - 12R^4Z^2 + 8R^2Z^4 + 2.182R^4$$
$$+ 4.791R^2Z^2 - 1.882R^2 - 0.509Z^2]. \quad (50)$$

This gives a Doublet-type equilibrium as shown in Fig. 9. The virtual casing on Γ can now be replaced by a more convenient casing l carrying a surface current J which gives the same containing field $\mathbf{B}^m$ inside the plasma:

$$B_\tau^m(s) = \oint_l J(l)b_\tau(s|l)\,dl \equiv C \cdot J, \quad (51)$$

where C is the integral operator and

$$b_\tau(s|l) = \mathbf{b}(s|l) \cdot (\hat{e}_1 \times \hat{n}(s)). \quad (52)$$

This is a Fredholm equation of the first kind for J and is incorrectly posed for numerical solution. Thus for currents J and $J_1 = J + \cos\omega s$ we get the same field $B_\tau^m(s)$ to any desired accuracy for sufficiently large ω. This sensitivity to high order multipole fields is eliminated by the Tikhonov (1963) regulari-

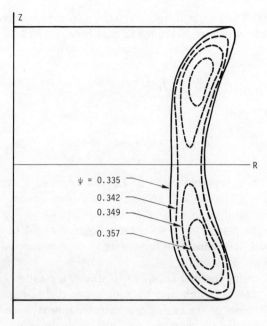

FIG. 9. A simple analytic doublet equilibrium from Eq. (50). From Zakharov (1973).

zation method which minimizes the functional

$$M(J, \alpha) = \oint_\Gamma (C \cdot J - B)^2 \, ds + \alpha \oint_l \left[k_1(l) \left(\frac{dJ}{dl} \right)^2 + k_2(l) J(l)^2 \right] dl, \tag{53}$$

where α is the regularization parameter chosen from a sequence which leads to the desired smooth solution of Eq. (50). The solution has the minimum norm, as defined by the second integral in Eq. (52), and the factors k_1, k_2 can be chosen to be 1. The Euler–Lagrange equation minimizing $M(J, \alpha)$ becomes

$$\alpha \, d^2 J/dl^2 - \alpha J - \oint_l \bar{b}(l \mid l') J(l') \, dl' = -\bar{B}(s) \tag{54}$$

where

$$\bar{b}(s/l) = \oint_\Gamma b_\tau(\xi \mid s) b_\tau(\xi \mid l) \, d\xi, \qquad \bar{B}(s) = \oint_\Gamma b_\tau(\xi \mid s) B_\tau^m(\xi) \, d\xi. \tag{55}$$

This equation is readily solved for $J = J_\alpha(l)$ by replacing the integro-differential operator by a suitable finite-difference form and inverting the resulting matrix. The regularizing parameter, α, can be reduced toward zero by a number of sequences (Gordonova and Morozov, 1973) related to the discrepancy,

$$D_\alpha = \oint_\Gamma (C \cdot J_\alpha - B_\tau)^2 \, ds.$$

Zakharov (1973) used

$$D_0 = 2 \times 10^{-6} \oint_\Gamma B_\tau^2 \, ds$$

as the maximum acceptable value of D_α and adjusted α according to the sequence, $\alpha_{n+1} = \alpha_n (D_0/2D_\alpha)^{1/2}$. In Fig. 10 is shown a more suitable casing for the Doublet equilibrium of Fig. 9 and this equilibrium has a separatrix quite close to the plasma. The solution procedure could have failed if the new casing l had been chosen so far out as to require concentrated currents or singularities inside or on l.

The virtual casing method allows us to specify a plasma shape or choose a single initial configuration and then to move the casing to a preferred position. The current in the casing can then be approximated by multipole windings and the slightly modified equilibrium recomputed. Suzuki (1974) has carried out all three phases of this program and calculates the multipole

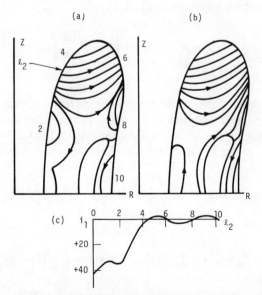

FIG. 10. (a) Maintaining magnetic field for the equilibrium of Fig. 9. (b) Vacuum equilibrium field. (c) Distribution of current J on outer casing, l. From Zakharov (1973).

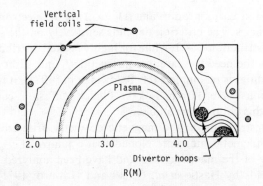

FIG. 11. Shell-less Tokamak equilibrium with plasma surface at the separatrix. From Suzuki et al. (1974).

equilibria from the integral formulation of Eq. (27). The toroidal plasma currents are replaced by ring currents on the mesh and the iteration becomes

$$\psi_{n+1} = -\int dR'\, dZ'\, \psi_0(R, Z \mid R'Z')\, 4\pi R' J_\phi(\psi_n)\, \delta(\psi_n), \tag{56}$$

where ψ_0 is given by Eq. (49), the flux of a unit current. A typical result (Suzuki, 1974) is shown in Fig. 11, a Tokamak equilibrium, leaning on an outer separatrix and supported by multipole coils. The configuration has $\beta_\rho = 1$, 3.3 MA of plasma current, and 1 MA in the main divertor hoop.

III. Anisotropic Pressure Equilibria

A. Mirror Machine Scaling Laws from Kinetic Theory

The equilibrium of a mirror-trapped plasma depends upon the details of individual particle motion in several ways, and it is therefore essential to establish the connection between fluid and kinetic theory. The equilibrium distribution, $f_i = f(\varepsilon, \mu, J, \sigma)$, of ions in a mirror machine is a function of the constants of particle motion: energy, $\varepsilon = \tfrac{1}{2}mv^2$, magnetic moment, $\mu = mv_\perp^2/2B$, longitudinal invariant, $J = \oint v_\parallel\, dl$, and the sign of the particle velocity, $\sigma = \pm 1$ (Taylor, 1964). As is well known, the loss cone for ions with $\varepsilon > \mu B_{\max} \equiv \mu R_M B_{\min}$, leads to high frequency ($\omega \sim \omega_{pi}$ or Ω_i) velocity-space instabilities which must be stabilized by making the plasma length, L_0, less than the shortest unstable wavelength. Post and Rosenbluth (1965, 1966) give a criterion for the most dangerous mode at high density,

$$L_0 < 100 a_i (1 + \Omega_e^2/\omega_{pe}^2)\, G(f_i), \tag{57}$$

where a_i is the ion Larmor radius and (Ω_e, ω_{pe}) are the electron cyclotron and plasma frequencies. The criterion depends sensitively on the shape of f_i: for a collisional distribution $G \sim 1$, while for highly peaked distributions, $G \sim 0.1$. This illustrates the need for care in constructing f_i and, although the theory of such instabilities is incomplete (cf. Baldwin et al., 1971), it will be assumed that distributions taken from Fokker–Planck diffusion calculations (see Killeen et al., this volume) are stable.

Stringent restrictions on the magnetic geometry arise from the need to conserve the magnetic moment. Nonadiabatic jumps in μ arise from the inhomogeneity of the magnetic field and have been analyzed in simple cusp and mirror fields by Hastie et al. (1969) and Howard (1971). Fortunately, small jumps in μ lead to stable "superadiabatic" oscillations about fixed points in phase space (Rosenbluth, 1972). Large jumps cause particles to find the loss cone in a few bounces, and a practical limit on the maximum ion energy for adiabatic containment is given by Foote (1972), based on extensive orbit calculations,

$$W_{max} = 8.5 \times 10^{-4} (Z^2/A) B_M^2 L^2 / R_M (R_M - 1)^3. \tag{58}$$

The magnetic field is fitted by

$$B = B_m/R_M [1 + (Z^2/L^2 + r^2/L_r^2)(R_m - 1)], \tag{59}$$

and $L^{-2} = L_z^{-2} + L_r^{-2}/(R_m - 1)$, and A, Z are atomic weight and number.

B. The Longitudinal Invariant

The Baseball magnetic well—originated by Larkin (1965) as the tennis ball coil—has two symmetry planes at right angles containing lines of force which lie entirely within these symmetry planes. Scalar functions, B, J, etc., are symmetric or antisymmetric under the symmetry operations, S_1, S_2, S_3, in polar coordinates (r, θ, z)

$$\begin{aligned} S_1: (r, \theta, z) &\to (r, -\theta, z), \\ S_2: (r, \pi/2 + \theta, z) &\to (r, \pi/2 - \theta, z), \\ S_3: (r, \pi/4 + \theta, z) &\to (r, \pi/4 - \theta, -z). \end{aligned} \tag{60}$$

Because of the symmetries, the drift surfaces $\psi(r, \theta, z, v = \mu/\varepsilon)$, on which $J = $ const., can be expanded in a Fourier series

$$\psi = \psi_0(r, z, v) + \sum_{n=1}^{\infty} \psi_{2n}(r, z, v) \cos 2n\theta, \tag{61}$$

where $\psi_{2n}(r, -z, v) = (-1)^n \psi_{2n}(r, z, v)$. Near the axis, $r = 0$, an expansion in the aspect ratio gives

$$\psi \simeq \psi_0 + \psi_2 \cos 2\theta. \tag{62}$$

The drift surfaces are therefore circular in the midplane, independent of v, and so all particles lie on the same "omnigenous" drift surfaces, $\psi(r, \theta, z)$. In an azimuthally symmetric (θ-independent) mirror machine such as a stuffed cusp (Hartman, 1967) or a Mirror-Levitron (Anderson et al., 1969) the drift surfaces are determined by the constant canonical angular momentum surfaces, $p_\theta = mv_\theta + eA_\theta/c$, which are just the flux surfaces, $\psi(r, z)$, and the result is trivial. Omnigenity of the drift surfaces is a symmetry property, applicable even at high beta, and it is fortunate that Baseball has just enough symmetry. Numerical calculations of J show that omnigenity applies in Baseball throughout the plasma volume (Hall et al., 1974) (see Fig. 12). The value of this concept for mirror equilibria is that the distribution function is separable

$$f = \omega(\psi(J)) F(\varepsilon, \mu, B_{\max}(\psi)). \tag{63}$$

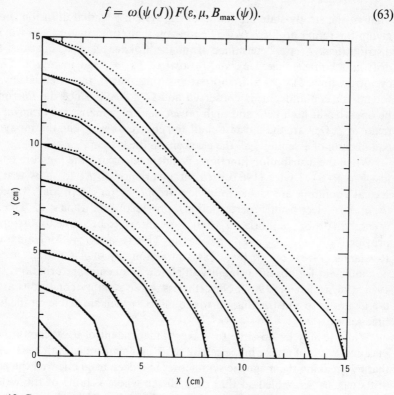

FIG. 12. Constant-J surfaces for particles bouncing in Baseball II. Dotted line: Near the mirror peak; solid line: near the minimum of B. From Hall et al. (1974).

The pressure tensor, $\mathsf{P} = P_\perp \mathsf{I} + (P_\parallel - P_\perp)\hat{\mathbf{b}}\hat{\mathbf{b}}$, becomes very simply

$$\begin{bmatrix} P_\parallel \\ P_\perp \end{bmatrix} = \sum_\sigma \int_0^\infty \varepsilon^{3/2}\,d\varepsilon \int_0^{B^{-1}} \frac{dvB}{[2(1-vB)]^{1/2}} f(\varepsilon,v,J) \begin{bmatrix} 2(1-vB) \\ vB \end{bmatrix}$$

$$= \omega(\psi) \begin{bmatrix} \hat{P}_\parallel(B,\psi) \\ \hat{P}_\perp(B,\psi) \end{bmatrix}. \tag{64}$$

The tensor P is a known function of (B,ψ) if F is obtained from the Fokker–Planck equation. It remains to solve the guiding center fluid equations (1), (2), and (3) for (ψ, B). This class of mirror equilibria is then self-consistent on all of the progressively shorter time-scales of the guiding center theory if they are ordered as follows

$$T \gg \tau_r \gg \tau_{II} \gg \omega_{DI}^{-1} \gg \omega_{bI}^{-1} \gg \Omega_I^{-1} \gg \omega_{pI}^{-1} \gg \omega_{pe}^{-1} \sim \Omega_e^{-1}. \tag{65}$$

That is, the steady-state lifetime T of the system $\gg$ radial diffusion time, τ_r, giving the radial profile $\omega(\psi)$. The ion–ion scattering time, τ_{II}, on which the distributions, F, are determined from the Fokker–Planck equations $\gg$ ion drift and bounce times, $\omega_{DI}^{-1}, \omega_{bI}^{-1}$, so that J is a good invariant. The ion cyclotron time, Ω_I^{-1}, is still shorter, ensuring that μ, the essential invariant for mirror containment, is conserved and $\Omega_I^{-1} \gg \omega_{pI}^{-1}$ since the mirror is to be operated at high beta and high power density. Finally, the electron time-scales $\omega_{pe}^{-1}, \Omega_e^{-1}$ are the shortest of all and ensure stability and thermodynamic equilibrium of essentially all the electrons in the system.

When the distribution function f is independent of J the pressure tensor is just $\mathsf{P} = \mathsf{P}(B)$. Taylor (1963) used the invariance of μ, ε to show that these special equilibria are stable to all low frequency ($\omega \ll \Omega_I$), long wavelength ($k_\perp a_i \ll 1$) electrostatic perturbations provided $\partial f(\varepsilon,\mu)/\partial \varepsilon < 0$. The more general flux tube equilibria, Eq. (63), were shown to be stable to interchanges provided $(\partial f/\partial \varepsilon)|_{\text{const.}\mu, J} < 0$ by Taylor (1964) and by McNamara and Rowlands (1963), but in these cases monotonicity of f is neither necessary nor sufficient for absolute stability to all low frequency perturbations (Grad, 1966, 1967a; Taylor, 1967). Nevertheless, these stability conditions are very useful guides to constructing mirror equilibria which are stable on the longest time-scales.

While it may be possible to make f independent of J by careful neutral injection this is of no use in the case of weak magnetic wells. Hall has suggested that by breaking the magnetic symmetries in a deep magnetic well the particle drifts can be scrambled so that they fill the whole volume of the well. The invariant J is till conserved but is independent of position, $J = J(v)$, the very opposite of omnigenity, and the highly stable $\mathsf{P}(B)$ equilibria are generated in a simple way, independent of the neutral injection process.

C. Mirror Energetics and the MIRICLE Machine

The Baseball and 2X field configuration are described in detail in Section III, E, but the economics of the mirror machine lead us to introduce a modified mirror, the MIRICLE machine (Hall and McNamara, 1974). A great deal of energy is invested in making high temperature plasma ($T_i \sim 100$ keV), and so a mirror reactor is essentially an energy amplifier. The amplification factor, Q = Thermonuclear power produced/Injected beam power, is given in its simplest form by

$$Q = \tfrac{1}{4} n^2 (\langle \sigma v \rangle_{\text{DT}} / I W_{\text{I}}) W_{\text{F}}, \tag{66}$$

where $\langle \sigma v \rangle_{\text{DT}}$ is the average DT fusion reaction rate, $W_{\text{F}} = 22.4$ MeV, the nuclear energy per reaction, I the beam current, and W_i the beam energy. The maximum value of Q in a bare mirror is $Q = 5 \log R$ but Fowler and Rankin (1962) showed that electron cooling and the effective reduction of the mirror ratio for ions by the ambipolar potential, Φ, to

$$R_{\text{eff}} = R/(1 + e\Phi/W_i) \tag{67}$$

reduce this to $Q \sim \log R_{\text{eff}}$. The electrons remain cool, $T_e \sim T_i/7.0$, and the ion lifetime maximizes near the ion–electron energy exchange time, $\tau_{ie} \sim \tfrac{1}{3}\tau_{II}$,

$$1/\tau \simeq 1/\tau_{II} \log R_{\text{eff}} + 1/\tau_{Ie}. \tag{68}$$

Every electron lost has to climb over an ambipolar potential $\Phi \sim 10 T_e$ but only one of the hottest electrons can escape with each ion. If the ends of the mirror are connected by a toroidal link in which the density n_T is allowed to rise to about 5% of the density in the well, n_w, then an electron only has to climb a potential of $\Phi \sim T_e \ln(n_w/n_T) \sim 3 T_e$ to get into the link. Since the toroidal link necessarily has a lower beta limit for MHD stability than the well, this link must be drained several times per ion lifetime, τ, to hold down the plasma pressure. However, a considerable saving of electron energy is made when compared with the bare mirror; the electrons heat up more, the cooling time $\tau_{Ie} \to 3\tau_{II}$ ceases to dominate the ion lifetime and $Q \sim 2 \log R$ is much closer to the theoretical limit of $5 \log R$. The device is known by the acronym MIRICLE for *MIR*rored *I*ons, *CL*osed *E*lectrons. The simplest realization of this concept would be the Mirror-Levitron whose equilibria are described in Section III, E, 1.

D. Guiding Center Equilibria

1. *Classical Magnetostatics of Tensor Pressure Plasmas*

It remains to show how the fluid equations (1)–(3) for tensor pressure equilibria may be reduced, depending on the symmetries of the configurations, to one or more scalar equations. The parallel component of Eq. (1) reflects the conservation of (ε, μ) and relates $p_\perp$ to $p_\parallel$ by

$$p_\perp = -B^2(\partial/\partial B)(p_\parallel/B). \tag{69}$$

The guiding center fluid equations are satisfied by any $f(\varepsilon, \mu, J)$ inserted in Eq. (64) though all the computations are restricted to the separable profiles of Eqs. (63). A simple approximation to the normal mode Fokker–Planck distribution is

$$f = \omega(\psi)(\mu B_{\max}/\varepsilon - 1)e^{-\varepsilon} \tag{70}$$

giving

$$p_\parallel = \omega(\psi)(B_{\max} - B)^{5/2}/B \equiv \omega \hat{p}_\parallel. \tag{71}$$

Other simple profiles, due to Taylor (1967), are

$$f = (\mu B_0 - \varepsilon)^{m-3/2} g(\mu), \qquad p_\parallel = cB(B_0 - B)^m. \tag{72}$$

The natural coordinates in which to describe the equilibrium quantities are (ψ, θ, B), where θ is an angle around the drift surfaces ψ, and B is a multi-valued coordinate along the field lines. The perpendicular component of Eq. (1) becomes

$$\mathbf{B} \times \nabla \times [(1+\eta)\mathbf{B}] + 4\pi(\partial p_\parallel/\partial \psi)\nabla\psi = 0, \tag{73}$$

where

$$\eta = 4\pi B^{-2}(p_\perp - p_\parallel) = -4\pi B^{-1} \partial p_\parallel/\partial B.$$

In configurations with toroidal symmetry the drift surfaces and flux surfaces coincide and the field $\mathbf{B}$ can be found in polar coordinates (r, θ, z) in terms of the stream function: $B_r = -r^{-1} \partial\psi/\partial z$, $B_z = r^{-1} \partial\psi/\partial r$. Equation (73) reduces to the tensor pressure extension of Eq. (3) (Grad and Rubin, 1958),

$$\Delta^*\psi + \sigma^{-1}(\nabla\psi \cdot \nabla\sigma) + \frac{r^2}{\sigma}\frac{\partial p_\parallel}{\partial \psi} + \frac{1}{\sigma^2}\frac{\partial}{\partial \psi}\left(\frac{\sigma^2}{2} r^2 B_\theta^2(\psi)\right) = 0, \tag{74}$$

where $\sigma = (1 + \eta)$. The toroidal field, B_θ, must be given as a function of ψ from the external coils and, because of the symmetry, is not modified by the plasma magnetization. In a low-beta plasma the equation can be simplified somewhat by assuming $\sigma r B_\theta \to$ const. and, for a $P(B)$ equilibrium, the third term also vanishes to leave, as the simplest tensor pressure equilibrium equation,

$$\Delta^*\psi + \sigma^{-1} \nabla\psi \cdot \nabla\sigma = 0. \tag{75}$$

Hall and McNamara (1975) have shown that the Eqs. (1)–(3) can be cast in the classical magnetostatic form

$$\mathbf{B} = \mathbf{H} + 4\pi\mathbf{M}, \quad \nabla \times \mathbf{H} = 0, \quad \nabla \cdot \mathbf{B} = 0, \tag{76}$$

where $\mathbf{M}(\mathbf{B})$ is a volume magnetization, proportional to the amount of plasma in the system. The simplest case of nonaxisymmetric equilibria is for $\omega = 1$, $P = P(B)$ when the last term of (73) vanishes and the field is given entirely by a single scalar, Ω,

$$\mathbf{B} = -\nabla\Omega - \eta\mathbf{B}. \tag{77}$$

The local magnetization vector, $4\pi\mathbf{M}_\mu = -\eta\mathbf{B}$, arises from the μ dependence of f and vanishes when the plasma pressure is isotropic. The magnetic potential, Ω, is found from $\nabla \cdot \mathbf{B} = 0$ which gives the Poisson equation

$$\nabla^2 \Omega = 4\pi\nabla \cdot \mathbf{M} \tag{78}$$

or, in this case,

$$\nabla \cdot (\sigma^{-1} \nabla\Omega) = 0. \tag{79}$$

It is interesting to note that for the collisional pressure profiles of Eq. (71), $\mathbf{M}$ is diamagnetic but that the Taylor profiles can become paramagnetic (Hall, 1972).

The more general flux tube equilibria give rise to a nonlocal drift magnetization $\mathbf{M}_J$ which requires a second, independent potential, Γ, to be calculated from the magnetic differential equation (MDE)

$$\nabla \cdot (\Gamma\mathbf{B}) = \hat{p}_\parallel. \tag{80}$$

The MDE can be cast in Lagrangian form as an integral equation along the field lines, l,

$$\nabla \cdot (\Gamma\mathbf{B}) \equiv \mathbf{B} \cdot \nabla\Gamma \equiv B\, \partial\Gamma/\partial\rho = \hat{p}_\parallel.$$

Therefore,
$$\tag{81}$$

$$\Gamma(\alpha, \beta, L) = \Gamma_0(\alpha, \beta, 0) + \int_0^l (dl/B)\, \hat{p}_\parallel,$$

where $\mathbf{B} = \nabla\alpha \times \nabla\beta$ defines the field line coordinates (α, β). The boundary condition for Eq. (80) is that $\Gamma_0 \equiv 0$ on an initial surface, $\mathscr{S}_0$, lying outside the mirror but intersecting all the field lines passing through the plasma. The final value of Γ at the end surface, $\mathscr{S}_c$, of the plasma is then

$$\overline{\Gamma}(\mathscr{S}_0) = \oint (dl/B)\,\hat{p}_\| \tag{82}$$

which actually describes the average drift surfaces ψ for the plasma. The form of $\mathbf{M}_J$, derived by Hall and McNamara (1975), is

$$\mathbf{M}_J = \xi \nabla \gamma, \tag{83}$$

where

$$\xi(\omega) = \int_0^\omega \overline{\Gamma}\, d\omega, \qquad \gamma = \Gamma/\overline{\Gamma} - 1/2. \tag{84}$$

This is the true magnetization which vanishes outside the plasma, vanishes if $f \neq f(J)$, and has the symmetries of the vacuum field, such as S_1, S_2, and antisymmetry under S_3 for the Baseball configuration. It is easy to see that the integrability condition that the parallel current, $j_\| = \hat{b} \cdot \nabla \times \mathbf{M}_J$, should vanish at the plasma surface is satisfied if $\omega = \omega(\overline{\Gamma})$. This profile is, of course, given by a radial diffusion calculation as a function of r along some line intersecting all the drift surfaces. The functions $\omega, \overline{\Gamma}, \xi$ are all related by the one-dimensional transformations

$$\omega(\Gamma) = \omega(\overline{\Gamma}(r)), \qquad \overline{\Gamma} = \overline{\Gamma}(\omega), \qquad \xi = \xi(\omega). \tag{85}$$

Since ω drops smoothly to zero at the plasma edge, this is a more useful basic coordinate than $\overline{\Gamma}$ and is calculated in three dimensions from the MDE

$$\nabla \cdot (\omega \mathbf{B}) = 0, \qquad \omega(\mathscr{S}_0) = \omega(\overline{\Gamma}). \tag{86}$$

The functions $(\overline{\Gamma}, \xi)$ are then constructed in three dimensions by interpolation in the one-dimensional relation (85).

2. MHD Stability of Anisotropic Pressure Plasma

It is obvious that Eqs. (74) and (75) change character when σ passes through zero, leading to loss of equilibrium (Grad, 1967a). Taylor and Hastie (1965) first derived this from the energy principle as a condition for stability to the "firehose" mode, along with a similar condition for this "mirror" mode.

Hall (1972) expresses these in the instructive form

$$H = B/\sigma > 0, \tag{87}$$

$$dH/dB = (d/dB)(B/\sigma) > 0. \tag{88}$$

These conditions are purely algebraic for $P(\psi, B)$ equilibria and do not even require solution of the equilibrium problem. The second condition can be integrated directly for a $P(B)$ equilibrium to give

$$\tilde{\beta}_{\max} \equiv 8\pi\rho/B_{\max}^2 = (1 - 1/R^2). \tag{89}$$

However, Hall (1972) has investigated the beta limits corresponding to the profiles (71), (72) and finds the actual limits to be somewhat lower. Similar results are given by Killeen et al. in this volume. The best results ($\tilde{\beta}_{\max} \sim 0.4$ at $R = 5$) are given by the Taylor profiles which have the curious property of being paramagnetic ($P_\parallel > P_\perp$) at the bottom of the well for $B < (M+1)^{-1}$. The collisional distribution gives $\tilde{\beta}_{\max} < 0.2$ at $R = 2$ and has been observed in 2XII by Molvik et al. (1974). The simplex algorithm has been employed by Hall et al. (1975) for optimizing the distribution function with respect to stability and economics.

The conditions (87), (88) are necessary for MHD stability and in the case of $P(B)$ equilibria they are also sufficient. Many sufficient conditions for stability to kink and interchange modes in general mirror equilibria are given in the literature (Grad, 1969; Hedrick et al., 1973). A simple sufficient condition for interchange stability of the separable equilibria is given by Hall and McNamara (1974):

$$D\bar{\Gamma}/D\omega \equiv \partial\bar{\Gamma}/\partial\omega + (\partial\bar{\Gamma}/\partial B)(\partial B/\partial\omega) \geq 0, \tag{90}$$

and a monotonically decreasing $\omega(r_0)$ will satisfy this.

E. NUMERICAL METHODS FOR GUIDING CENTER EQUILIBRIA

1. *Azimuthally Symmetric Equilibria*

Killeen and Whiteman (1966) solved the flux equation (75) by a straightforward ADI scheme for Taylor $P(B)$ pressures in a stuffed cusp mirror. A toroidal magnetic well is formed by equal and opposite currents in two circular coils, radius 1.0, at $z = \pm 1.0$, and a current along the polar axis, $r = 0$. A typical result is shown in Fig. 13 where $P_\parallel = CB(1.25 - B)^3$ in the well. The mirror condition (88) is the more restrictive and gives $C_{\max} = 0.83$, corresponding to a mild well digging by the plasma, mainly at the well bottom

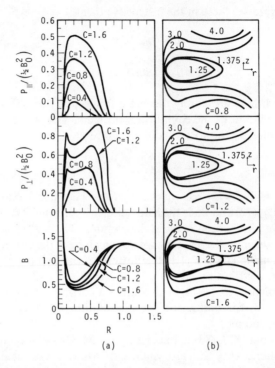

FIG. 13. Anisotropic $P(B)$ equilibria in a stuffed cusp. $P_\parallel = CB(1.25 - B)^3$. (a) Magnetic field, $p_\perp, p_\parallel$. (b) Contours of constant B in the R–Z plane. Cases with $C > 0.83$ are unstable. From Killeen and Whiteman (1966).

where most of the plasma is situated. The code was able to find unstable equilibria because σ did not have a zero anywhere.

Fisher and Killeen (1971) solved the flux equation (124), with $\sigma r B_\theta = $ const., for a number of pressure profiles in a Mirror-Levitron. This field is produced by a toroidal field B_θ, the field of a single "floating" circular coil at ($r = 1$, $z = 0$), and a small multipole field from six external windings. The resulting flux surfaces and B contours are shown in Fig. 14, and, while the well depth is too small to be of practical value, this is representative of the toroidally linked MIRICLE machine (Hall and McNamara, 1974). Fisher and Killeen used pressures of the form

$$p_\parallel \equiv \omega(\psi)\hat{p}_\parallel(B) = p_0(\psi - \psi_1)(\psi - \psi_2)(B_1/B)(1 - B/B_1)^{5/2} \quad (91)$$

for $\psi_1 \leqslant \psi \leqslant \psi_2$, $B \leqslant B_1$, and zero otherwise. They found that for isotropic pressures, $\hat{p}_\parallel \equiv 1$, the maximum value of $\beta_\perp = 8\pi P_\perp / B_{\min}^2$ was $\beta_\perp = 2.3$, while for an anisotropic pressure of the form (91), with $B_1 = 30$, the mirror mode

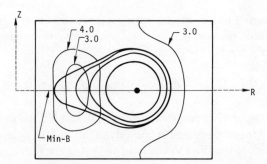

FIG. 14. Flux surfaces and B surfaces in a Mirror-Levitron. From Fisher and Killeen (1971).

limits $\beta_\perp$ to $\beta_\perp = 0.27$. Unfortunately, they made no attempt to evaluate other criteria for ballooning modes in the link, and the first limit is undoubtedly far too high.

2. P(B) *Equilibria in Baseball Wells*

Anderson and Killeen (1972) have solved the potential equation (79) for Baseball- and Ioffe-type magnetic wells. The plasma potential is separated from the vacuum potential of the coils by setting $\Omega = \Omega_p + \Omega_v$ so that the boundary condition is $\Omega_p = 0$ on some (relatively) distant boundary. The ADI methods discussed in Section II, B are unstable in three dimensions and so the Douglas–Gunn (1964) modification is used by converting Eq. (79) to a diffusion equation in which time plays the role of iteration parameter:

$$\partial \Omega_p / \partial t = \nabla^2 \Omega_p - \sigma^{-1} \nabla \sigma \cdot \nabla \Omega_p + \sigma^{-1} \nabla \sigma \cdot \mathbf{B}_v. \tag{92}$$

The equation is solved in cylindrical coordinates (r, θ, z) with the axis along the center of the coil system. Representing the central difference operators, equivalent to the differential operators of (92), by $\Delta_r, \Delta_\theta, \Delta_z$ and the time-step by $\rho = \Delta t^{-1}$, Eq. (92) becomes

$$\rho(\Omega^{n+1} - \Omega^n) = (\Delta_r + \Delta_\theta + \Delta_z) \cdot \Omega + g. \tag{93}$$

The Douglas–Gunn iteration cycle comprises the three steps:

$$\begin{aligned}
(\rho - \tfrac{1}{2}\Delta r) \cdot \Omega^{n+1} &= (\rho + \tfrac{1}{2}\Delta_r + \Delta_\theta + \Delta_z) \cdot \Omega^n + g^n, \\
(\rho - \tfrac{1}{2}\Delta_\theta) \cdot \Omega^{n+2} &= \rho \Omega^{n+1} - \tfrac{1}{2}\Delta_\theta \Omega^n, \\
(\rho - \tfrac{1}{2}\Delta_z) \cdot \Omega^{n+3} &= \rho \Omega^{n+2} - \tfrac{1}{2}\Delta_z \cdot \Omega^n.
\end{aligned} \tag{94}$$

Short wavelength Fourier modes turn each second-order operator into an algebraic operator such as, $\frac{1}{2}\Delta_r \to -d_r(k)$. It is easy to calculate the amplification factor, E_k, for Eq. (94) and show

$$E_k = (\rho - d_r)(\rho - d_\theta)(\rho - d_z)/(\rho + d_r)(\rho + d_\theta)(\rho + d_z), \tag{95}$$

and therefore $|E_k| \leq 1$, demonstrating absolute stability to short wavelength perturbations. Since the method is quasi-dynamic and the potential form gives the opposite sign of the $(\nabla \sigma \cdot \nabla)$ operator to that in the flux formulation, (74),

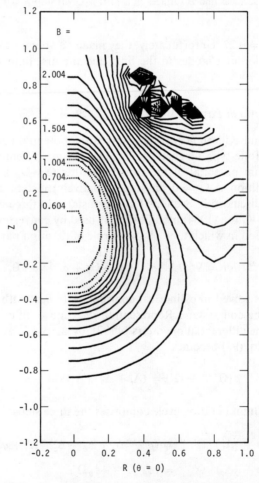

FIG. 15. Baseball II $P(B)$ equilibrium near the mirror stability limit. From Anderson and Killeen (1971).

the equation actually does become hyperbolic as the mirror mode instability threshold is reached. The method diverges and the code cannot find unstable equilibria. A typical high-beta result is shown in Fig. 15 for $P_\parallel = 0.2B(1.8 - B)^4$, in a Baseball II coil, with the vacuum field normalized to 1.0 at the minimum. The plasma diamagnetism has increased the mirror ratio from its vacuum value of 2.0 to about 3.3, a result often used in scaling mirror fields. The central $\beta_\perp$ is 1.0, close to the maximum of 1.04 for this profile, and numerical instability is beginning, as evidenced by the ripples in the contours.

3. Magnetic Differential Equations

The flux tube equilibria require the solution of the two MDE's, (80) and (86), for the drift magnetization potential, Γ, and the pressure surfaces, $\omega(\bar{\Gamma})$. The typical MDE, (80), being a first-order equation, has curious properties as can be seen from the Lagrangian equivalent equation (81):

1. The solutions are local to field lines on which Γ_0 or $\hat{p}_\parallel \neq 0$.
2. The solutions are multivalued on closed field lines.
3. If $\hat{p}_\parallel \equiv 0$ there are two classes of solutions corresponding to $\Gamma_0 = \alpha$ or β on $\mathscr{S}_0$.

In Baseball geometry it is most convenient to use a regular cartesian mesh $(x, y, z = i\Delta x, j\Delta y, k\Delta z, i = 1, M, j = 1, N, h = 1, K)$ spanning one-eighth of the configuration and bounded by the symmetry planes, $x = 0, y = 0, z = 0$, and the outermost edges of the plasma. A typical pressure surface is shown in Fig. 16 (Boyd et al., 1975) together with the grid on which the solution was computed. The magnetic field B is given and, together with Γ, $\hat{p}_\parallel$, this requires about 200,000 words of memory on 32^3 mesh points. Even on this mesh the field lines can cross more than one grid cell in one step in the z-direction, and so the Courant–Friedrichs condition cannot be satisfied everywhere for an explicit integration scheme. The interior points of the mesh are therefore connected by a conservative ADI scheme, the first step of which is

$$\delta_x(\Gamma B_x)^{k+1/2} + (2\Gamma B_z/\Delta z)^{k+1/2} = \tfrac{1}{4}(p_\parallel^{k+1} + 3p_\parallel^{k}) - \delta_y(\Gamma B_y)^k + (2\Gamma B_z/\Delta z)^k, \tag{96}$$

where δ_x is the centered difference operator,

$$\delta_x Q = (Q_{i+1,j,k} - Q_{i-1,j,k})/2\Delta x,$$

and the fields at the half-step are $B_x^{k+1/2} = \tfrac{1}{2}(B_x^{k+1} + B_x^{k})$. Unnecessary indices have been suppressed. Since $B_x(x = 0) \equiv 0$ and $B_y(y = 0) \equiv 0$, this scheme is not sufficiently accurate in the symmetry planes and the following

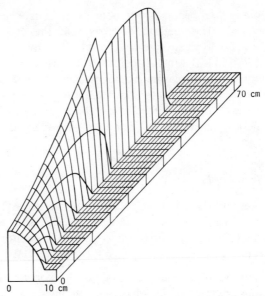

FIG. 16. Isometric plot of a flux tube $\omega = \omega(\overline{\Gamma})$ in 2XIIB showing fanning. The cross section in the midplane is made noncircular by the quadrupole vacuum field distorting the drift surfaces. From Boyd et al. (1975).

implicit nonconservative scheme is used (on the plane $y = 0$):

$$\tfrac{1}{2}B_x^{k+1/2}\,\delta_x\Gamma_{i1}^{k+1/2} + (2B_z^{k+1/2}/\Delta z)\,\Gamma_{i1}^{k+1/2}$$
$$= (2B_z^{k+1/2}/\Delta z)\,\Gamma_{i1}^k + p_\|^{k+1/4} - \tfrac{1}{2}B_x^{k+1/2}\,\delta_x\Gamma_{i1}^k, \tag{97}$$

which also does not involve interior values of Γ. The interior points immediately adjacent to the symmetry planes are on a boundary for the interior scheme and are connected to the symmetry plane by the nonconservative scheme (on the $y = 0$ plane):

$$(B_x^{k+1/2}/2\Delta x)(\Gamma_{i+1}^{k+1/2} - \Gamma_{i-1,2}^{k+1/2}) + (B_z^{k+1/2}/\Delta z)\,\Gamma_{i2}^{k+1/2}$$
$$= (B_z^{k+1/2}/\Delta z)\,\Gamma_{i2}^k + p_\|^{k+1/4} - (B_y/2\Delta_y)(\Gamma_{i3}^k - \Gamma_{i1}^k). \tag{98}$$

This connects the interior values, $\Gamma_{i2}^{k+1/2}$, explicitly to the values Γ_{i1}^k on the $y = 0$ symmetry plane and, as a boundary condition, to the value of $\Gamma_{12}^{k+1/2}$ on the $x = 0$ plane.

The external boundary conditions on this scheme must be carefully specified since the surface $\mathscr{S}_0$ is the plane $z = 0$ and any part of the planes $x = M\Delta x$, $y = N\Delta y$ on which field lines enter the solution volume. The particular MDE's of the equilibrium theory require the simple conditions

$\Gamma = 0$, $\omega = 0$ on entering field lines. On field lines leaving the volume no boundary condition is required, and the difference equations are closed with the second-order-accurate approximation

$$\delta_x Q = (1/2\Delta x)(3Q_M - 4Q_{M-1} + Q_{M-2}). \tag{99}$$

These difference equations are in tridiagonal form and are solved by the usual Gaussian elimination process. The second half of the ADI scheme uses the same set of equations, advanced half a step, and with x, y interchanged. The final result, Γ^{k+1}, is accurate to second order in $\Delta x, \Delta y, \Delta z$ and agrees to within about 1% in the Baseball field with a Lagrangian calculation, which follows individual field lines with an eighth-order-accurate scheme, and is orders of magnitude faster.

The local stability of the ADI scheme is well known but the global properties of the scheme can also be analyzed very easily. Equation (96) is just a difference form of the first-order, ordinary differential equation

$$(\partial/\partial x)(\Gamma B_x) + \Gamma B_z/\Delta = R(x), \tag{100}$$

with the exact solution, corresponding to one half-step of the ADI scheme,

$$\Gamma = \Gamma_0 + \exp\left(-\int_0^x B_z\, dx'/B_x \Delta\right) \int_0^x \exp\left(\int_0^{x'} B_z\, dx''/B_x \Delta\right) R(x')\, dx'. \tag{101}$$

If $R(x)$ is finite between 0 and x_1 and drops abruptly to zero at x_1, the solution Γ will only drop exponentially to zero beyond x_1. The ADI scheme described is therefore not very good at propagating discontinuous functions, even though the original MDE constrains the solution to a bundle of field lines. This is the reason why the pressure profile $\omega(\overline{\Gamma})$ is used as the fundamental coordinate in the equilibrium computations.

This observation is widely applicable to ADI schemes: A set of nonlinear partial differential equations, such as those of magnetohydrodynamics, would be reduced to a set of *linear, ordinary* differential equations containing many parameters for the numerical scheme. All the tools of modern analysis can be applied to the set of equations. The nonlinear flux, nV, for example, may be approximated by

$$(nV) = \alpha_1 n^k V^k + \alpha_2 n^{k+1} V^k + \alpha_3 n^k V^{k+1}, \quad \alpha_1 + \alpha_2 + \alpha_3 = 1, \tag{102}$$

where k indicates the time level of the computed quantity. The choice of the parameters α_i can be guided by the global analysis of the integration scheme demonstrated here and can help with the treatment of shocks and boundary conditions.

The difference equations (96)–(99) have been coded at Livermore for the CDC 7600 computers using the vector extensions of FORTRAN which will be available across the USA to users of the CTR Computing Center at Livermore. The three-dimensional scalar arrays are stored in large core on the 7600 and brought into small core one or two planes at a time. The coefficients for one step of the ADI scheme are computed by vector statements of the form A = GAMMA*BY. The compiler translates these into a call to a hand coded routine which multiplies the arrays GAMMA, BY together, element by element, at the maximum speed of which the machine is capable. The syntax of the language extensions is a little awkward for addressing parts of a vector array. This is overcome by constructing a set of problem oriented macros, called CLICHES by the compiler, such as the one below for taking the x derivative of a function:

```
    CLICHE SUBI (ONDISC) (2)
C···FIRST VECTOR ARGUMENT [1] = ELEMENTS (I + 1, J) – (I – 1, J)
C···OF SECOND
    [1](M; M) = [2](M + 1; M + 1) – [2](M – 1; M – 1)
    ENDCLICHE
C···EXAMPLE OF USE FOR SPECIFIC VECTOR ARRAYS DQDX, Q
    SUBI [DQDX] = [Q](I + 1) – [Q](I – 1)
```

The two arguments of the macro are given in [ ] after the macro name and all other symbols are just commentary to make the coding clear. Use of the vector extensions can gain a factor of two or more in execution speed over FORTRAN. One penalty is that guard cells are required around the array boundaries to avoid taking derivatives like $D_M j = (Q_{1,j+1} - Q_{M,j})$ across rows of a two-dimensional calculation. A great deal of coding is required to repair the effects of the vectorized operations on the boundaries. Most of this is done with the aid of calls to hand-coded routines which, on the 7600, are able to index rows and columns of the arrays. Although the coding is hard to read this is certainly simpler than writing in machine language, and the technique is easily implemented on other computers.

The second half of the ADI scheme is done by physically transposing the data in memory, and interchanging B_x, B_y, so that the vector operations can address the columns in the y-direction sequentially. The logic of the coding is convoluted but the same integrator is used for both steps of the ADI.

4. *Flux Tube Equilibria in 2XII*

These equilibrium computations require the vacuum magnetic field $\mathbf{B}_v$ of the 2XII coils, the pressure profiles $\hat{p}_\parallel(B)$, $p_\perp(B)$, and the radial distribution

$\omega(x)$ to be given initially. The MDE (80) is solved for Γ at low beta using the vacuum field by integrating from a plane $Z = 70$ cm, just outside the mirror throat, to the midplane $Z = 0$. The average drift surfaces $\bar{\Gamma}$ are then found, using the S_3 symmetry, from

$$\bar{\Gamma}(x, y, z = 0) = \Gamma(x, y, 0) + \Gamma(y, x, 0).$$

A spline fit is then made to $\bar{\Gamma} = \bar{\Gamma}(x, 0, 0)$ and an interpolation table at equal intervals $(0(0.01)1)$ of $\omega = (1 - x^2/a^2)^2$ is constructed for $\bar{\Gamma}(\omega)$ and $\xi(\omega)$. This simple model for ω has zero value and first derivative at the plasma edge, $x = a$, for the benefit of the MDE solver. The interpolation table is first used in the midplane to find $\omega = \omega(\bar{\Gamma}(x, y, 0))$ which gives the boundary condition for the MDE, $\nabla \cdot (\omega \mathbf{B}) = 0$, for propagating the pressure surfaces throughout the mesh to $z = L$. The functions $\bar{\Gamma}, \xi$ are then constructed in three dimensions from the interpolation table and $\omega(x, y, z)$. The drift magnetization $\mathbf{M}_J$ is constructed from ξ and $\gamma = \Gamma/\bar{\Gamma} - 1/2$ and stored on disk where it can be examined interactively by diagnostic codes. The physically relevant components of all such fields are in the directions of the induction, $\hat{b} = \mathbf{B}/B$, the curvature, $\mathbf{k} = -\hat{b} \cdot \nabla \hat{b}$, and the binormal $\hat{n} = \hat{b} \times \hat{k}$, $\hat{k} = \mathbf{k}/k$. The drift magnetization is shown in the midplane and the symmetry planes in Figs. 17 and 18. The same set of equally spaced contour heights is used throughout these pictures with zero lying between the light and heavy contours.

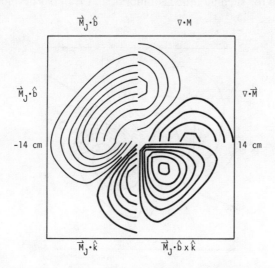

FIG. 17. Components of drift magnetization, $\mathbf{M}_J$, and $\nabla \cdot (\mathbf{M}_\mu + \mathbf{M}_J)$ in 2XIIB at the midplane, $Z = 0$.

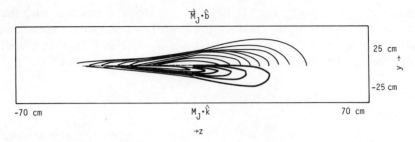

FIG. 18. Components of drift magnetization in 2XIIB as seen in the symmetry planes. The uppermost contour is at the edge of the plasma where $M = 0$.

The pressure tensor, $P(x, y, z) = \omega(\bar{\Gamma}) P(B)$, and hence the μ-magnetization, $\mathbf{M}_\mu$, are also constructed and stored. The divergence of the total magnetization is calculated and the Poisson equation for the magnetic potential Ω is solved using the integral solution. This gives the correct boundary condition, $\Omega(\infty) = 0$, but a fast Poisson solver is being developed at the time of writing. This gives the plasma magnetic field $\mathbf{H}_p$, which is displayed in Figs. 19 and 20, and the total plasma magnetic field, $\mathbf{B}_p = \mathbf{H}_p + 4\pi\mathbf{M}$, displayed in Figs. 21 and 22. The 2XII plasma is long and thin, 140 cm × 30 cm, and in this approximation $\mathbf{H}_p \sim -4\pi(\mathbf{M}_J - (\mathbf{M}_J \cdot \hat{b})\hat{b})$. This is apparent from Figs. 14–22, and the final low-beta plasma field has a small perpendicular component in the region of weakest field curvature. This may have an effect on the maximum beta for such flux tube equilibria, and Hall has developed new trial

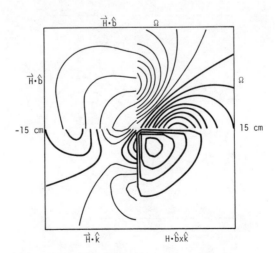

FIG. 19. Plasma potential Ω and magnetic field $\mathbf{H} = -\nabla\Omega$ in midplane, $Z = 0$, of 2XIIB.

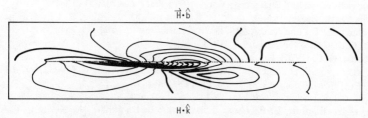

FIG. 20. Magnetic field **H** of low-beta plasma in 2XIIB as seen in the symmetry planes $X = 0$ for $Z > 0$, $Y = 0$ for $Z < 0$. Solid contours are $H > 0$ and the changes at $Z = 0$ demonstrate the symmetries of the field components under S_3.

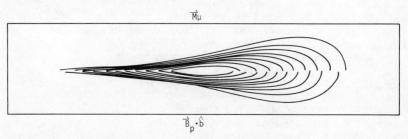

FIG. 21. Gyromagnetization and total parallel magnetic field in 2XIIB.

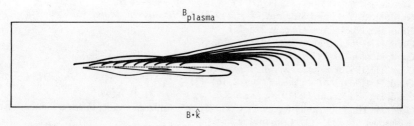

FIG. 22. Total plasma field in 2XIIB and the perpendicular component in the symmetry planes $X = 0$, $Y = 0$. Observe that $\mathbf{H} \simeq -4\pi \mathbf{M}_J$ and total perpendicular plasma field is $O(r/L)$ smaller than $\mathbf{M}_\mu = O(\beta)\mathbf{B}_{\text{vac}}$.

functions for the energy principle which utilize this plasma field. Further numerical work is required to iterate the process described for calculating these equilibria at finite beta and to evaluate Hall's stability limits.

REFERENCES

Anderson, D. V., and Killeen, J. (1972). *J. Comput. Phys.* **10**, 133–150.
Anderson, O. A., Birdsall, D. H., Hartman, C. W., Lauer, E. J., and Furth, H. P. (1969). *IAEA Novosibirsk Conf.* **1**, 443–447.

Artsimovich, L. A., and Shafranov, V. D. (1972). *JETP Lett.* **15**, 51–54.
Baldwin, D. E., Beasley, C. O., Berk, H. L., Farr, W. M., Harding, R. C., McCune, J. E., Pearlstein, L. D., and Sen, A. (1971). *IAEA Madison Conf.* **2**, 735–755.
Becker, G. (1974). *Nucl. Fusion* **14**, 319–321.
Betancourt, O. L. (1974). Courant Inst. Report MF-67 COO-3077-49.
Blank, A. A., Grad, H., and Weitzner, H. (1969). *IAEA Novosibirsk Conf.* **2**, 607–617.
Bloch, E. (1969a). Courant Inst. Report NYO-1840-116.
Bloch, E. (1969b). *Phys. Fluids* **12**, 129–132.
Bodin, H. A. B., Butt, E. P., Crow, J. E., Irons, F. E., Junker, J., Newton, A. A., and Robinson, D. C. (1971). *IAEA Madison Conf.* **1**, 225–250.
Boyd, J. K., McNamara, B., Hall, L. S., and Wilgus, C. A. (1975). *In* "Conference on Numerical Simulation of Plasma," pp. 127–130. Courant Inst.
Callen, J. D., and Dory, R. A. (1972). *Phys. Fluids* **15**, 1523–1528.
Chu, M. S., Dobrott, D., Jensen, T. H., and Tamano, T. (1974). *Phys. Fluids* **17**, 1183–1187.
Dobrott, D., Chu, M. S., and Ohkawa, T. (1973). *Phys. Fluids* **16**, 1870–1875.
Douglas, J., and Gunn, J. (1964). *Numer. Math.* **6**, 428–435.
Feneberg, W., and Lackner, K. (1973). *Nucl. Fusion* **13**, 549–556.
Fisher, S. (1971). *Phys. Fluids* **14**, 962–966.
Fisher, S., and Killeen, J. (1971). *Phys. Fluids* **14**, 1240–1246.
Foote, J. H. (1972). *Plasma Phys.* **14**, 543–554.
Fowler, T. K., and Rankin, M. (1962). *Plasma Phys.* **4**, 311–320.
Friedberg, J. P. (1971). *IAEA Madison Conf.* **3**, 215–222.
Friedman, N. (1970). Courant Inst. Report NYO-1480-161.
Gajewski, R. (1972). *Phys. Fluids* **15**, 70–74.
Gibson, A., and Taylor, J. B. (1967). *Phys. Fluids* **10**, 2653–2659.
Gordonova, V. I., and Morozov, V. A. (1973). *USSR Comput. Math. Math. Phys.* **13**, 1–9.
Grad, H. (1966). *Phys. Fluids* **9**, 225–251.
Grad, H. (1967a). *Phys. Fluids* **10**, 137–154.
Grad, H. (1967b). *IAEA Madison Conf.* **3**, 229–239.
Grad, H. (1967c). *Proc. Symp. Appl. Math.* **18**, 162–248.
Grad, H. (1969). *IAEA Culham Conf.* **2**, 161–175.
Grad, H., and Rubin, H. (1958). *IAEA Salzburg Congr.* **31**, 190–197.
Grad, H., Hu, P. N., and Stevens, D. C. (1975). *In* "Conference on Numerical Simulation of Plasmas," pp. 135–137. Courant Inst.
Hall, L. S. (1972). *Phys. Fluids* **15**, 882–890.
Hall, L. S., and McNamara, B. (1974). *Bull. Amer. Phys. Soc.* **19**, 872.
Hall, L. S., and McNamara, B. (1975). *Phys. Fluids* **18**, 552–565.
Hall, L. S., and Walker, H. (1975). UCRL-51820.
Hall, L. S., McNamara, B., Boyd, J. K., Finan, C. A., Fuss, D., and Wilgus, C. A. (1974). *IAEA Tokyo Conf.* Paper D1-2 or UCRL-75993.
Hartman, C. W. (1967). *Phys. Fluids* **10**, 1685–1688.
Hastie, R. J., Hobbs, G. D., and Taylor, J. B. (1969). *IAEA Novosibirsk Conf.* **1**, 389–401.
Hedrick, C. L., Guest, G. E., and Nelson, D. B. (1973). ORNL-TM-4076.
Hernegger, F., and Mashke, E. K. (1974). *Nucl. Fusion* **14**, 119–121.
Hockney, R. (1967). *Methods Comput. Phys.* **9**, 136–211.
Howard, J. E. (1971). *Phys. Fluids* **14**, 2378–2384.
Kerner, W., Pfirsch, D., and Tasso, H. (1972). *Nucl. Fusion* **12**, 433–435.
Killeen, J., and Whiteman, K. J. (1966). *Phys. Fluids* **9**, 1846–1852.
Kruskal, M. D., and Kulsrud, R. M. (1958). *Phys. Fluids* **1**, 265–274.
Larkin, F. M. (1965). U.K.A.E.A. Culham Report.

Laval, G., Mashke, E. K., and Pellat, R. (1970). IAEA Trieste Report IC/70/110.
McNamara, B., and Rowlands, G. (1963). *Proc. Inst. Mech. Eng.* **178**, 47–50.
Marder, B., and Weitzner, H. (1970). *Plasma Phys.* **12**, 435–445.
Mashke, E. K. (1973). *Plasma Phys.* **15**, 535–541.
Matsui, T., Tanaka, Y., and Okuda, T., (1973). *Nucl. Fusion* **13**, 671–675.
Molvick, A. W., Coensgen, F. H., Cummins, W. F., Nexsen, W. E., and Simonen, T. C. (1974). *Phys. Rev. Lett.* **32**, 1107–1110.
Morse, P. M., and Feshback, H. (1953). *Methods Theor. Phys.* **2**, 1247–1250.
Mukhovatov, V. S., and Shafrahov, V. D. (1971). *Nucl. Fusion* **11**, 605–633.
Ohkawa, T. (1968). *Kakuyugo-Kenkyu* **2**, 557–564.
Ohkawa, T., and Jensen, T. H. (1970). *Plasma Phys.* **12**, 789–797.
Okabayashi, M., and Sheffield, G. (1974). *Nuc. Fusion* **14**, 263–265.
Post, R. F., and Rosenbluth, M. N. (1966). *Phys. Fluids* **9**, 730–749.
Richtmyer, R. D., and Morton, K. W. (1967). "Difference Methods for Initial Value Problems." Wiley (Interscience), New York.
Robinson, D. C. (1969). *Plasma Phys.* **11**, 893–897.
Rosenbluth, M. N. (1972). *Phys. Rev. Lett.* **29**, 408–410.
Rosenbluth, M. N., and Post, R. F. (1965). *Phys. Fluids* **8**, 547–550.
Stevens, D. C. (1974). *Phys. Fluids* **17**, 222–226.
Suydam, B. G. (1958). *IAEA Geneva Conf.* **31**, 157–159.
Suzuki, Y. (1973). *Nucl. Fusion* **13**, 369–372.
Suzuki, Y. (1974). *Nucl. Fusion* **14**, 346–352.
Suzuki, Y., Kameari, A., Ninomiya, H., Masuzaki, M., and Tayana, H. (1974). *IAEA Tokyo Conf.* Paper CN-33/A11-2.
Taylor, J. B. (1963). *Phys. Fluids* **6**, 1529–1536.
Taylor, J. B. (1964). *Phys. Fluids* **7**, 767–773.
Taylor, J. B. (1967). *Phys. Fluids* **10**, 1357–1358.
Taylor, J. B. (1974) *IAEA Tokyo Conf.* Paper CN-33.
Taylor, J. B., and Hastie, R. J. (1965). *Phys. Fluids* **8**, 323–332.
Tikhonov, A. N. (1963). *Dokl. Akad. Nauk SSSR* **151**, 501–504.
Woltjer, J. (1958). *Proc. Nat. Acad. Sci. U.S.* **44**, 489–497.
Zakharov, L. E. (1973). *Nuc. Fusion* **13**, 595–602.
Zwicker, H., Wilhelm, R., and Krause, H. (1971). *IAEA Madison Conf.* **1**, 251–258.

Computation of the Magnetohydrodynamic Spectrum in Axisymmetric Toroidal Confinement Systems

RAY C. GRIMM, JOHN M. GREENE, AND JOHN L. JOHNSON*

PLASMA PHYSICS LABORATORY
PRINCETON UNIVERSITY, PRINCETON, NEW JERSEY

I. Introduction 253
  A. Purpose 253
  B. Possible Approaches to the Problem 255
  C. Outline of This Discussion 257
II. Formulation of the Problem 257
  A. Equilibrium 257
  B. Coordinates 258
  C. Mapping 259
  D. Projections 261
III. Representation of the Normal-Mode Equations 262
  A. Lagrangian 262
  B. Decomposition of Expansion Functions 263
  C. Matrix Elements 264
  D. Computation of the Spectrum and Associated Eigenmodes . . . 271
IV. Application 272
V. Discussion 276
  A. Geometry 276
  B. Resistive Models 276
  C. Generalized Pressure Models 277
  D. Nonlinear Extensions 277
References 278

I. Introduction

A. PURPOSE

IT IS POSSIBLE WITH present-day computers to obtain detailed information about the magnetohydrodynamic behavior of a confined plasma. Our purpose

---

* On loan from Westinghouse Research Laboratories, Pittsburgh, Pennsylvania.

here is to discuss special computational techniques for the study of the linearized motion of plasma around its equilibrium state in a general axisymmetric toroidal configuration. We adopt a simple ideal magnetohydrodynamic model since it should be clear how generalizations to more realistic descriptions could be effected. We emphasize the two most obvious phenomena one should obtain from such a study: the existence of possible unstable modes and the nature of the complete spectrum.

1. *Instabilities*

Much effort has been expended in determining the stability properties of equilibrium configurations (Greene and Johnson, 1965). Due to the complexity of the problem, we have often been satisfied to know whether or not an unstable mode can occur without obtaining information concerning growth rates or the nature of the unstable perturbation. Current interest in Tokamaks with noncircular cross sections (Grad and Hogan, 1970; Artsimovich and Shafranov, 1972; Jensen *et al.*, 1975) provides considerable motivation for a treatment that provides more of this information. The problem has usually been attacked by making a separation into localized and kink modes. The localized modes have been studied analytically to obtain stability criteria in terms of equilibrium properties on a single magnetic surface. Kink modes, where driving forces throughout the plasma volume must be considered, are usually treated numerically using simplified models.

a. Localized Interchanges. Very general stability criteria have been determined for modes that are singular near a particular magnetic surface (Mercier, 1962; Greene and Johnson, 1962). Since small nonideal effects such as resistivity are important only in this singular region, it is possible to determine the properties of nonideal modes analytically (Glasser *et al.*, 1975). Evaluation of analytic stability criteria (Chance *et al.*, 1975) usefully complements any numerical stability calculation since the highly localized modes are most difficult to represent accurately using a finite mesh.

b. Kink Modes. Nonlocalized modes have always required serious consideration because they have been observed experimentally to limit confinement (Sinclair *et al.*, 1965; von Goeler *et al.*, 1974). Evaluation of stability criteria for such modes is difficult. Therefore, almost all work on them has involved assuming a one-dimensional model in which plasma pressure gradients can be ignored (Shafranov, 1970; Robinson, 1971). Several useful analytic beginnings have been made. The stability of kink modes in a straight column with an elliptic cross section was investigated by Laval *et al.* (1974) and Dewar *et al.* (1974). Some progress was made by Frieman *et al.* (1973)

to determine the effect of toroidicity on kink modes by employing expansion techniques. Obviously, a rigorous numerical study of these modes in realistic configurations is desirable.

2. *Spectra*

The complexity and variety of properties possessed by the normal modes of the ideal magnetohydrodynamic equations makes the determination of the complete spectrum an interesting and rewarding exercise (Grad, 1973). Considerable work has been done toward understanding the spectrum in a diffuse linear pinch, a one-dimensional model where all equilibrium properties depend only on the radius (Appert *et al.*, 1974b; Tataronis and Grossmann, 1972; Goedbloed, 1974). Generalization to toroidal configurations is vastly more difficult. Knowledge of the mathematical properties of the spectrum in such systems illuminates general tendencies and provides a framework for the interpretation of results of individual calculations. Since geometric effects can distort the characteristic modes considerably, it is reasonable and, indeed, desirable to treat the instability problem as part of a more general treatment of the complete spectrum (Dewar *et al.*, 1975).

A possible additional application of an understanding of the spectrum has to do with heating a plasma by externally applying radio frequency radiation. Knowledge of mode structure and wave propagation in a torus could facilitate the choice of the excitation to insure that the driven modes can deposit energy well inside the plasma.

B. POSSIBLE APPROACHES TO THE PROBLEM

1. *Time Dependent*

One obvious approach is to treat an initial-value problem in which the time development of a small perturbation from an equilibrium is followed. This appealing technique has been implemented with much success (Sykes and Wesson, 1974; Bateman *et al.*, 1974). It is useful for study of the fastest-growing instability as, in principle, this mode should quickly dominate everything else. Care must be taken when there are two modes with nearby growth rates since the process of projection onto the most unstable mode depends on the separation of the eigenvalues. When an explicit finite-difference scheme is employed, the problem is aggravated if the accompanying oscillatory modes have a high frequency since then the numerical stability criterion demands a correspondingly small time-step (Roberts and Potter, 1970). It is worthwhile pursuing a dynamic technique as it should provide insight into problems of plasma simulation which must be faced when the nonlinear interaction of the different modes is considered.

2. Variational Approaches

The other major technique utilizes the Lagrangian, associated with linearized perturbations about an equilibrium configuration, to formulate an eigenvalue problem. A number of choices must be made before this approach can be implemented.

As was noted earlier, if one is satisfied with obtaining stability criteria and is willing to forego accurate determination of growth rates or the shape of the unstable perturbations, considerable analytical simplification is effected by replacing the kinetic energy term in the Lagrangian with a convenient normalization and extremizing the potential energy functional. This is especially useful for analytic work.

The Lagrangian can be made stationary by a solution of the associated Euler–Lagrange equations, a set of partial-differential equations that can be solved numerically. This method is most useful for one-dimensional stability calculations (Goedbloed and Hagebeuk, 1972; Furth et al., 1973) and has been adapted for two-dimensional systems (Laval et al., 1974). Baker and Mann (1972) have used a similar approach successfully for studying a toroidal stabilized pinch. Introducing truncated Fourier series in the surface variables, they analytically reduced the problem to a set of ordinary differential equations with which they constructed a quadratic form and tested stability.

The alternative approach involves the use of a set of carefully selected expansion functions which enables the problem to be reduced to the minimization of an algebraic quadratic form (Mikhlin, 1965) with respect to certain variational parameters. Such an approach has been used by Kerner and Tasso (1975). In the most general Rayleigh–Ritz procedure, trial functions depending nonlinearly on the variational parameters can be efficiently employed to determine the lowest eigenvalue accurately (Schwartz, 1963). However, the computation of the higher eigenvalues may then be difficult to carry out. The less sophisticated Galerkin procedure, where the parameters enter only linearly, provides a practical numerical approach. Here, the perturbations are represented by a finite subset of a complete set of functions, $\xi = \sum a_m \Phi_m$, with the Φ_m's independent of the expansion parameters. Then the variational calculation is reduced to the determination of eigenvalues and eigenfunctions of the matrix eigenvalue problem

$$\sum_m [\omega^2 \langle \Phi_{m'}^* | K | \Phi_m \rangle - \langle \Phi_{m'}^* | \delta W | \Phi_m \rangle] a_m = 0, \qquad (1)$$

where δW is the change in potential energy associated with a displacement ξ from an equilibrium configuration, $\omega^2 K$ is the kinetic energy, and the brackets denote integration over the volume. Here, the vector nature of the ξ's and Φ's is denoted symbolically; actually the different vector components must be treated independently. This procedure preserves the virtues of the Lagrangian system, including the feature that the lowest eigenvalue is always approached from above as the accuracy is increased. Thus, we cannot find an instability in a stable configuration. Further, since the numerical approximations obtained from this approach involve only square-integrable functions, difficulties associated with singular perturbations are avoided. At the same time, good representation of such modes can be obtained with a suitable choice of expansion functions (Appert et al., 1975; Dewar et al., 1975).

The Galerkin method is particularly convenient if the functions are expanded in separated variables. It can use either global functions, where the support of the elements of the trial functions is the whole space, or finite elements, where the support is local. Freidberg and Marder (1973) have successfully used the global function technique. Ohta et al. (1972), Boyd et al. (1973), and Appert et al. (1974a) have studied the finite element approach. In our work, we adopt a combination of finite elements to represent variation normal to the magnetic surfaces and Fourier series for behavior in them. We believe this is the most efficient and convenient representation (Dewar et al., 1975). We choose that set of finite elements with lowest continuity properties consistent with finiteness of the various integrals. Other authors (Takeda et al., 1972; Appert et al., 1975) have found this to be most satisfactory.

C. OUTLINE OF THIS DISCUSSION

In the next section, we describe our formalism, specifying the equilibrium, adopting a natural coordinate system, and choosing the projections of the displacement vectors. Then, in the following section, we describe our treatment of the normal mode equations. We adopt the Galerkin method to make the Lagrangian stationary. We describe our decomposition of the expansion functions and write down the matrix that must be diagonalized. We give some results for a specific application in Section IV and discuss the accuracy and usefulness of our representation and how this specific code fits into a larger program utilizing other magnetohydrodynamic models in Section V.

II. Formulation of the Problem

A. EQUILIBRIUM

A major part of any discussion of the magnetohydrodynamic properties of a toroidal configuration consists of the evaluation of an equilibrium. For axisymmetric systems this problem reduces to one of solving the well-known differential equation

$$X\frac{\partial}{\partial X}\frac{1}{X}\frac{\partial \Psi}{\partial X} + \frac{\partial^2 \Psi}{\partial Z^2} = 2\pi X J_\phi = -4\pi^2\left(X^2 \frac{dp(\Psi)}{d\Psi} + \frac{R^2 B_0^2}{2}\frac{dg^2(\Psi)}{d\Psi}\right) \quad (2)$$

for a reasonable prescription of the pressure $p(\Psi)$, the toroidal field function $g(\Psi) \equiv XB_\phi/RB_0$, and the current in external conductors. Here, X, ϕ, Z are cylindrical coordinates (see Fig. 1), R is the radius of the magnetic axis, B_0 is the externally imposed toroidal field at R, and Ψ is the poloidal magnetic flux inside the surface. Solution of this problem, especially for systems in which there is a vacuum region between the plasma surface and the external coils or conducting boundary, is worthy of a paper in its own right (McNamara, this volume). We have an efficient program that utilizes boundary conditions at infinity and thus does not rely on the imposition of a prescribed magnetic surface (Chance et al., 1975). Once the equilibrium configuration has been determined, we express the magnetic field in the form

$$\mathbf{B} = B_0[f(\psi)\nabla\phi \times \nabla\psi + Rg(\psi)\nabla\phi], \quad (3)$$

with ψ a magnetic surface label related to the poloidal flux through $\Psi = 2\pi B_0 \int f\, d\psi$.

B. Coordinates

Considerable thought should be given to the choice of the coordinate system adopted for the calculation. From a numerical point of view, an improper choice could introduce the need for considerable interpolation with its concomitant introduction of error. Most importantly, we know from the history of variational calculations that the eigenfunctions are most easily represented computationally by the wise choice of a natural coordinate system. Although it is unlikely that in any complicated geometry a coordinate system can be chosen such that the normal-mode eigenfunctions can be expressed entirely in terms of functions of separated variables, the choice of coordinate system can considerably influence the number of terms in such an expansion required to obtain a good representation.

It seems obvious that ϕ should be chosen as one of the coordinates to utilize the symmetry of the system.

The physical characteristics of the plasma dictate the choice of the second coordinate. Since current can flow freely along field lines and fluid cannot cross them, it is clear that plasma behavior is quite anisotropic. Mathematically, the ideal fluid equations are higher order in derivatives within magnetic surfaces than in derivatives across them. To represent this with good numerical accuracy, it seems necessary to use ψ as our second coordinate.

Since the operator $\mathbf{B} \cdot \mathbf{V}$ occurs frequently in the fluid equations, it is reasonable to select the third coordinate Θ so that the magnetic field lines are straight. This is done by making the Jacobian

$$\mathscr{J} \equiv (\nabla\psi \times \nabla\Theta \cdot \nabla\phi)^{-1} = vX^2/2\pi R \tag{4}$$

with

$$v \equiv (R/2\pi) \int_p d\tau/X^2, \tag{5}$$

the integral being taken over the plasma volume $\psi \leq 1$. Then

$$\mathbf{B} \cdot \mathbf{V} = (2\pi R f B_0/vX^2)(\partial/\partial\Theta + q\,\partial/\partial\phi) \tag{6}$$

with $q \equiv vg/2\pi f$, the usual safety factor. With this Jacobian, the function ψ measures the toroidal flux associated with the externally imposed field inside a surface, normalized to its value at the plasma–vacuum interface. The coordinate Θ divides this same flux into equal intervals.

This ψ, Θ, ϕ coordinate system introduces some complication because it is not orthogonal (see Fig. 1). We think the additional care this entails is more than compensated by making the calculations conform well with the physics. One particularly nice feature is that the plasma–vacuum interface is a coordinate surface. This makes it easy to use Green's function techniques to express the extremized contribution to the potential energy of the system from the vacuum region outside the plasma in terms of the components of an arbitrary displacement of this surface. Although it should be easy to generalize the model to a system with internal separatrices, as in the Doublet devices, considerable effort will have to be expended before one can treat exactly a Tokamak in which the plasma extends right up to a divertor.

C. Mapping

The use of this coordinate system entails the problem of converting the knowledge of $\Psi(X, Z)$, given by the equilibrium calculation, to that of $X(\psi, \Theta)$ and $Z(\psi, \Theta)$. Obviously this must be done accurately.

To make the transformation it is useful to introduce a local r, θ, ϕ coordinate system (see Fig. 1), centered on the magnetic axis, such that

$$X = R - r\cos\theta, \tag{7}$$

$$Z = r\sin\theta. \tag{8}$$

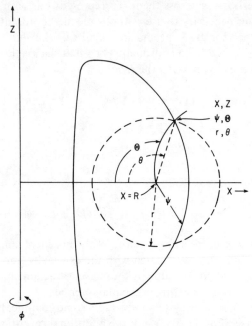

FIG. 1. Coordinate systems. Three different sets of coordinates are used in this work. The usual X, ϕ, Z cylindrical coordinates are most useful for the numerical solution of Eq. (2) to determine equilibrium configurations. The ψ, Θ, ϕ system is constructed as described in the text to incorporate as much physics as possible into the metric. The well-known r, θ, ϕ system facilitates construction of the operational system from the knowledge of the equilibrium in the cylindrical one.

Equating the expression for the line elements in the two coordinate systems, we find that, on a constant-Ψ surface,

$$\partial\Theta/\partial\theta|_\Psi = -4\pi^2 R B_0 \, fr/vX [\partial\Psi/\partial X \cos\theta - \partial\Psi/\partial Z \sin\theta]. \tag{9}$$

Similarly, using the chain rule for differentiation to determine $\partial X/\partial\Theta$ and $\partial Z/\partial\Theta$ at constant Ψ in terms of $\partial\Psi/\partial Z$ and $\partial\Psi/\partial X$, we obtain

$$\partial X/\partial\theta|_\Psi = -r \, \partial\Psi/\partial Z/[\partial\Psi/\partial X \cos\theta - \partial\Psi/\partial Z \sin\theta], \tag{10}$$

$$\partial Z/\partial\theta|_\Psi = r \, \partial\Psi/\partial X/[\partial\Psi/\partial X \cos\theta - \partial\Psi/\partial Z \sin\theta]. \tag{11}$$

Finally, evaluating the integral in $\psi = (R/2\pi v)\int d\tau/X^2$ in the r, θ, ϕ coordinate system, making an integration by parts in the integral with respect to $d\tan\theta$, and using Eq. (8), we can determine $\psi(\Psi) = \Upsilon(2\pi)$, where

$$\partial\Upsilon(\theta)/\partial\theta|_\Psi = -rRZ \, \partial\Psi/\partial Z/vX[\partial\Psi/\partial X \cos\theta - \partial\Psi/\partial Z \sin\theta]. \tag{12}$$

These four equations are integrated using a Newton–Cotes predictor–corrector algorithm employing cubic spline fits to the equilibrium values of $\Psi(X, Z)$. Since $\psi = 1$ on the plasma boundary, Eq. (12) determines v as well as $\psi(\Psi)$. Similarly, Eq. (9) determines $\Theta(\theta)$ and thus $f(\psi)$ on each surface since $\Theta = 2\pi$ when $\theta = 2\pi$. Using Eqs. (10) and (11) to determine $X(\psi, \Theta)$ and $Z(\psi, \Theta)$, it is easy to evaluate the metric tensor elements, since

$$|\nabla\psi|^2 = (X^2/\mathscr{J}^2)[(\partial X/\partial\Theta)^2 + (\partial Z/\partial\Theta)^2], \tag{13}$$

$$\nabla\psi \cdot \nabla\Theta = -(X^2/\mathscr{J}^2)(\partial X/\partial\psi\, \partial X/\partial\Theta + \partial Z/\partial\psi\, \partial Z/\partial\Theta), \tag{14}$$

and

$$|\nabla\Theta|^2 = (X^2/\mathscr{J}^2)[(\partial X/\partial\psi)^2 + (\partial Z/\partial\psi)^2]$$
$$= [X^2/\mathscr{J}^2 + (\nabla\psi \cdot \nabla\Theta)^2]/|\nabla\psi|^2 \tag{15}$$

at each mesh point. This completes the problem of expressing the equilibrium in the operational coordinate system.

D. Projections

The next problem to be considered is the decomposition of the displacement vector in the plasma region. Keeping in mind the fact that the Lagrangian is diagonalized by the normal modes of the physical system, it seems essential to choose components of the perturbations that represent the polarizations of the various modes, at least in the low pressure, long wavelength limit that is reasonable for present-day devices. This is also likely to be the most troublesome regime since the range between the highest and lowest eigenvalues is most exaggerated here. Indeed, for typical parameters, the eigenvalue ω^2 ranges over eight orders of magnitude. The lowest frequency, or sound, modes consist of flow along the magnetic field lines. The next branch, that of shear-Alfvén waves, has nearly divergence-free motion perpendicular to the field. The fast magnetosonic mode is primarily due to this perpendicular compressibility.

The sound waves are well treated by the projection

$$\boldsymbol{\xi} = (\mathscr{J}\xi_\psi/gR^2B_0)\nabla\Theta \times \mathbf{B} + i(\mathscr{J}\xi_s/gR^2B_0)\mathbf{B} \times \nabla\psi + i(\tau/B_0)\mathbf{B}, \tag{16}$$

while, at least in the long wavelength limit, the shear-Alfvén and fast compressional branches are decoupled by the transformation

$$\begin{aligned}\xi_\psi &= \delta - 2\pi i\, \partial\zeta/\partial\Theta, \\ \xi_s &= 2\pi\, \partial\zeta/\partial\psi,\end{aligned} \tag{17}$$

so that

$$\nabla \cdot \boldsymbol{\xi} = \frac{1}{RX^2} \frac{\partial X^2 \delta}{\partial \psi} + \frac{vf \nabla \psi \cdot \nabla \Theta}{2\pi R^3 g} \frac{\partial \delta}{\partial \phi} + \frac{2\pi i}{RX^2} \left(\frac{\partial X^2}{\partial \Theta} \frac{\partial \zeta}{\partial \psi} - \frac{\partial X^2}{\partial \psi} \frac{\partial \zeta}{\partial \Theta} \right)$$
$$- \frac{ivf}{R^3 g} \left(|\nabla \psi|^2 \frac{\partial^2 \zeta}{\partial \phi\, \partial \psi} + \nabla \psi \cdot \nabla \Theta \frac{\partial^2 \zeta}{\partial \phi\, \partial \Theta} \right) + \frac{2\pi i Rf}{vX^2} \left(\frac{\partial \tau}{\partial \Theta} + q \frac{\partial \tau}{\partial \phi} \right). \tag{18}$$

III. Representation of the Normal-Mode Equations

A. Lagrangian

The problem of determining the spectrum can be posed as that of finding estimates of the eigenvalues ω^2 and eigenfunctions $\boldsymbol{\xi}$ that make the Lagrangian

$$L = \omega^2 K(\boldsymbol{\xi}^*, \boldsymbol{\xi}) - \delta W(\boldsymbol{\xi}^*, \boldsymbol{\xi}) \tag{19}$$

stationary with respect to variations of $\boldsymbol{\xi}$ (Bernstein et al., 1958). Here, $\mathrm{Re}[\boldsymbol{\xi}(\mathbf{r}) \exp(-i\omega t)]$ is the displacement of a fluid element from its equilibrium position $\mathbf{r}$ and $\omega^2 K$ and δW are the kinetic and potential energy functionals

$$2K = \int_p d\tau\, \rho |\boldsymbol{\xi}|^2, \tag{20}$$

$$2\delta W = \int_p d\tau (|\mathbf{Q} - \mathbf{B}(\boldsymbol{\xi} \cdot \nabla p)/B^2|^2 + [(\mathbf{J} \cdot \mathbf{B})/B^2] \mathbf{B} \times \boldsymbol{\xi}^* \cdot \mathbf{Q} - 2\boldsymbol{\xi} \cdot \nabla p\, \boldsymbol{\xi}^* \cdot \boldsymbol{\kappa}$$
$$+ \gamma p |\nabla \cdot \boldsymbol{\xi}|^2) + 2\int_s d\sigma |\boldsymbol{\xi} \cdot \mathbf{n}|^2 \mathbf{n} \cdot [\![\kappa B^2]\!] + \int_v d\tau |\nabla \times \mathbf{A}|^2, \tag{21}$$

with ρ the plasma density, $\mathbf{Q} \equiv \nabla \times (\boldsymbol{\xi} \times \mathbf{B})$ the perturbed magnetic field, $\boldsymbol{\kappa} = (\mathbf{B}/B) \cdot \nabla(\mathbf{B}/B)$ the local magnetic field line curvature, γ the ratio of specific heats, $[\![\ ]\!]$ denoting the jump in crossing the plasma–vacuum interface, and $\mathbf{A}$ the vector potential for the perturbed magnetic field in the vacuum region.

The admissible variational functions are those for which the displacement has a finite kinetic energy norm and the normal component of the perturbed magnetic field is continuous at the plasma–vacuum interface and vanishes at the vacuum wall. To keep the kinetic energy finite, ξ_ψ must vanish at the magnetic axis. Since δ and ζ are defined uniquely only up to an arbitrary function of Θ, this can be achieved consistently by making $\delta = \zeta = 0$ at this point.

The surface term δW_s contributes only when a surface current is present on the plasma–vacuum interface. Since such localized currents are seldom realistic, the purpose of incorporating this term into the formalism would be to facilitate comparison with special simplified models. Numerical evaluation of this term is straightforward, and the results enter the Lagrangian in the same place as the contributions from the vacuum region. Henceforth, we omit this term.

B. Decomposition of Expansion Functions

As noted in connection with Eq. (1), approximating ξ by a linear superposition of M linearly independent expansion functions leads to a matrix eigenvalue problem whose solution yields approximations $\omega^{(M)2}$, $a_m^{(M)}$ to ω^2 and a_m. We assume, without proof, that $\xi^{(M)}$ converges to a solution of Eq. (1) in the limit as $M \to \infty$. This is a reasonable assumption since it is known that convergence will certainly be achieved when δW is positive definite (Mikhlin, 1971), which it is when the plasma is stable. Note that Rayleigh's principle implies that, if $|\xi - \xi^{(M)}|$ is $O(\varepsilon)$, then $\omega^2 - \omega^{(M)2}$ is $O(\varepsilon^2)$ where $\varepsilon \to 0$ as $M \to \infty$.

It is convenient to use a Fourier series in Θ and ϕ,

$$\Phi(\mathbf{r}) = \sum_{l,n} \Phi_{l,n}(\psi) \exp[i(l\Theta - n\phi)]. \tag{22}$$

The different terms in n decouple and we can drop this subscript accordingly. Obviously, we cannot expect decoupling in Θ. It is useful to note, however, that most configurations of physical interest possess symmetry about some constant Z-plane. In this case, the real and imaginary terms do not couple.

We adopt a tent-function expansion to represent the ψ behavior of the δ_l's and the ζ_l's,

$$\Phi_{l,m}^{(M)} = M[(\psi - \psi_{m-1}) H(\psi - \psi_{m-1}) H(\psi_m - \psi) \\ - (\psi - \psi_{m+1}) H(\psi - \psi_m) H(\psi_{m+1} - \psi)], \tag{23}$$

and a step-function expansion for the τ_l's,

$$\Phi_{l,m}^{(M)} = H(\psi - \psi_{m-1}) H(\psi_m - \psi). \tag{24}$$

The Heaviside functions $H(p) \equiv (p + |p|)/2p$ make the $\Phi_{l,m}^{(M)}$'s vanish except in the intervals (ψ_{m-1}, ψ_{m+1}) in Eq. (23) or (ψ_{m-1}, ψ_m) in Eq. (24), where $\psi_m \equiv m/M$. This could easily be generalized to unequally spaced meshes.

It should be noted that the expansion functions we use for the τ_l's are not continuous. This is reasonable since no ψ-derivatives of this component enter the Lagrangian and because with this choice the τ_l's can more closely couple with $\mathbf{V} \cdot \boldsymbol{\xi}_\perp$ to separate the slow wave from the other branches. The use of discontinuous expansion functions has been studied in detail for the special case of a diffuse linear pinch (Appert *et al.*, 1975).

C. Matrix Elements

Here we describe the contribution to the eigenvalue matrix, Eq. (1), from the plasma region and then discuss the formalism in the vacuum.

1. *Plasma*

The evaluation of the matrix elements of Eq. (1) is straightforward but somewhat tedious. We have carried through the algebra analytically and have checked the results by means of the MACSYMA symbolic programming facility (Rosen *et al.*, 1974). Here we present the results in integral form. The specific matrix elements are evaluated by inserting the expansions of Eqs. (22) through (24). Thus, treating δ, ζ and their ψ-derivatives δ', ζ' as independent,

$$K = \frac{\omega^2 v^3}{2R} \sum_{l'l} \int_0^1 d\psi \int_0^{2\pi} d\Theta \exp[i(l - l')\Theta]$$

$$\times (\zeta_{l'}^* \zeta_{l'}^{*\prime} \delta_{l'}^* \delta_{l'}^{*\prime} \tau_{l'}^*) \begin{bmatrix} \Gamma & \Sigma & \Upsilon & 0 & 0 \\ \Sigma^\dagger & \Xi & \Lambda & 0 & 0 \\ \Upsilon^\dagger & \Lambda^\dagger & \Pi & 0 & 0 \\ 0 & 0 & 0 & 0 & 0 \\ 0 & 0 & 0 & 0 & \Omega \end{bmatrix} \begin{bmatrix} \zeta_l \\ \zeta_l' \\ \delta_l \\ \delta_l' \\ \tau_l \end{bmatrix}, \qquad (25)$$

and

$$\delta W_p = \frac{v B_0^2}{2R} \sum_{l'l} \int_0^1 d\psi \int_0^{2\pi} d\Theta \exp[i(l - l')\Theta]$$

$$\times (\zeta_{l'}^* \zeta_{l'}^{*\prime} \delta_{l'}^* \delta_{l'}^{*\prime} \tau_{l'}^*) \begin{bmatrix} A & P & U & X & Z \\ P^\dagger & B & Q & V & Y \\ U^\dagger & Q^\dagger & C & R & W \\ X^\dagger & V^\dagger & R^\dagger & D & S \\ Z^\dagger & Y^\dagger & W^\dagger & S^\dagger & E \end{bmatrix} \begin{bmatrix} \zeta_l \\ \zeta_l' \\ \delta_l \\ \delta_l' \\ \tau_l \end{bmatrix}, \qquad (26)$$

with

$$\Gamma = (ll'\rho X^4/R^6 g^2)[R^2 g^2 |\nabla\Theta|^2 + f^2(\nabla\psi \cdot \nabla\Theta)^2],$$

$$\Xi = \rho X^6 B^2 |\nabla\psi|^2/R^6 g^2 B_0^2,$$

$$\Pi = \Gamma/4\pi^2 ll',$$

$$\Omega = \rho X^2 B^2/v^2 B_0^2,$$

$$\Sigma = -il'\rho X^6 B^2 \nabla\psi \cdot \nabla\Theta/R^6 g^2 B_0^2,$$

$$\Lambda = -\Sigma/2\pi l',$$

$$\Upsilon = \Gamma/2\pi l;$$

$$A = \frac{4\pi^2 ll'}{R^2}\left[\frac{(l-nq)(l'-nq)f^2}{R^2 g^2}[R^2 g^2 |\nabla\Theta|^2 + f^2(\nabla\psi \cdot \nabla\Theta)^2]\right.$$
$$+ f'^2 |\nabla\psi|^2 + \frac{f'}{f}\left(\frac{X^2 p'}{B_0^2} + R^2 gg'\right)$$
$$- ff' \frac{\partial}{\partial\Theta}\nabla\psi \cdot \nabla\Theta - \frac{f^4}{R^2 g^2}\nabla\psi \cdot \nabla\Theta \frac{\partial^2}{\partial\Theta^2}\nabla\psi \cdot \nabla\Theta$$
$$\left.+ \frac{p'}{B_0^2}\frac{\partial X^2}{\partial\psi} + \frac{\gamma p}{X^2 B_0^2}\left(\frac{\partial X^2}{\partial\psi}\right)^2 + \frac{n^2 q^2 f^4 X^2 (B^2 + \gamma p)(\nabla\psi \cdot \nabla\Theta)^2}{R^4 g^4 B_0^2}\right],$$

$$B = \frac{4\pi^2}{R^4 g^2}\left[(l-nq)(l'-nq)\frac{X^2 B^2 f^2}{B_0^2}|\nabla\psi|^2 + \frac{n^2 q^2 f^4 X^2 (B^2 + \gamma p)|\nabla\psi|^4}{R^2 g^2 B_0^2}\right.$$
$$\left.- f^4 |\nabla\psi|^2 \frac{\partial^2}{\partial\Theta^2}|\nabla\psi|^2 + \frac{R^2 g^2 \gamma p}{X^2 B_0^2}\left(\frac{\partial X^2}{\partial\Theta}\right)^2\right],$$

$$C = A/4\pi^2 ll',$$

$$D = X^2(B^2 + \gamma p)/R^2 B_0^2,$$

$$E = 4\pi^2 (l-nq)(l'-nq) R^2 f^2 \gamma p/X^2 v^2 B_0^2,$$

$$P = -\frac{4\pi^2 l'}{R^4 g^2}\left\{i\left((l-nq)(l'-nq)\frac{f^2 X^2 B^2}{B_0^2}\nabla\psi \cdot \nabla\Theta\right.\right.$$
$$+ \frac{n^2 q^2 f^4 X^2 (B^2 + \gamma p)|\nabla\psi|^2 \nabla\psi \cdot \nabla\Theta}{R^2 g^2 B_0^2}$$
$$\left.- f^4 \nabla\psi \cdot \nabla\Theta \frac{\partial^2}{\partial\Theta^2}|\nabla\psi|^2 - \frac{R^2 g^2 \gamma p}{X^2 B_0^2}\frac{\partial X^2}{\partial\psi}\frac{\partial X^2}{\partial\Theta}\right)$$
$$- l\left[R^2 g^2\left(\frac{X^2 p'}{B_0^2} + R^2 gg'\right) + ff'|\nabla\psi|^2\right]$$

$$-f^4 \nabla\psi \cdot \nabla\Theta \frac{\partial}{\partial\Theta}|\nabla\psi|^2 + f^4|\nabla\psi|^2 \frac{\partial}{\partial\Theta}\nabla\psi \cdot \nabla\Theta\Big]$$

$$+ nq\Big[\frac{X^2B^2}{B_0^2}\Big(\frac{X^2p'}{B_0^2} + R^2gg' + ff'|\nabla\psi|^2\Big)$$

$$+ \frac{\gamma pf^2}{B_0^2}\Big(|\nabla\psi|^2 \frac{\partial X^2}{\partial\psi} + \nabla\psi \cdot \nabla\Theta \frac{\partial X^2}{\partial\Theta}\Big)\Big]\Big\},$$

$$Q = \frac{2\pi}{R^4g^2}\Big\{i\Big((l-nq)(l'-nq)\frac{f^2X^2B^2}{B_0^2}\nabla\psi \cdot \nabla\Theta$$

$$+ \frac{n^2q^2f^4X^2(B^2+\gamma p)|\nabla\psi|^2 \nabla\psi \cdot \nabla\Theta}{R^2g^2B_0^2}$$

$$- f^4 \nabla\psi \cdot \nabla\Theta \frac{\partial^2}{\partial\Theta^2}|\nabla\psi|^2 - \frac{R^2g^2\gamma p}{X^2B_0^2}\frac{\partial X^2}{\partial\psi}\frac{\partial X^2}{\partial\Theta}\Big)$$

$$+ l'\Big[R^2g^2\Big(\frac{X^2p'}{B_0^2} + R^2gg' + ff'|\nabla\psi|^2\Big)$$

$$- f^4 \nabla\psi \cdot \nabla\Theta \frac{\partial}{\partial\Theta}|\nabla\psi|^2 + f^4|\nabla\psi|^2 \frac{\partial}{\partial\Theta}\nabla\psi \cdot \nabla\Theta\Big]$$

$$- nq\Big[\frac{X^2B^2}{B_0^2}\Big(\frac{X^2p'}{B_0^2} + R^2gg' + ff'|\nabla\psi|^2\Big)$$

$$+ \frac{\gamma pf^2}{B_0^2}\Big(|\nabla\psi|^2 \frac{\partial X^2}{\partial\psi} + \nabla\psi \cdot \nabla\Theta \frac{\partial X^2}{\partial\Theta}\Big)\Big]\Big\},$$

$$R = \frac{inqf^2X^2(B^2+\gamma p) \nabla\psi \cdot \nabla\Theta}{R^4g^2B_0^2}$$

$$+ \frac{1}{R^2}\Big(\frac{X^2p'}{B_0^2} + R^2gg' + ff'|\nabla\psi|^2 - f^2\frac{\partial}{\partial\Theta}\nabla\psi \cdot \nabla\Theta + \frac{\gamma p}{B_0^2}\frac{\partial X^2}{\partial\psi}\Big),$$

$$S = -2\pi(l-nq)\gamma pf/vB_0^2,$$

$$U = A/2\pi l,$$

$$V = -\frac{2\pi i}{R^2}\Big(f^2\frac{\partial}{\partial\Theta}|\nabla\psi|^2 + \frac{\gamma p}{B_0^2}\frac{\partial X^2}{\partial\Theta}\Big) - \frac{2\pi nqf^2X^2(B^2+\gamma p)|\nabla\psi|^2}{R^4g^2B_0^2},$$

$$W = -2\pi(l-nq)\frac{\gamma pf}{vB_0^2}\Big(\frac{inqf^2 \nabla\psi \cdot \nabla\Theta}{R^2g^2} + \frac{1}{X^2}\frac{\partial X^2}{\partial\psi}\Big),$$

$$X = 2\pi l'R,$$

$$Y = 4\pi^2(l - nq)\frac{\gamma p f}{v B_0{}^2}\left(\frac{i}{X^2}\frac{\partial X^2}{\partial \Theta} + \frac{nqf^2|\nabla\psi|^2}{R^2 g^2}\right),$$

$$Z = 2\pi l' W.$$

Here the daggers denote Hermitian conjugates.

2. Vacuum

It is clear from Eqs. (19), (20), and (21) that the vacuum contribution to the Lagrangian should be obtained by extremizing δW_v with respect to **A**, keeping the normal component of the perturbed field $\mathbf{n} \cdot \nabla \times \mathbf{A}$ equal to $\mathbf{n} \cdot \mathbf{Q}$ on the interface and $\mathbf{n} \cdot \nabla \times \mathbf{A} = 0$ on the vacuum wall. Lüst and Martensen (1960) have shown that this contribution can be expressed in terms of surface integrals over the plasma–vacuum interface

$$2\delta W_v = -\int_0^{2\pi}\int_0^{2\pi} \mathscr{J}\chi^* \mathbf{Q} \cdot \nabla\psi \, d\Theta \, d\phi$$

$$+ \sum_{i=1,2} \mathscr{L}_i^{-1}\left|\int_0^{2\pi}\int_0^{2\pi} \mathscr{J}\xi \cdot \nabla\psi \, \mathbf{B} \cdot \mathscr{D}_i \, d\Theta \, d\phi\right|^2. \tag{27}$$

Here, χ^* is a single-valued scalar potential that satisfies $\nabla^2\chi^* = 0$, $\mathbf{n} \cdot \nabla\chi^* = \mathbf{n} \cdot \nabla \times \mathbf{A}^*$ at the boundaries, and the $\mathscr{D}_i$'s are two orthogonal vectors such that $\nabla \times \mathscr{D}_i = \nabla \cdot \mathscr{D}_i = 0$ in the vacuum region, $\mathscr{D}_i \cdot \mathbf{n} = 0$ on the plasma–vacuum interface and on the vacuum wall, and $\oint \mathscr{D}_i \cdot \mathbf{dl}_k \propto \delta_{ik}$ where the contours ($k = 1, 2$) are closed curves going respectively the long or short way around the plasma. The induction matrix elements are given by $\mathscr{L}_i = 4\pi \int d\tau \mathscr{D}_i^2$. In writing this, we have assumed that the vacuum region is triply connected. If there were coils in this region, we would have introduced more $\mathscr{D}_i$'s to account for the additional flux constraints. The last term in Eq. (27) arises from the condition that the perturbed magnetic field cannot be given completely in terms of a single-valued scalar potential. It obviously contributes only to axisymmetric modes, $n = 0$.

The first problem is to evaluate $\chi^*(\Theta)$ on the plasma–vacuum interface in terms of $\mathbf{n} \cdot \nabla\chi^*$. This is accomplished by using Green's theorem. Then

$$X_s\chi^*(X_s, Z_s) = -\frac{1}{2\pi}\oint X_t dl_t[\chi^*(X_t, Z_t)\mathbf{n} \cdot \nabla_t \mathscr{G}(X_t, Z_t | X_s, Z_s)$$

$$- \mathscr{G}(X_t, Z_t | X_s, Z_s)\mathbf{n} \cdot \nabla_t \chi^*(X_t, Z_t)] \tag{28}$$

where we take the principal part of the integral and evaluate the contour on a line of constant ϕ over both the interface and the wall. Here, $\mathscr{G}(X_t, Z_t | X_s, Z_s)$

satisfies

$$\nabla_t^2 \mathcal{G}(X_t, Z_t | X_s, Z_s) \exp(-in\phi) = -4\pi\delta(X_s - X_t)\delta(Z_s - Z_t)\exp(-in\phi),$$

$\mathbf{n} \cdot \nabla_t$ denotes the normal derivative pointing outward from the vacuum, and $X_t dl_t = -\oint_t d\Theta_t |\nabla\psi|_t$ on the interface and $X_t dl_t = -\oint_t d\Xi_t |\nabla\Upsilon|_t$ on the wall, Υ = const. Here Υ and Ξ are defined with the same Jacobian as in the plasma region. A convenient solution of this equation, which vanishes at infinity and thus necessitates integration of Eq. (28) over the vacuum wall, is

$$\mathcal{G}(X_t, Z_t | X_s, Z_s) = 2\pi^{1/2}\Gamma(\tfrac{1}{2} - n)(X_s/r)P^n_{-1/2}(w), \tag{29}$$

where

$$r \equiv [(X_s^2 - X_t^2)^2 + (Z_s - Z_t)^4 + 2(X_s^2 + X_t^2)(Z_s - Z_t)^2]^{1/4}, \tag{30}$$

and

$$w \equiv [X_s^2 + X_t^2 + (Z_s - Z_t)^2]/r^2.$$

This function is evaluated by expressing the Legendre functions $P_{\pm 1/2}(w)$ in terms of elliptic integrals, determined by Hastings' approximations, and using recurrence relations to obtain the $P^n_{-1/2}(w)$'s. It is useful to note that for $n=0$ this function is the Green's function for an axisymmetric filament source passing through the point X_s, Z_s in the cross section. It is not difficult to show that, if toroidal curvature is eliminated by assuming that $X - R \ll 1$, $\mathcal{G}(X_s, Z_s | X_t, Z_t) = 2K_0(2\pi n r_{s,t}/L)$, where in the argument of the Bessel function L is the periodicity length and $r_{s,t} \equiv [(X_s - X_t)^2 + (Z_s - Z_t)^2]^{1/2}$. If $n = 0$ or $L \to \infty$, $\mathcal{G}$ can be reduced to $-2\ln r_{s,t}$ since the boundary conditions are not essential in this formalism. Marder (1974) used this limit to study systems with arbitrary cross sections.

Before considering the solution of Eq. (28), we turn to the second term in Eq. (27). It is useful to choose

$$\mathcal{D}_1 = \nabla\phi/2\pi \tag{31}$$

as one of the vectors introduced to satisfy the flux constraints in Eq. (27). Then, using the techniques that lead to Eq. (12), we can write

$$\mathcal{L}_1 = 2\oint (Z/X)(\partial X/\partial\theta)\,d\theta \Big|_{\text{interface}}^{\text{wall}}. \tag{32}$$

We can express the poloidal vector $\mathcal{D}_2$ in terms of a scalar α,

$$\mathcal{D}_2 = \nabla\phi \times \nabla\alpha, \tag{33}$$

where $\mathbf{V} \cdot (1/X^2) \mathbf{V}\alpha = 0$, and $\alpha = 1$ on the plasma–vacuum interface and $\alpha = 0$ on the wall. We obtain an equation for $\mathbf{n} \cdot \mathbf{V}_t \alpha(X_t, Z_t)$, analogous to the one for $\chi^*(X_s, Z_s)$,

$$\alpha(X_s, Z_s) = -\frac{X_s}{2\pi} \oint \frac{dl_t}{X_t} [\alpha(X_t, Z_t) \mathbf{n} \cdot \mathbf{V}_t \mathcal{G}(X_t, Z_t | X_s, Z_s)$$
$$- \mathcal{G}(X_t, Z_t | X_s, Z_s) \mathbf{n} \cdot \mathbf{V}_t \alpha(X_t, Z_t)]. \quad (34)$$

Again we take the principal part of the integral and evaluate the contour on a line of constant ϕ over both the interface and the wall. Here $\mathcal{G}(X_t, Z_t | X_s, Z_s)$, which satisfies

$$\mathbf{V}_t \cdot (1/X_t^2) \mathbf{V}_t \mathcal{G}(X_t, Z_t | X_s, Z_s) = -(4\pi/X_s^2) \delta(X_s - X_t) \delta(Z_s - Z_t),$$

can be chosen to be

$$\mathcal{G}(X_t, Z_t | X_s, Z_s) = -4\pi(X_t/r) P^1_{-1/2}(w) \quad (35)$$

with r and w given in Eq. (30). Then

$$\mathscr{L}_2 = \frac{4\pi v}{R} \int_0^{2\pi} \mathbf{V}\psi \cdot \mathbf{V}\alpha \, d\Theta \quad (36)$$

integrated over the plasma interface.

We can now reduce the integral equations, Eqs. (28) and (34), to matrix equations by introducing Fourier decomposition in Θ. From Eq. (28), expressing $\chi(1, \Theta) = \Sigma_l \eta_l \exp il\Theta$ on the interface, and $\chi(\Upsilon_w, \Xi) = \Sigma_l \mu_l \exp il\Xi$ on the wall, we obtain a set of coupled equations

$$\sum_{l'} \{[(8\pi^3 R/v) \delta_{l,l'} - A_{l,l'}] \eta_{l'}^* - (iB_0/q) B_{l,l'} \xi_{l'}^* - C_{l,l'} \mu_{l'}^*\} = 0,$$

$$\sum_{l'} \{[(8\pi^3 R/v) \delta_{l,l'} - D_{l,l'}] \mu_{l'}^* - E_{l,l'} \eta_{l'}^* - (iB_0/q) F_{l,l'} \xi_{l'}^*\} = 0,$$

that determine these Fourier components in terms of $\xi_l^*(1)$. Here

$$A_{l,l'} \equiv \int d\Theta_s \, d\Theta_t (X_t^2/X_s) \mathbf{V}\psi \cdot \mathbf{V}_t \mathcal{G}(X_t, Z_t | X_s, Z_s) \exp[i(l\Theta_s - l'\Theta_t)],$$

$$B_{l,l'} \equiv (l' - nq) \int d\Theta_s \, d\Theta_t (1/X_s) \mathcal{G}(X_t, Z_t | X_s, Z_s) \exp[i(l\Theta_s - l'\Theta_t)],$$

$$C_{l,l'} \equiv \int d\Theta_s \, d\Xi_t (X_t^2/X_s) \mathbf{V}\Upsilon \cdot \mathbf{V}_t \mathcal{G}(X_t, Z_t | X_s, Z_s) \exp[i(l\Theta_s - l'\Xi_t)],$$

$$D_{l,l'} \equiv \int d\Xi_s \, d\Xi_t (X_t^2/X_s) \nabla \Upsilon \cdot \nabla_t \mathscr{G}(X_t, Z_t | X_s, Z_s) \exp[i(l\Xi_s - l'\Xi_t)],$$

$$E_{l,l'} \equiv \int d\Xi_s \, d\Theta_t (X_t^2/X_s) \nabla \psi \cdot \nabla_t \mathscr{G}(X_t, Z_t | X_s, Z_s) \exp[i(l\Xi_s - l'\Theta_t)],$$

$$F_{l,l'} \equiv (l' - nq) \int d\Xi_s \, d\Theta_t (1/X_s) \mathscr{G}(X_t, Z_t | X_s, Z_s) \exp[i(l\Xi_s - l'\Theta_t)],$$

with due care being taken in evaluation over integrable singularities. Similarly, we obtain a set of equations for the components of $\mathbf{n} \cdot \nabla \alpha$, defined by

$$\nabla \psi \cdot \nabla \alpha(1, \Theta) = \Sigma_l \lambda_l \exp il\Theta \quad \text{on the interface}$$

and

$$\nabla \psi \cdot \nabla \alpha(\Upsilon_w, \Xi) = \Sigma_l \nu_l \exp il\Xi \quad \text{on the wall};$$

$$\Sigma_l [P_{l'l} \lambda_l + Q_{l'l} \nu_l] = T_{l'},$$

$$\Sigma_l [R_{l'l} \lambda_l + S_{l'l} \nu_l] = U_{l'}.$$

Here

$$P_{l'l} \equiv \int d\Theta_s \, d\Theta_t \, X_s \hat{\mathscr{G}}(X_t, Z_t | X_s, Z_s) \exp[i(l\Theta_t - l'\Theta_s)],$$

$$Q_{l'l} = R_{ll'}^* \equiv \int d\Theta_s \, d\Xi_t \, X_s \hat{\mathscr{G}}(X_t, Z_t | X_s, Z_s) \exp[i(l\Xi_t - l'\Theta_s)],$$

$$S_{l'l} \equiv \int d\Xi_s \, d\Xi_t \, X_s \hat{\mathscr{G}}(X_t, Z_t | X_s, Z_s) \exp[i(l\Xi_t - l'\Xi_s)],$$

$$T_{l'} \equiv \int d\Theta_s \, d\Theta_t \, X_s \nabla \psi \cdot \nabla_t \hat{\mathscr{G}}(X_t, Z_t | X_s, Z_s) \exp(-il'\Theta_s) - 4\pi^2 R \delta_{l,0}/v,$$

$$U_{l'} \equiv \int d\Xi_s \, d\Theta_t \, X_s \nabla \psi \cdot \nabla_t \hat{\mathscr{G}}(X_t, Z_t | X_s, Z_s) \exp(-il'\Xi_s).$$

We solve these equations implicitly so that, defining the matrices

$$G \equiv \frac{v}{8\pi^3 R} \left(I - \frac{v}{8\pi^3 R} D\right)^{-1} E,$$

$$H \equiv \frac{1}{4\pi^2 R} \left(I - \frac{v}{8\pi^3 R} D\right)^{-1} F,$$

$$K \equiv \frac{v}{4\pi^2 R} \left[I - \frac{v}{8\pi^3 R}(A + CG)\right]^{-1} \left(B + \frac{v}{2\pi} CH\right),$$

$$V \equiv (P - QS^{-1}R)^{-1}(T - QS^{-1}U),$$

we obtain for the contribution from the vacuum region

$$\delta W_v = (vB_0^2/2R) \sum_{l'l} (\zeta_{l'}^* \zeta_{l'}^{*\prime} \delta_{l'}^* \delta_{l'}^{*\prime} \tau_{l'}^*) \begin{bmatrix} \mathscr{A} & 0 & \mathscr{B} & 0 & 0 \\ 0 & 0 & 0 & 0 & 0 \\ \mathscr{B}^\dagger & 0 & \mathscr{C} & 0 & 0 \\ 0 & 0 & 0 & 0 & 0 \\ 0 & 0 & 0 & 0 & 0 \end{bmatrix} \begin{bmatrix} \zeta_l \\ \zeta_l' \\ \delta_l \\ \delta_l' \\ \tau_l \end{bmatrix} \qquad (37)$$

with

$$\mathscr{C} = \frac{(l-nq)}{q^2} K_{l',l} + \frac{v}{R\mathscr{L}_1} \delta_{l,0} \delta_{l',0} \delta_{n,0} + \frac{f^2 V_l V_{l'}}{8\pi R^2 V_0} \delta_{n,0},$$

$$\mathscr{B} = 2\pi l' \mathscr{C}$$

$$\mathscr{A} = 4\pi^2 ll' \mathscr{C},$$

with $\mathscr{L}_1$ given by Eq. (32).

D. Computation of the Spectrum and Associated Eigenmodes

Substituting the appropriate finite element expansions of Eqs. (23) and (24) into Eqs. (25), (26), and (37), we construct numerically matrix elements of the eigenvalue problem, Eq. (1). The complex eigenvectors of this equation have length $3M(2L+1)$ where the Fourier series are truncated at $l = \pm L$. The corresponding matrices can thus be very large, but are sparse since the overlap integral between linear piecewise finite elements m' and m vanishes except when $m' = m$ or $m \pm 1$. Additional zero contributions arise because of the form chosen for the τ's, since only elements with $m' = m$ survive. In the calculations performed to date, it has not been necessary to use this latter sparseness, and we treat the matrices as block tridiagonal in the index m, with $3(2L+1)(9LM - 6L + 5M - 3)$ nonzero elements. It should be noted that this expression is linear in M.

Numerical evaluation of the integrals involved in the expressions for the matrix elements in Eqs. (25), (26), and (37) is straightforward and easily carried out with simple quadrature formulas such as Simpson's and Filon's algorithms. When differentiation of the kernels is required, cubic spline interpolation is employed to keep the truncation error consistent with the accumulated errors in the quadrature formulas. Because of the large number of terms it is expedient to utilize the symmetries involved, particularly in performing the Fourier sine and cosine integrals, to conserve both storage and execution time. For example, when there is reflection symmetry around $Z = 0$, the line $\Theta = 0$ can be chosen to lie on this axis and the matrix elements are real.

Diagonalization of the matrix is easily carried out directly by computation of $K^{-1/2}$ (note that K is positive definite) and solution of the associated equation

$$K^{-1/2} \delta W K^{-1/2} - \omega^2 I = 0. \tag{38}$$

Standard routines are available in most library packages, but some massaging proves valuable to take advantage of the diagonally banded structure. Due to the basically quadratic convergence offered by Rayleigh's principle, our experience has been that the frequencies of the global unstable modes are well represented on relatively coarse meshes (small values of M). For the higher eigenvalues (where a variational principle exists in a subspace orthogonal to the lower modes), and particularly for modes near the marginal point, considerably larger values of M are required. It is worth remarking that, with fixed word length W, machine roundoff will set a limit on the number of accurate digits in the computed eigenvalues, given very roughly by $N = W - R$ where $R \approx \log_{10} |\omega_{\max}^2 / \omega_{\min}^2|$. Since increasing the number of expansion functions results in extending the range of frequencies (the fast magnetosonic branch has an accumulation point at $\omega^2 = \infty$), such considerations will eventually determine the practical accuracy of the results.

IV. Application

In any large program of numerical work, it is important to verify that the results are accurate. A major purpose of our work to date has been to investigate such questions in order to determine the usefulness of the representation.

A convenient representation for studying this consists of a straight cylindrical plasma column, with an elliptic cross section, confined in an axial magnetic field B_z, carrying a uniform axial current J_z, and embedded in a vacuum. To work with this model, we set $x = X - R$, $y = Z$, and $z = R\phi$, let R go to infinity, and then introduce a periodicity length L. Then, in Eq. (3),

$$f(\psi) = b^2 a^2 L J_z / 4\pi B_0 (b^2 + a^2)$$

is a constant and

$$B_0^2 [g^2(\psi)]' = -[p'(\psi) + 2\pi f B_0 J_z / L].$$

From Eq. (2), we take the magnetic surfaces to be similar ellipses

$$\psi = x^2/a^2 + y^2/b^2. \tag{39}$$

The ψ, Θ, z coordinate system associated with Eq. (4) is

$$x = a\psi^{1/2} \cos \Theta, \tag{40}$$

$$y = b\psi^{1/2} \sin \Theta. \tag{41}$$

The special case where $p(\psi)/B_0^2 \sim (a/L)^2 \ll 1$, which provides a good approximation of typical Tokamak configurations, is amenable to analytic solution (Dewar et al., 1974).

The most salient features concerning the stability of this system are illustrated in Fig. 2, which shows the values of ω^2 for the lowest even modes as a function of nq for several values of ellipticity. The numerical results are in excellent agreement with the analytic ones for constant plasma density (solid lines). For nonconstant density, convergence occurs with $M \sim 5$ (Dewar et al., 1975). As has been shown analytically, the l value of the most unstable mode varies as a function of nq—for fixed ellipticity, higher values of l are required for convergence of the representation when q is increased. Similar conclusions follow for fixed q as the ellipticity is increased, as indicated in Fig. 3 where we

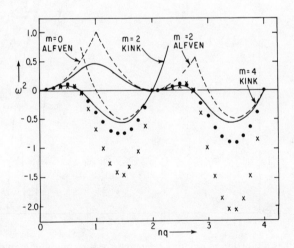

FIG. 2. The lowest even branches of the $m = 0, 2,$ and 4 modes for the uniform mass density case in a long, low pressure, constant current, elliptic plasma column, embedded in a large axial field, for $a/b = 1.0$ (dashed lines) and 0.5 (solid lines). Also shown are numerical results with a parabolic mass density profile such that $\rho(1)/\rho(0) = 0.25$ (dots) and a Gaussian profile such that $\rho(1)/\rho(0) = 0.044$ (crosses). The unit of ω^2 is normalized in terms of the average density. From Dewar et al. (1974).

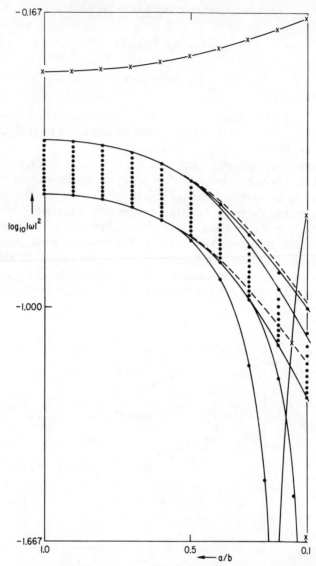

FIG. 3. $|\omega^2|$ as a function of a/b for a long, low pressure, constant current, elliptic configuration with $q = 1.5$ and a density distribution $\rho = 0.8 + 0.4\psi$. The upper curve is the unstable $l = 2$ kink, crosses denoting unstable eigenvalues. The inner solid curves mark the boundaries of the shear-Alfvén continuum as calculated by truncating the Fourier harmonics at $l = \pm 6$. The other solid lines denote discrete modes emerging from the continuum. In particular, the $l = 4$ kink becomes unstable at $a/b = 0.21$ and the $l = 6$ kink at $a/b = 0.12$. The $l = 2$ and 4 kink results are also obtained from calculations truncated at $l = \pm 4$ but the $l = 6$ mode is missing. The dashed curves show that the calculated continuum boundaries are displaced upward by such an approximation.

show modes from the lower part of the spectrum as a function of a/b. For $q = 1.5$, the dominant unstable mode has $l = 2$; as a/b is decreased additional modes with higher l values are driven unstable and more Fourier harmonics are required to represent this effect accurately. Further, for large ellipticities, higher values of l are required for convergence of the limits of the continua. There are no simple general rules for determining where to truncate the representation. As is the case in most numerical work, insight gained from simple analytical models is invaluable as an alternative to the trial-and-error approach.

Continuous spectra are characterized by eigenfunctions having singular behavior somewhere inside the plasma. An analysis of the nature of such functions (Dewar et al., 1975) provided analytic expressions for the ratio of the jumps in the logarithmic derivatives of the different Fourier components of a mode in crossing such a singular surface. Such a mode is shown in Fig. 4

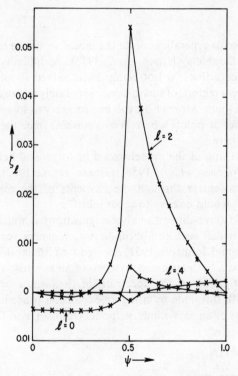

FIG. 4. The $l = 0, 2,$ and 4 Fourier components of ζ for a stable mode in the shear-Alfvén continuum for a long, low pressure, elliptic plasma column with $a/b = 0.5, q = 1.5,$ and $\rho(\psi) = 1.2 - 0.4\psi$, showing the logarithmic singularity predicted analytically to occur at $\psi_s = 0.51$. Agreement between estimates of the jump in the derivative across this surface of the various components, as evaluated computationally and analytically, agree to within 20% if six finite elements are used and to within 7% if $M = 10$. From Dewar et al. (1975).

for the low pressure, $a/b = 0.5$, $q = 1.5$, parabolic density model. The singularity occurs analytically at $\psi_s = 0.51$. The efficacy of the finite-element representation is demonstrated by the good agreement with analysis achieved with a coarse mesh.

V. Discussion

The formalism described in this article is applicable for studying the spectrum of any closed axisymmetric equilibrium configuration in which there is only one closed magnetic field line (the magnetic axis) in the region occupied by a plasma that is described by an ideal, scalar-pressure, magnetohydrodynamic model. Here we briefly comment on several directions in which the work can and should be extended.

A. Geometry

An obvious useful generalization of the model would be to multiple-region systems such as Doublets (Jensen et al., 1975). A difficulty associated with functions of the coordinate ψ not being single valued should be handled by treating the different regions of such a device separately and providing matching conditions to join them. More effort will be necessary to solve the fundamental problem that arises at points where $|\nabla\psi|$ vanishes, since the coordinate Θ is not well defined there.

A second extension of the model would be to general helically symmetric configurations (Johnson et al., 1958; Greene et al., 1962). Since the only essential difference enters through the elements of the metric tensor, this useful application would be easy to accomplish.

Application to three-dimensional configurations is much more difficult. A basic problem is that such equilibria do not, in general, exist (Grad, 1967). However, Greene and Johnson (1961) showed that Stellarator equilibria exist to all orders, and can be determined, in terms of an expansion in the rotational transform. Investigation of the spectrum of Stellarator or Scyllac devices could be done with this code by using an expansion procedure to reduce the calculation to that of an axisymmetric model (Johnson and Greene, 1961).

B. Resistive Models

Although stability with respect to ideal magnetohydrodynamic modes is essential for magnetic confinement of a plasma, it may not be sufficient. As an example of how this work can be extended to the study of more general problems, it is useful to consider tearing modes, where the finite resistivity of the plasma allows the magnetic field lines to break along resonant surfaces, on which the equilibrium field lines are closed. These unstable modes could cause significant problems since they can have growth rates that scale as fractional powers of the resistive diffusion time. They have received extensive analytic study (Glasser *et al.*, 1975) and expressions are available to determine the growth rates, given sufficient information concerning the behavior of the $\omega^2 = 0$ modes in the ideal regions on either side of the resistive layer. It should be straightforward to obtain this information by modifying the present program.

C. Generalized Pressure Models

The simple scalar-pressure model that is used in the derivation of the energy principle may not provide a good description of flow along the magnetic field lines. Current interest in two-component plasma devices, in which a directed beam of high energy particles is injected into the plasma to provide heating, makes the study of more realistic models desirable. An energy principle using a double-adiabatic pressure model in which both perpendicular and parallel pressure components are treated separately was obtained by Bernstein *et al.* (1958). Kruskal and Oberman (1958) derived a variational principle using a guiding center model in which flow perpendicular to the field is represented by a displacement vector, but a distribution function is retained to describe the parallel motion (see also Connor and Hastie, 1974). The techniques employed in this article should be extended to utilize these models.

D. Nonlinear Extensions

Knowledge of the linear perturbations about an equilibrium configuration is useful. However, it is essential that the nonlinear behavior of the unstable modes be investigated. Since the most important of these instabilities are nearly axisymmetric or helically symmetric, it is reasonable to solve the time-dependent equations numerically on a two-dimensional model to determine the time evolution. Considerable progress has been made in following the nonlinear kink mode (White *et al.*, 1975). It is our belief that considerable care should be taken in the formulation of such programs in the selection of the coordinate system and the projection of the vectors so that the results are not dominated by numerical errors. Techniques have been developed for following the time evolution in low pressure models (Winsor *et al.*, 1970; Grimm and Johnson, 1975; Potter, this volume). This work should be extended.

The three-dimensional problem is much more difficult. Some progress has been made (Brackbill, this volume) in following the evolution of the approximately isotropic modes in high-pressure configurations where the frequencies of the fast waves and the instability growth rates are not disparate. The lack of static equilibria, the formation of magnetic islands, and related phenomena make it essential in any such calculation to utilize a formalism capable of treating these problems. Otherwise, there will be no way of ascertaining to what extent the results are related to real physics or to numerology.

To summarize this work, in this paper we have shown that the Galerkin method utilizing finite elements provides a useful formalism for obtaining eigenvalues of a two-dimensional system described by a partial differential equation. We have discussed the necessity of carefully selecting the coordinate system and projections so as to make the model represent the physics and not be dominated by numerical difficulties, and have described an explicit model that accomplishes this. The success we have had indicates that this approach should be applicable to a wide variety of problems.

ACKNOWLEDGEMENTS

Many people deserve recognition when such an extensive program is undertaken. In particular, we must acknowledge the contributions of Dr. M. S. Chance, who has made significant application of the program, particularly for carrying out the calculations presented in Fig. 2. Dr. R. L. Dewar provided considerable help in the formulation and development of the program. Dr. B. Rosen rendered valuable assistance to the implementation of the model, especially the mapping procedure. We are indebted to Drs. D. A. Baker, J. P. Freidberg L. W. Mann, B. M. Marder, F. Troyon, E. Rebhan, and K. Lackner for helpful discussions and suggestions which influenced the direction of this work significantly and to many members of the Princeton University Plasma Physics Laboratory staff, in particular, Drs. P. H. Rutherford, C. R. Oberman, and E. A. Frieman, for their enthusiastic interest and support. Mr. R. F. Kluge contributed significantly to the programming. Finally, we must express our appreciation to Ms. E. F. Engelbart for capably typing this manuscript.

The work was supported by the U.S. Atomic Energy Commission, Contract AT(11-1)-3073 with Princeton University. It used computer facilities provided in part by National Science Foundation Grant NSF-GP 579, and some supported by Advanced Research Projects Agency, Grant DA-ARO-D-31-124-73-G 138. We are indebted to the Los Alamos Scientific Laboratory for several library subroutines.

REFERENCES

Appert, K., Berger, D., Gruber, R., Troyon, F., and Rappaz, J. (1974a). *Z. Angew. Math. Phys.* **25**, 229–240.
Appert, K., Gruber, R., and Vaclavik, J. (1974b). *Phys. Fluids* **17**, 1471–1472.
Appert, K., Berger, D., Gruber, R., and Rappaz, J. (1975). *J. Comput. Phys.* **18**, 284–299.

Artsimovich, L. A., and Shafranov, V. D. (1972). *Pis'ma Zh. Eksp. Teor. Fiz.* **15**, 72–76; *Sov. Phys.—JETP Lett.* **15**, 51–54 (1972).
Baker, D. A., and Mann, L. W. (1972). *Bull. Amer. Phys. Soc.* [2] **17**, 847.
Bateman, G., Schneider, W., and Grossmann, W. (1974). *Nucl. Fusion* **14**, 669–683.
Bernstein, I. B., Frieman, E. A., Kruskal, M. D., and Kulsrud, R. M. (1958). *Proc. Roy. Soc., Ser. A* **244**, 17–40.
Boyd, T. J. M., Gardner, G. A., and Gardner, L. R. T. (1973). *Nucl. Fusion* **13**, 764–766.
Chance, M. S., Dewar, R. L., Glasser, A. H., Greene, J. M., Grimm, R. C., Jardin, S. C., Johnson, J. L., Rosen, B., Sheffield, G. V., and Weimer, K. E. (1975). In "Plasma Physics and Controlled Nuclear Fusion Research 1974," Vol. 1, pp. 463–472. IAEA, Vienna.
Connor, J. W., and Hastie, R. J. (1974). *Phys. Rev. Lett.* **33**, 202–205.
Dewar, R. L., Grimm, R. C., Johnson, J. L., Frieman, E. A., Greene, J. M., and Rutherford, P. H. (1974). *Phys. Fluids* **17**, 930–938.
Dewar, R. L., Greene, J. M., Grimm, R. C., and Johnson, J. L. (1975). *J. Comput. Phys.* **18**, 132–153.
Friedberg, J. P., and Marder, B. M. (1973). *Phys. Fluids* **16**, 247–253.
Frieman, E. A., Greene, J. M., Johnson, J. L., and Weimer, K. E. (1973). *Phys. Fluids* **16**, 1108–1125.
Furth, H. P., Rutherford, P.H., and Selberg, H. (1973). *Phys. Fluids* **16**, 1054–1063.
Glasser, A. H., Greene, J. M., and Johnson, J. L. (1975). *Phys. Fluids* **18**, 875–888.
Goedbloed, J. P. (1974). *Bull. Amer. Phys. Soc.* [2] **19**, 941.
Goedbloed, J. P., and Hagebeuk, H. J. L. (1972). *Phys. Fluids* **15**, 1090–1101.
Grad, H. (1967). *Phys. Fluids* **10**, 137–154.
Grad, H. (1973). *Proc. Nat. Acad. Sci. U.S.* **70**, 3277–3281.
Grad, H., and Hogan, J. (1970). *Phys. Rev. Lett.* **24**, 1337–1340.
Greene, J. M., and Johnson, J. L. (1961). *Phys. Fluids* **4**, 875–890.
Greene, J. M., and Johnson, J. L. (1962). *Phys. Fluids* **5**, 510–517.
Greene, J. M., and Johnson, J. L. (1965). *Advan. Theor. Phys.* **1**, 195–244.
Greene, J. M., Johnson, J. L., Kruskal, M. D., and Wilets, L. (1962). *Phys. Fluids* **5**, 1063–1069.
Grimm, R. C., and Johnson, J. L. (1975). *J. Comput. Phys.* **17**, 192–208.
Jensen, T. H., Fisher, R. K., Hsieh, C. L., Mahdavi, M. A., Vanek, V., and Ohkawa, T. (1975). *Phys. Rev. Lett.* **34**, 257–260.
Johnson, J. L., and Greene, J. M. (1961). *Phys. Fluids* **4**, 1417–1426.
Johnson, J. L., Oberman, C. R., Kulsrud, R. M., and Frieman, E. A. (1958). *Phys. Fluids* **1**, 281–296.
Kerner, W., and Tasso, H. (1975). In "Plasma Physics and Controlled Nuclear Fusion Research 1974," Vol. 1, pp. 475–484. IAEA, Vienna.
Kruskal, M. D., and Oberman, C. R. (1958). *Phys. Fluids* **1**, 275–280.
Laval, G., Pellat, R., and Soule, J. S. (1974). *Phys Fluids* **17**, 835–845.
Lüst, R., and Martensen, E. (1960). *Z. Naturforsch. A* **15**, 706–713.
Marder, B. M. (1974). *Phys Fluids* **17**, 634–639.
Mercier, C. (1962). *Nucl. Fusion, Suppl. Pt.* **2**, 801–808.
Mikhlin, S. G. (1965). "The Problem of the Minimum of a Quadratic Functional," Sects. 11–13. Holden-Day, San Francisco, California.
Mikhlin, S. G. (1971). "The Numerical Performance of Variational Methods," Chapter 7. Walters-Noordhoff, Gröningen, The Netherlands.
Ohta, M., Shimomura, Y., and Takeda, T. (1972). *Nucl. Fusion* **12**, 271–272.
Roberts, K. V., and Potter, D. E. (1970). *Methods Comput. Phys.* **9**, 339–420.
Robinson, D. C. (1971). *Plasma Phys.* **13**, 439–462.

Rosen, B., Greene, J. M., Grimm, R. C., and Johnson, J. L. (1974). *Bull. Amer. Phys. Soc.* [2] **19**, 940.
Schwartz, C. (1963). *Methods Comput. Phys.* **2**, 241–266.
Shafranov, V. D. (1970). *Zh. Tekh. Fiz.* **40**, 241–253; *Sov. Phys.—Tech. Phys.* **15**, 175–183 (1970).
Sinclair, R. M., Yoshikawa, S., Harries, W. L., Young, K. M., Weimer, K. E., and Johnson, J. L. (1965). *Phys. Fluids* **8**, 118–133.
Sykes, A. and Wesson, J. A. (1974). *Nucl. Fusion* **14**, 645–648.
Takeda, T., Shimomura, Y., Ohta, M., and Yoshikawa, M. (1972). *Phys. Fluids* **15**, 2193–2201.
Tataronis, J., and Grossman, W. (1972). *Bull. Amer. Phys. Soc.* [2] **17**, 847.
von Goeler, S., Stodiek, W., and Sauthoff, N. (1974). *Phys. Rev. Lett.* **33**, 1201–1203.
White, R., Monticello, D., Rosenbluth, M. N., Strauss, H., and Kadomtsev, B. B. (1975). *In* "Plasma Physics and Controlled Nuclear Fusion Research 1974," Vol. 1, pp. 495–503. IAEA, Vienna.
Winsor, N. K., Johnson, J. L., and Dawson, J. M. (1970). *J. Comput. Phys.* **6**, 430–448.

Collective Transport in Plasmas

JOHN M. DAWSON

PHYSICS DEPARTMENT
UNIVERSITY OF CALIFORNIA AT LOS ANGELES
LOS ANGELES, CALIFORNIA

HIDEO OKUDA

PLASMA PHYSICS LABORATORY
PRINCETON UNIVERSITY
PRINCETON, NEW JERSEY

and

BERNARD ROSEN

DEPARTMENT OF PHYSICS
STEVENS INSTITUTE OF TECHNOLOGY
HOBOKEN, NEW JERSEY

I. Introduction	282
II. The Simulation Model	283
A. Introduction to the Model	283
B. Determination of Fields and Forces	284
C. The Particle Advancement Method	286
D. A Two and One-Half Dimensional Model with Magnetic Mirroring	287
E. Some Coding Developments	288
III. Elementary Theory of Convective Diffusion in a Uniform Thermal Plasma	289
A. Diffusion in Two Dimensions	289
B. Diffusion for Three Dimensions	292
IV. The Simulation of Plasma Diffusion across a Magnetic Field (Uniform Thermal Plasma)	295
A. Two-Dimensional Plasma	295
B. Two-Dimensional Electron Diffusion	299
C. Three-Dimensional Transport	301
V. Simulation of Diffusion in Nonuniform Plasmas	310
A. Convective Diffusion in an Inhomogeneous Plasma	310
B. Energy Transport across Magnetic Field by Plasma Waves	314
C. Neoclassical Diffusion in a Toroidal Magnetic Field	319
References	325

I. Introduction

FROM THE EARLIEST DAYS of the investigation of magnetically confined plasmas, it has often been observed that their transport across the field was often orders of magnitudes greater than could be accounted for by binary encounters (Bohm, 1949). This has generally been attributed to turbulence created by plasma instabilities (Hockney, 1966b) or by the process involved in its production and heating (Harries, 1970). It has been thought that if a stable plasma could be created and if the processes of generation were gentle enough, the transport would decrease to that predicted by binary encounter theory.

This article will report on recent studies of plasma and heat transport across magnetic fields by the use of computer simulation. These studies indicate that the problems of anomalous transport go much deeper than had been expected, for they show that collective processes can dominate the transport even for plasmas in thermal equilibrium. They thus point up serious weaknesses in existing theories of plasma transport. At the same time they exhibit many of the features found in experiments. Actual quantitative comparison exists in only a few isolated cases (Okuda *et al.*, 1972; Okuda and Dawson, 1973a; Tamano *et al.*, 1973; Armentrout *et al.*, 1974). This is largely due to the complexities of most plasma experiments and to the limitations on present computers with regard to size and speed. Nevertheless, these studies give strong clues as to what is going on in real plasma; coordinated experiments and simulations designed to study isolated phenomenon should prove extremely fruitful. These studies also serve as powerful guides to theory. They are generally more precise than most experiments and can be diagnosed much better. They yield more detailed information on the relevant processes and one can isolate the mechanisms by turning off the fields due to select Fourier modes. Already some theoretical approximations have been shown to be in error while other simple approximations have been shown to yield results which fit the simulations with a good degree of accuracy (Okuda *et al.*, 1974; Chu *et al.*, 1975).

The magnitude of the enhanced transport is sensitive to many things; it is sensitive to the magnetic field strength, to the plasma density, to the magnetic geometry. Under some conditions, transport due to binary encounters dominates; this is particularly true at low magnetic fields and with ergodic field lines. As the magnetic field rises and particle confinement improves, the transport due to collisions decreases and that due to slow convective motion and to waves (which are not inhibited by the magnetic field) tends to dominate. These effects can even dominate by orders of magnitude in thermal plasmas (i.e., only thermal levels of fluctuations are excited) in high magnetic fields. Of course if these modes are excited above the thermal level, then their dominance can become exceedingly large so that it is little wonder that so many experiments witness large anomalous transport.

This paper will report on these computer studies of collective transport. In order to put the results in perspective, some theoretical framework is required; a brief discussion of this will be given for the uniform plasma case in Section III. For nonuniform plasmas, some theory will be given along with the discussion of the simulations; however, it is impossible to give an adequate treatment in this short section, and the interested reader should consult the references for more complete detail. Section II will be devoted to a review of some of the computational techniques used and the diagnostic methods employed. The remainder of the article will be devoted to summarizing the major results which have been achieved to date. The subjects covered will include two-, two and one-half-, and three-dimensional models of transport in thermal plasmas and studies of transport in nonuniform plasmas using two and one-half dimensional models. Both studies of plasma and heat transport will be discussed, although by far the major effort has been devoted to convective transport of plasma. Some discussions will be given of electron transport produced by high frequency waves.

II. The Simulation Model (Kruer *et al.*, 1973)

A. Introduction to the Model

In all of the so-called "particle-pushing" codes some method must be used to ascertain the forces which accelerate the plasma particles. At present this is most often done by deriving the electric and magnetic field strengths from the charge and current densities using Maxwell's field equations. Both the fields and sources are known at the grid points of the spatial mesh. The method used to accelerate the particles is tied to that used for the field determination in order that energy and momentum be conserved as well as is possible. In general, some compromise in either the conservation of energy or of momentum must be made (Langdon, 1970).

The numerical experiments described in this paper involve time-varying electric fields, while the magnetic field strength is static. It is assumed throughout that only the irrotational part of the electric field is present. This field is determined by Poisson's equation

$$\text{div}(E) = 4\pi\rho,$$
$$\text{curl}(E) = 0, \tag{1}$$

where ρ is the electric charge density. The magnetic fields generated by the particle currents are negligible. The charge density at the grid points is determined from the particle positions by the subtracted dipole scheme (SUDS) in the manner described below. The electric force on any given particle, which in this work is an extended charge with a Gaussian-shaped charge distribution, is calculated in turn by means of Fourier transforms. Finally the plasma particles are accelerated in the combined electric and magnetic fields by means of an algorithm that combines leapfrog and implicit features.

The remainder of this section has four parts; in the first we describe the dipole expansion scheme and its more efficient, if slightly less accurate, form—the SUDS algorithm. In the second the particle acceleration scheme is discussed, and in the third a model which can take care of mirroring magnetic field is introduced. Finally, some aspects of code optimization are covered.

B. Determination of Fields and Forces

If $n(r)$ denotes the number density of particles per unit cell and $F(r)$ describes the charge distribution of any single charge, then the charge density at r is given by

$$\rho(\mathbf{r}) = \int n(\mathbf{r}')F(\mathbf{r}-\mathbf{r}')d^N\mathbf{r}', \qquad (2)$$

where N is the number of space dimensions. In terms of the Fourier transforms of ρ, F, and n, which we denote as $\tilde{\rho}$, $\tilde{F}$, and $\tilde{n}$, respectively,

$$\tilde{\rho}(\mathbf{k}) = \tilde{F}(\mathbf{k})\tilde{n}(\mathbf{k}). \qquad (3)$$

The distinguishing feature of the dipole method is the means for assigning the number density at the grid points. Basically the idea is to represent $\rho(\mathbf{r})$ by a multipole expansion. Thus, if $\mathbf{R}_G$ represents the grid point nearest the particle position $\mathbf{r}_i$ then

$$\rho(\mathbf{r}) = \sum_i \sum_{l=0}^{\infty} (1/l!)((\mathbf{r}_i - \mathbf{R}_G) \cdot \nabla_{R_G})^l F(\mathbf{r} - \mathbf{R}_G). \qquad (4)$$

In doing this, we have taken the number density to be a sum of delta functions and then expanded $F(\mathbf{r} - \mathbf{r}_i)$ about $\mathbf{R}_G$. In the dipole expansion scheme we truncate this expansion after the second term so that

$$\rho(\mathbf{r}) = \sum_{R_G} \rho_{\text{NGP}}(\mathbf{R}_G)F(\mathbf{r} - \mathbf{R}_G) + \boldsymbol{\rho}_D(\mathbf{R}_G) \cdot \nabla_{R_G} F(\mathbf{r} - \mathbf{R}_G). \qquad (5)$$

At the grid point R_G we have then introduced a charge density

$$\rho_{\text{NGP}}(R_G) \equiv \sum_i{}' 1$$

and a dipole moment

$$\rho_D(R_G) \equiv {\sum_i}' (\mathbf{r}_i - \mathbf{R}_G)$$

where the sums are over those particles whose nearest grid point (NGP) is $\mathbf{R}_G$.

(The designation ρ_{NGP} points up a feature of this method, viz., that the transition to less accurate but quicker methods of charge calculation can be made by mere omission of portions of the code.)

In a similar manner the electric force, $\mathscr{F}_i$, on the ith extended particle can be approximated as

$$\mathscr{F}_i = \int F_M(\mathbf{r})[1 + (\mathbf{r}_i - \mathbf{R}_G) \cdot \nabla_{R_G}] E(\mathbf{R}_G - \mathbf{r}) d^N\mathbf{r}, \tag{6}$$

where $F_M(x) = F(-x)$.

In this original version of the dipole scheme, the charge density due to a particle was represented at the nearest grid point by several quantities (in particular the monopole and dipole moments), rather than by a single quantity at several grid points as in the CIC and the PIC schemes. There were fewer indexing operations required in charge assignments, but the drawback was the number of Fourier transforms required. One transform was required for the charge density and one for each component of the dipole moment, one for each component of the electric field and one for each component of the tensor $\mathbf{VE}$.

The subtracted dipole scheme (SUDS) replaces the gradient operations indicated above by finite differencing. This reduces considerably the extent of the computations. More explicitly, we have

$$\sum_{R_G} \rho_D(\mathbf{R}_G) \cdot \nabla_{R_G} F(\mathbf{r} - \mathbf{R}_G) \Rightarrow$$

$$\sum_{R_G} \sum_{\alpha=1}^{N} [\rho_D(R_G)_\alpha/2|\Delta X_\alpha|](F(\mathbf{r} - \mathbf{R}_G - \Delta\mathbf{X}_\alpha) - F(r - \mathbf{R}_G + \Delta\mathbf{X}_\alpha)),$$

where $\Delta\mathbf{X}_\alpha$ is the cell spacing in the α direction. In other words, the extended dipole moment at R_G is replaced by extended charges of opposing polarities at adjacent grid points. Note that in two dimensions, the grid points among which the charge is shared lie on a cross rather than on the corners of a square, and that furthermore, the sharing of charges is done by distributing the accumulated dipole moments rather than being done one particle at a time. Figure 1 exhibits the nature of this approximation for a Gaussian-shaped charge distribution such as we have used.

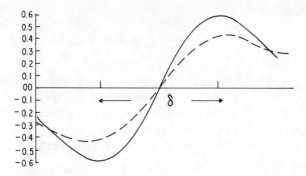

FIG. 1. Comparison of the derivative of a Gaussian (solid line) with the difference between two Gaussians (dashed line).

In summary the charge density in the SUDS scheme is given by[1]

$$\rho(\mathbf{r}) = \sum_{R_G} F(\mathbf{r} - \mathbf{R}_G)$$
$$\cdot \left\{ \rho_{\text{NGP}}(\mathbf{R}_G) + \sum_{\alpha=1}^{N} (1/2|\Delta \mathbf{X}_\alpha|) [\rho_D(\mathbf{R}_G - \Delta \mathbf{X}_\alpha) - \rho_D(\mathbf{R}_G + \Delta \mathbf{X}_\alpha)] \right\}. \quad (7)$$

The determination of $\rho(\mathbf{k})$ is made simply by multiplying the Fourier transform of F by that of the quantity in the curly brackets of the last equation. We can apply the same considerations to the determination of the electric force $\mathscr{F}$ on the extended particle. In this way, the gradient operations indicated in Eq. (6) can be converted to finite-differencing operations.

A recent study of this scheme has been made which indicates that its long wavelength behavior is approximately as accurate as that of the usual charge sharing methods (Chen and Okuda, 1975), insofar as grid effects are concerned.

C. THE PARTICLE ADVANCEMENT METHOD

The finite-difference method for integrating the equations of motion of the particles is the leapfrog scheme as modified by Hockney (1966a) and Buneman (1967) so as to apply to cases in which static magnetic fields are present. In brief the algorithm is

$$\frac{\mathbf{V}(T + DT) - \mathbf{V}(T)}{DT} = \frac{e\alpha}{m} \left[\mathbf{E}\left(T + \frac{DT}{2}\right) + \frac{\mathbf{V}(T + DT) + \mathbf{V}(T)}{2c} \times \mathbf{B}_0 \right] \quad (8a)$$

$$[\mathbf{r}(T + 3DT/2) - \mathbf{r}(T + DT/2)]/DT = \mathbf{V}(T + DT), \quad (8b)$$

[1] There is a sign error in the corresponding formula in Kruer et al. (1973).

where α is chosen so that the particle circles about the lines of magnetic force (B_0 is the magnetic field strength) with the correct gyrofrequency, $\omega_c \equiv eB_0/mc$, in the absence of electric forces. The quantity DT is the time-step. The quantity α is given by

$$\alpha = \tan(\omega_c DT/2)/(\omega_c DT/2). \tag{9}$$

It is relatively simple to make the method explicit by solving (8a) for $V(T + DT)$. The algorithm has the additional advantages of: (1) conserving kinetic energy in the absence of electric forces; (2) predicting that the average velocity approaches the drift value (given by $(\mathbf{E} + \mathbf{V} \times \mathbf{B}_0) = 0$) as the mass of the particles approaches zero.

D. A Two and One-Half Dimensional Model with Magnetic Mirroring (Kamimura and Dawson, 1975)

As we shall see in the discussion of the results, the free flow of charges along magnetic lines of force is a very important effect since it determines the lifetime of charge accumulation and of associated convective motions. Anything which inhibits this motion enhances the lifetime of the charge fluctuations and leads to enhanced diffusion. One such mechanism which occurs in almost all fusion devices is magnetic mirroring. We should like to include such effects in our models so as to study these effects. In order to do this we have constructed the model (Kamimura and Dawson, 1975) shown in Fig. 2. It is what is called a two and one-half dimensional model; it consists of finite size charged rods which are constrained to be normal to the x, y-plane and are allowed to have velocity components in the x-, y-, and z-directions (five-dimensional phase space). The magnetic field points in an arbitrary direction. A point on one of the charged rods executes cyclotron motion

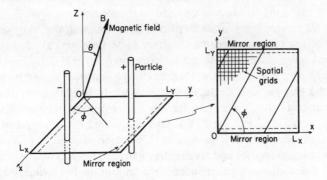

Fig. 2. Two and one-half dimensional electrostatic particle model with mirror.

about a line of force. When a particle enters the cell adjoining the y boundary, it is tested to determine whether or not its component of velocity parallel to the magnetic field exceeds some multiple of its velocity perpendicular to the magnetic field.

$$v_B \gtrless [(B_{max}/B_{min}) - 1]^{1/2} v_\perp = R v_\perp. \tag{10}$$

The quantity B_{max}/B_{min} is the mirror ratio; if $v_B > R v_\perp$ the particle is allowed to pass through the boundary and reenters the system on the opposite side; if $v_B < R v_\perp$ the sign of v_B is reversed. The value $R = 0$ corresponds to no mirror.

Two special diagnostics are used to view the results of these calculations. First, we may compute the x, y positions of the guiding centers of a large number of particles and plot their positions at various times. In this way we gain knowledge about how the particles move in the x, y-plane. Second, we may follow the full three-dimensional motion of the guiding center of a point on a rod. We follow that point which initially lies in the x, y-plane; we do not make use of the periodicity here, and thus keep track of the number of times the particle crosses the system. This point is then projected back along the magnetic field until it intersects the x, y-plane. The intersection point is plotted. This plot shows how the particle has moved across the field. When this is done for a large number of test particles, it gives information on how the plasma diffuses across the magnetic field. The diffusion across B is determined by computing the mean square displacement as a function of time of this B projection of the guiding centers of a large number of test particles.

E. Some Coding Developments (Rosen et al., 1972)

The determination of diffusion rates requires rather long runs with tens of thousands of iterations, so that there is a distinct advantage to optimizing the codes if resources are limited. In this section we wish to make mention of some of the programming techniques we used to reduce running time.

First, there are the obvious steps of reprogramming the most often-used portions of the code in machine language. The time for calculating Fourier transforms was reduced by using a form adapted for real data. Furthermore, the operation of converting the electric charge density Fourier transform to that of the electric field included the immediate formation of the correct components of the force needed in the leapfrog scheme (essentially $E_x + E_y \omega_c \cdot DT/2$, $E_y - E_x \omega_c DT/2$). This eliminated two multiplications and two additions per particle. The usual manipulation of including factors of the time-step in the velocities and forces eliminated additional multiplications.

Index calculations were reduced to a minimum by means of masking operations. These are particularly simple in the case where the number of cells in each direction is a power of 2. This masking of indices was used both in the distribution of the dipole moment in SUDS and in assigning particles to the correct cell. Two features of coding are hardware-dependent but are sufficiently useful to be mentioned. The gathering of data by the IBM 360-91 makes use of the following useful techniques: (1) The indices of the electric force array needed for the Ith $+ 1$ particle were calculated while the Ith particle was processed. (2) The x and y positions of the particles were loaded contiguously.

As a result of these methods the average time per particle per iteration was 20 μsec. This included time for the Fourier transforms and refers to a two-dimensional case in which the magnetic field was perpendicular to the X–Y plane.

In order to reduce computer storage requirements in the three-dimensional cases and in some of the larger two-dimensional cases only the potential due to the charge distribution was used. The forces were calculated from the potential by first and second differences.

III. Elementary Theory of Convective Diffusion in a Uniform Thermal Plasma

In order to understand the results of the numerical simulation, we require a theoretical picture of the processes taking place. While the theory presented here is rough, it points up the processes taking place and accounts reasonably well for the observations (Dawson *et al.*, 1971). It also touches on critical questions which we must ask of the computation and which a more refined theory must answer.

A. Diffusion in Two Dimensions

First we consider a two-dimensional plasma of charged rods parallel to B which we take to be in the z-direction, and we allow the rods to move only in the x, y-directions. We consider the system to be doubly periodic of size L^2, and the plasma is assumed to be a thermal plasma; i.e., there are only thermal fluctuations in it. As we shall see, the collective transport is due to convective shear flow of the plasma as illustrated in Fig. 3. We can Fourier analyze this motion and write

$$\mathbf{v}_T(\mathbf{r}) = \sum_\mathbf{k} \mathbf{v}_T(\mathbf{k}) e^{i\mathbf{k}\cdot\mathbf{r}}, \tag{11}$$

$$\mathbf{v}_T(\mathbf{k}) = \frac{-1}{L^2} \int \frac{\mathbf{k} \times (\mathbf{k} \times \mathbf{v}(\mathbf{r}))}{k^2} e^{-i\mathbf{k}\cdot\mathbf{r}} d^2r, \tag{12}$$

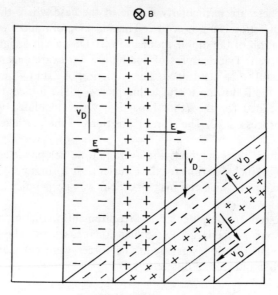

FIG. 3. Convective motion.

where $\mathbf{v}_T \cdot \mathbf{k} = 0$ and $\mathbf{v}_T$ is the transverse or shearing flow of the plasma. Associated with this motion is an electric field such that

$$[\mathbf{E}(\mathbf{k}) \times \mathbf{B}] c/B^2 = \mathbf{v}_T(\mathbf{k}). \tag{13}$$

There is an energy associated with the flow[2]

$$W_v(k) = \tfrac{1}{2}\rho \langle v_T^2(k) \rangle L^2, \tag{14}$$

$$\rho = \sum_i n_i \cdot m_i \quad \text{summed over species}$$

and also with the electric field

$$W_E(\mathbf{k}) = \langle \mathbf{E}^2(\mathbf{k}) \rangle L^2/8\pi. \tag{15}$$

Adding these two contributions, we get for the total energy of the disturbance,

$$\begin{aligned} W &= [\rho \langle v_T^2(k) \rangle/2 + \langle E^2(k) \rangle/8\pi] L^2 \\ &= (\langle E^2(k) \rangle/8\pi)(1 + 4\pi\rho c^2/B^2) L^2. \end{aligned} \tag{16}$$

[2] The total mass flow is considered here, and v is to be mass flow velocity.

Now according to equilibrium statistical mechanics, each Fourier mode should have energy $T/2$ (T is the temperature measured in energy units). Thus we have

$$(\langle \mathbf{E}^2(k)\rangle/8\pi)[1 + (4\pi\rho c^2/B^2)]L^2 = T/2 \tag{17}$$

or[3]

$$\langle \mathbf{E}^2(k)\rangle/8\pi = T/2L^2[1 + (4\pi\rho c^2/B^2)]. \tag{18}$$

The mean square flow velocity is

$$\langle \mathbf{v}_T^2(\mathbf{k})\rangle = 4\pi T c^2/L^2 B^2[1 + (4\pi\rho c^2/B^2)]. \tag{19}$$

We now estimate the diffusion caused by this flow as follows. The flow causes the particles to execute a random walk in space. As time progresses, the flow will change in a random manner as the eddies grow and decay, and there will be some correlation time or coherence time for each $\mathbf{k}$; call this time $\tau(\mathbf{k})$. During a coherence time, the particle is displaced a distance $\mathbf{v}_T(k)\tau(\mathbf{k})$, and its mean square displacement is $\langle \mathbf{v}_T^2(\mathbf{k})\tau^2(\mathbf{k})\rangle \simeq \langle \mathbf{v}_T^2(k)\rangle \tau^2(k)$.

During a time t the particle will make $t/\tau(\mathbf{k})$ such random steps, and thus its mean square displacement will be

$$\langle \Delta \mathbf{r}^2(k)\rangle = \langle \mathbf{v}_T^2(\mathbf{k})\rangle \tau(\mathbf{k})t. \tag{20}$$

Summing over all modes gives the mean square displacement as

$$\langle \Delta \mathbf{r}^2\rangle = \sum_{\mathbf{k}} \langle v_T^2(k)\rangle \tau(\mathbf{k})t. \tag{21}$$

Substituting in $\langle \mathbf{v}_T^2(k)\rangle$ from (19) gives

$$\langle \Delta \mathbf{r}^2\rangle = \frac{4\pi Tc^2 t}{L^2 B^2[1 + (4\pi\rho c^2/B^2)]}\sum \tau(k). \tag{22}$$

Converting from a sum to an integral and using the fact that the density of modes is $L^2 k\,dk/2\pi$ results in

$$\langle \Delta \mathbf{r}^2\rangle = \frac{tT}{2\pi\rho[(B^2/4\pi\rho c^2) + 1]}\int_{k_{\min}}^{k_{\max}} \tau(k)k\,dk = Dt, \tag{23}$$

[3] A more rigorous treatment which includes the energy associated with pressure variations gives

$$\langle \mathbf{E}^2(k)\rangle/8\pi = T/2L^2[1 + (4\pi\rho c^2/B^2)](1 + k^2\lambda_D^2),$$

which gives correction for $k\lambda_D > 1$, where λ_D is the Debye length.

where k_{min} is the minimum value of k for which we can apply the thoery, i.e., $k_{min} = 2\pi/L$ and k_{max} is the maximum value of k ($k_{max} = \text{Min}\{1/\lambda_D, 1/\rho_c\}$ where λ_D is the Debye length and ρ_c is the cyclotron radius), and D is the diffusion coefficient.

To complete the treatment we need a method of finding $\tau(k)$. Now the convective motion which is causing the diffusion is also destroying the existing convective flow; the shearing motion in one mode is tearing up that due to another mode. This is the classic picture of turbulence in which the eddies destroy one another. We might therefore try using the assumption that the lifetime is determined by the diffusion, or

$$\tau(k) = 1/k^2 D. \tag{24}$$

Making this substitution gives

$$D^2 = \{T/2\pi\rho[1 + (B^2/4\pi\rho c^2)]\} \ln k_{max}/k_{min}. \tag{25}$$

This whole treatment can be given somewhat more rigorously in terms of fluctuation dissipation theory. Such a treatment is given by Okuda and Dawson (1973a).

The above treatment is clearly rough, and there are a number of points which should be clarified. First, the correlation time used in the theory should be that seen by a particle, not by a stationary observer. We shall see from the simulation results that the correlation times for the electric field as seen by a stationary observer can be much longer than $1/k^2 D$, which means that the electric field pattern can remain relatively stationary while the particles diffuse through them. This will be discussed in greater detail at the time those results are presented. Second, the short wavelength modes should not be destroyed by motion produced by the long wavelength modes; they should simply be carried along by this motion (Dupree, 1974). Thus D should be a function of k.

Despite these shortcomings, the above treatment appears to fit the results of the simulation rather well. However, more extensive theoretical and computational treatments are called for to resolve these questions.

B. Diffusion for Three Dimensions

Diffusion in a three-dimensional plasma is much more complex but can be attacked in much the same way (Okuda and Dawson, 1972). In three dimensions we consider a system which is triply periodic of period L^3. We must consider two different situations. First, the magnetic field lines may be parallel to one of the coordinate axes; in this case they will close on themselves on transiting across the system and one has closed flux lines. The other possibility is that the B field makes an arbitrary angle to the coordinate axes.

In this case, in general, the field lines will fill the volume ergodically. However, here it is also possible, for certain choices of angles, that the field lines will close on themselves after a finite number of transits across the cube. We shall see that the diffusion depends critically on the arrangement (Okuda and Dawson, 1973c).

Similar situations occur in existing fusion devices so that the results may have some bearing on the operation of these devices.

1. *Closed Field Lines*

First let us consider the case of the magnetic field parallel to one of the coordinate axes, as shown in Fig. 4. In this case, the situation turns out to be very similar to the two-dimensional case just discussed. One gets charged flux tubes, and the associated $E \times B$ motion convects plasma across the field. Because the flux tubes are closed, any excess charge that occurs on one can only be dissipated by slow diffusion. The motion is perpendicular to B and is thus two-dimensional.

We may analyze this case in a manner similar to that used in the previous case. We Fourier analyze the shear flow associated with k's perpendicular to B; thus we write

$$\mathbf{v}_T(r) = \sum_k \mathbf{v}_T(\mathbf{k}) e^{i\mathbf{k}\cdot\mathbf{r}},$$

$$\mathbf{v}_T(k) = \frac{-1}{L^3} \int \frac{\mathbf{k} \times (\mathbf{k} \times \mathbf{v}(r)) e^{-i\mathbf{k}\cdot\mathbf{r}}}{k^2} d^3 r. \quad (26)$$

As before, we compute the energy per mode (both kinetic and electrical) associated with this motion and find

$$W = (\langle E^2(k)\rangle/8\pi)[1 + (4\pi\rho c^2/B^2)] L^3. \quad (27)$$

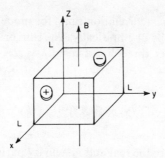

FIG. 4. Closed field line mode.

Equating this to $T/2$ gives

$$\langle E^2(k)\rangle/8\pi = T/2L^3[1+(4\pi\rho c^2/B^2)],$$
$$\langle v_T^2(k)\rangle = 4\pi Tc^2/L^3B^2[1+(4\pi\rho c^2/B^2)]. \tag{28}$$

The diffusion rate caused by this random flow is

$$\frac{\langle \Delta r^2\rangle}{t} = \sum_k \frac{4\pi Tc^2\tau(k)}{L^3B^2[1+(4\pi\rho c^2/B^2)]}$$
$$\approx \frac{T}{2\pi\rho[(B^2/4\pi\rho c^2)+1]L}\int_{k_{\min}}^{k_{\max}}\tau(k)k\,dk, \tag{29}$$

where $k_{\min}$ is the minimum value of k, $k_{\min}=2\pi/L$, and $k_{\max}$ is the maximum value for which the theory applies, $k_{\max}=\min(1/\lambda_D,1/\rho_c)$, and we only sum over k's perpendicular to B. Again assuming that the lifetime is determined by the turbulent diffusion we have

$$\tau(k) = 1/k^2 D,$$
$$D^2 = \{T/2\pi\rho L[1+(B^2/4\pi\rho c^2)]\}\ln(k_{\max}/k_{\min}). \tag{30}$$

2. *Ergodic Lines*

If the lines of force are ergodic then an accumulation of charge at one point can be quickly neutralized by a flow of charge along the lines of force. We may still compute the diffusion caused by the collective modes by using our earlier approach. If the electric field fluctuations are of very low frequency and essentially purely damped, then the random walk argument gives

$$\langle\Delta r^2\rangle/t = \sum_k \langle v_T^2(k)\rangle \tau(k) = \sum_k \langle |[\mathbf{E}(\mathbf{k})\times\mathbf{B}/B^2]c|^2\rangle \tau(k), \tag{31}$$

where $\tau(k)$ is the lifetime (correlation time) for mode k, and it must be determined either theoretically or empirically. The sum on k should now be carried out over all k, i.e., over three-dimensional k-space. If the important waves have a finite real frequency, then Eq. (31) must be replaced by (Chu *et al.*, 1975)

$$\langle\Delta r^2\rangle/t = \sum_k \langle |[\mathbf{E}(\mathbf{k})\times\mathbf{B}/B^2]c|^2\rangle \gamma(k)/[\omega_r^2(k)+\gamma^2(k)], \tag{32}$$

where $\omega_r(k)$ and $\gamma(k)$ are the real and imaginary parts of the frequency of the mode. In general, unless k is very nearly perpendicular to B, $\tau(\mathbf{k})$ is very short if one considers the plasma to be collisionless. In that case one finds that the collective modes contribute little to the transport if they are at the thermal level. However, if any effect occurs which tends to limit the motion of charges along the field lines, then the shorting of charge fluctuations will be inhibited, $\tau(k)$ will increase, and the importance of collective transport will be greater. Two situations have been investigated where this occurs; in the first of these collisions caused inhibitions of the flow of charge, while in the second magnetic mirroring was responsible. In both cases it was found that introducing the inhibition of charge flow caused a pronounced increase in the plasma diffusion. The theory for the collisional case can be found in Okuda *et al.* (1972). We will discuss the mirroring case here.

3. *Closing of Field Lines after an Integer of Crossings of the System*

A third case which can occur is that in which the field lines close after a number of passes across the system. In this case the situation is very similar to that encountered when the lines of force are parallel to one of the coordinate axes. We may expect that the convective transport can be obtained by replacing L in Eq. (30) of the previous treatment by the distance required for closure of the lines of force. This is indeed found to fit the simulations rather well (Okuda and Dawson, 1972).

IV. The Simulation of Plasma Diffusion across a Magnetic Field (Uniform Thermal Plasma)

A. Two-Dimensional Plasma

We first consider the case of a two-dimensional plasma consisting of charge rods aligned parallel to a uniform magnetic field. In this model, particles (rods) can move only perpendicularly to the magnetic field which we take to be in the z-direction. We investigate test particle diffusion for a thermal plasma and compute the guiding center diffusion using roughly 10% of the particles as test particles. By plotting the mean square displacement versus t and taking the slope at large times, we obtain the diffusion coefficient, $D = \lim_{t \to \infty} \langle \Delta r^2 \rangle / t$. An example of such a measurement is shown in Fig. 5. Some results from these simulations are shown in Figs. 6 and 7. These figures show plots of the diffusion coefficient versus ω_c^{-1} for two different size systems, for different numbers of particles per Debye square, and for two mass ratios. From these figures it is clear that there are three regions of diffusion. At low values of the magnetic field the diffusion follows the classical diffusion coefficient predicted from binary collision theory; it is proportional to B^{-2}.

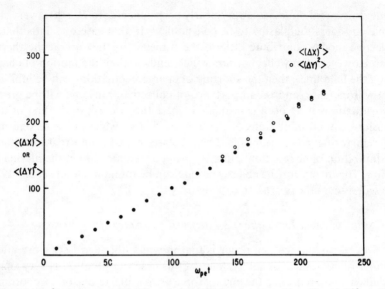

FIG. 5. $\langle \Delta r^2 \rangle$ versus time. Guiding center diffusion for ions: 64×64 grid, 2^{12} particles, $\lambda = 5, \rho e = 20, m_i/m_e = 1.25, n\lambda_p^2 = 25$.

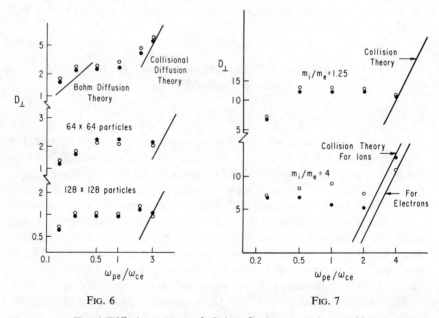

FIG. 6.

FIG. 7.

FIG. 6. Diffusion versus ω_c^{-1}. ●, ion; ○, electron; $m_i/m_e = 1.25$.

FIG. 7. Diffusion versus ω_c^{-1}. 128×128 grid, 64×64 particles; ●, ion; ○, electron.

As the field is increased, the diffusion rate deviates from that predicted by this theory and becomes almost independent of B. At still higher magnetic fields the diffusion rate appears to be proportional to B^{-1}.

Since it is the electric field fluctuations which cause the diffusion, a quantity of considerable interest is the correlation time for the electric field fluctuations given by

$$C(\tau) = \int_0^\infty E(k, t + \tau) E(k, t) \, dt. \tag{33}$$

Of particular interest is the low-frequency part of this. A plot of such correlation times versus k^2 is shown in Fig. 8 (Okuda *et al.*, 1974). (The ∞ upper limit must, of course, be replaced by the maximum value allowed by the simulation.) As can be seen the correlation time fits very well the law $\tau \propto k^{-2}$. In Fig. 8 there is also a plot of $\tau = (k^2 D)^{-1}$; this is the value τ would have if these fluctuations were destroyed by the diffusion of the particles. As can be seen, this time is much shorter than that actually found. Since in theory we assumed that the correlation time was $(k^2 D)^{-1}$, one may ask how is it possible for the theory to explain the diffusion and not predict the proper lifetimes for the field fluctuations. The answer is that the quantity which should enter the theory is the correlation time as seen by the diffusing particles, and not the intrinsic correlation time for the mode. We see that the correlation time as seen by

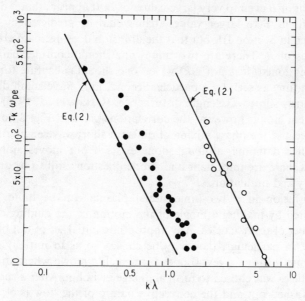

FIG. 8. Correlation function versus τ. Mode $(1, 0)$, $k\lambda_D = \pi/8$.

the particle, $(k^2 D)^{-1}$, can be much shorter than this intrinsic time. This means that for this case the electric field fluctuations exist for a long time while the particles diffuse through them.

It is interesting to compare this result with what one would obtain if the destruction of the electric field fluctuations were due to collisional diffusion of the particles

$$\tau^{-1} = k^2 r_e^2 v,$$

where r_e is the Larmor radius and v is the collision time. One can do a standard fluid analysis (Okuda and Dawson, 1973a) of the decay of the convective motion and one finds

$$\tau^{-1} = k^2 r_e^2 v/(1 + B^2/4\pi\rho c^2). \tag{34}$$

The dielectric term $(1 + B^2/4\pi\rho c^2)$ in the denominator enters because particle collisions dissipate only particle momentum, but there is also momentum stored in the electromagnetic field $(\mathbf{E} \times \mathbf{B}/4\pi c)$ which helps sustain the motion. If we assume that in our case we should replace $r_e^2 v$ by the turbulent diffusion coefficient D, then we find the predicted τ agrees quite well with the observations.

If one assumes that the correlation time as seen by the particles is scaled according to Eq. (34), then it is seen that the diffusion rate would remain independent of B even to very large values of B. It appears that it does remain independent of B to larger values than would be predicted by the simple theory given in Section III, but that the diffusion does decrease at sufficiently large values of B. There are two things one should consider here. First, as already mentioned, it is not $\tau(k)$ which one should substitute for τ, but the correlation time as seen by a particle. Second, long wavelength disturbances do not destroy short wavelength disturbances (Okuda et al., 1975) but simply convect them along; however, the correlations given in Fig. 8 do not reflect this fact because the phase mixing of different short wavelength disturbances will look like damping, although none exists if one moves along with the fluid. Thus there are important and subtle questions still to be answered by both theory and simulation.

The diffusion in a two-dimensional plasma can be quite graphically demonstrated by making a movie of the motion of the guiding centers of a large number of test particles. Any appropriate initial set of test particles can be chosen; in particular, they may be chosen so as to initially form some figure. Figure 9 shows a few frames from such a movie where the initial set of test particles was chosen to form an image of Bohm. After a short time the image is washed out and the convective nature of the flow is clearly shown (classical diffusion would spread the points evenly in all directions).

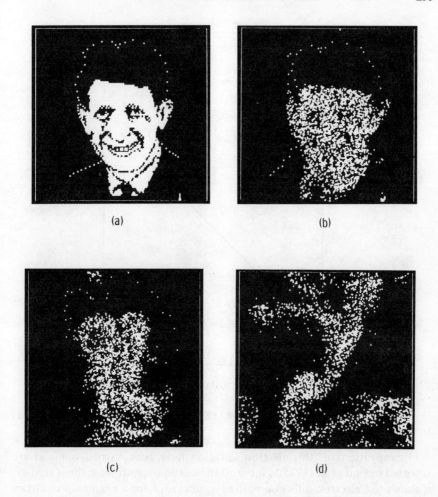

FIG. 9. The guiding center position for a set of test particles chosen so as to form an image of Bohm initially at times (a) $\omega_p t = 0$, (b) $\omega_{pe} t = 6$, (c) $\omega_{pe} t = 75$, (d) $\omega_{pe} t = 250$.

B. Two-Dimensional Electron Diffusion

A similar type of diffusion can occur for electrons alone. In this case, the electrons $E \times B$ drift in the electric field fluctuations associated with lower hybrid waves; the heavy ions move across the magnetic field in response to the electric field, but only at a slow rate. Despite the ion motion, the electrons execute significant $E \times B$ drift motion because of their high cyclotron frequency. This can lead to rapid electron diffusion and transport of heat through the mixing of hot and cold electrons.

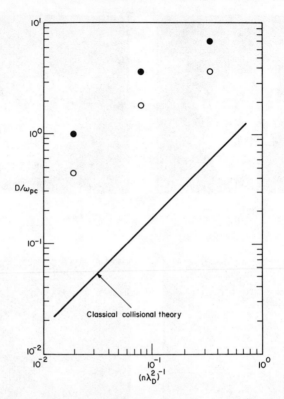

FIG. 10. D versus $n\lambda_D{}^2$ for electron. ●, Calculated from Eq. (30); ○, measured; $m_i = 100 m_e$; $\omega_{ce} = \omega_{pe}$; $T_e = T_i$.

Investigations of this electron diffusion have been carried out and are reported in Chu et al. (1975). Figure 10 shows some results from those studies; it shows test electron diffusion plotted against $n\lambda_D{}^2$ for a plasma of electrons and ions with a mass ratio of 100. Also shown on this plot is the curve one would get for collisional diffusion and that which one would get if he assumed infinitely massive ions and that the theory of Section III applied to electrons only. The latter curve fits better than collisional diffusion but the values are somewhat too high. This is because the long wavelength modes are oscillatory at the ion plasma frequency rather than purely damped; their contribution to diffusion is small. [See Eq. (40).] This enhanced electron diffusion also shows the dielectric effect given in Eq. (34), but one must use only the electron mass density and not the total plasma density. This is shown in Fig. 11.

This enhanced electron diffusion can lead to damping of lower hybrid waves. The motion associated with these waves is one in which the ion oscillates back and forth across the magnetic field in a direction parallel to k while the electrons execute an $E \times B$ motion perpendicular to k and B. The

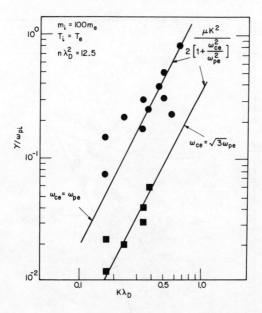

Fig. 11. The effect of electron dielectric on convective damping of lower hybrid waves.

turbulent shearing motion due to many such lower hybrid waves scrambles the $E \times B$ electron motion associated with any particular wave and thus dissipates the energy associated with this motion. Figure 12 shows an example of this damping; the curve labeled ion Landau damping is what one would expect from the ions alone (for all practical purposes they move in straight lines here), while the curve labeled $(k^2 D)^{-1}$ is what would be predicted for a fluid theory using an electron viscosity consistent with the diffusion observed in Fig. 10.

C. Three-Dimensional Transport

1. Closed Field Lines

We now turn to the question of convective transport in three dimensions. As was remarked earlier there are two situations which can be considered. For the first, the magnetic lines are parallel to one of the coordinate axes and thus close on themselves after one transit across a periodic system. The second situation has the field lines at an arbitrary angle to the axes, and they may or may not close on themselves after a discrete number of transits across the system, depending on the value of the angle. We now consider the first of these.

As mentioned in the theory section, this situation is quite similar to the purely two-dimensional one where one has charged flux tubes which can only

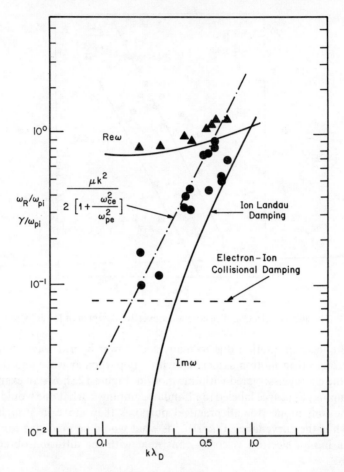

FIG. 12. Damping of lower hybrid waves by electron transport. $m_i = 100 m_e$; $T_i = T_e$; $\omega_{ce} = \omega_{pe}$; $n\lambda_D^2 = 3.125$; $\mu = 4D$.

lose their charge by slow diffusion. The $E \times B$ motion associated with these charges leads to convective motion and enhanced transport.

Such situations have been simulated on a three-dimensional finite size particle model, illustrated in Fig. 13. A sample of the results is shown in Fig. 14. As expected, the diffusion is very similar to that found in two dimensions; at low fields it follows the B^{-2} dependence of binary collision theory. As the field increases it flattens out and becomes almost independent of B. These results are in quite good agreement with the predictions of the theory section. At high field it presumably goes over into a B^{-1} dependence, as predicted by theory; however, these three-dimensional runs are expensive

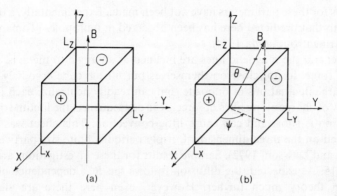

FIG. 13. (a) $(\theta, \psi) = (0, 0)$: closed field lines. (b) $(\theta, \psi) \neq (0, 0)$: nonclosed field lines.

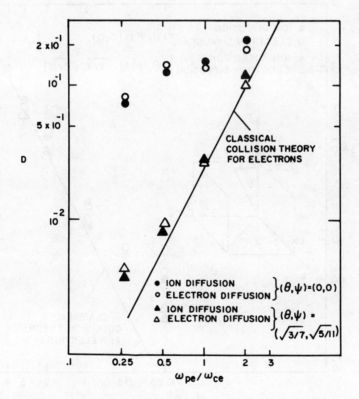

FIG. 14. Diffusion for the three-dimensional model with closed field lines. $n\lambda_D^3 = 170$; $m_i/m_e = 1.25$.

and runs for these parameters have not been made. Experimentally, a diffusion similar to that predicted here has been observed in multipoles (Tamano et al., 1973; Armentrout et al., 1974).

In general the magnetic lines are inclined at an angle to the axis, and the lines of force do not close on themselves but fill the cube ergodicly. In this case charge fluctuations of opposite sign can readily neutralize each other by free flow of charge along lines of force. As a consequence, the lifetime of charge fluctuation is short and can cause little convection. This effect was also investigated on the three-dimensional, triply periodic, finite size particle model (Okuda and Dawson, 1972). Sample results for these investigations are shown in Fig. 15. As expected, the diffusion follows the B^{-2} dependence of binary collision theory much further. However, even here there are significant deviations at high fields.

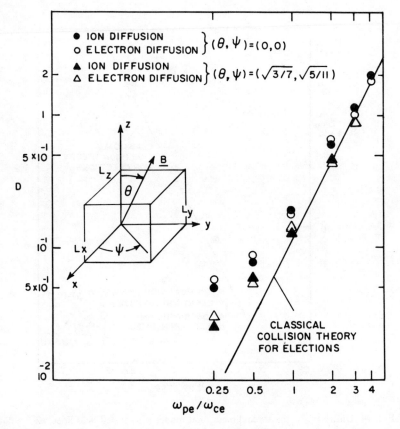

FIG. 15. Diffusion in the three-dimensional model with open field lines. $n\lambda_D^3 = 3.5$; $m_i/m_e = 1.25$.

2. Closure after a Finite Number of Passes across the System

When the magnetic field makes an angle with the coordinate axes it is also possible for the lines to close after a discrete number of passes across the system. We would expect that in this case the diffusion would behave much like that observed for field lines parallel to the coordinate axes; the theory that was developed earlier indicates that the only change should be that the diffusion should decrease as the square root of the distance between closures.

This effect was investigated by setting the angle ϕ (Fig. 13) equal to 45° and varying the angle θ (Okuda and Dawson, 1973c). For different values of θ the lines either close or are open. The line could close after one pass across the cube, after two passes, or after three, etc.; these correspond to rational iota for a Stellarator or Tokamak. On the other hand, they might not close at all, as would correspond to irrational values of iota. Figure 16 shows a plot

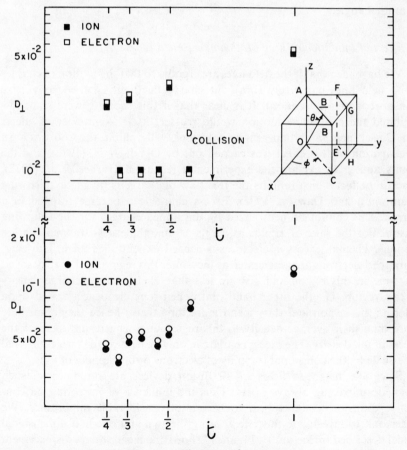

FIG. 16. Resonant diffusion. $n\lambda_D^3 = 55$; $\omega_{ce}/\omega_{pe} = 2$.

of the diffusion coefficient for displacements perpendicular to the $\phi = 45°$ direction versus i. When i takes on a rational value, p/q, the lines close on themselves after passing across the system q times. We see that the diffusion rate shows large peaks around $i = 1, 1/2, 1/3, 1/4$, and that the magnitude of the diffusion is roughly proportional to $i^{1/2}$, as predicted by theory. Experimentally, a similar phenomenon has been observed in the Wendelstein II (Grieger et al., 1971) Stellarator.

The lower part of the figure shows the diffusion rate parallel to the $\phi = 45°$ direction, i.e., parallel to magnetic surfaces. This diffusion is large and shows no striking variations with θ. This diffusion motion would be parallel to magnetic surfaces in Tokamaks or Stellarators. It corresponds to diffusive motion associated with charged magnetic surfaces and would be difficult to observe experimentally. Nevertheless, it could play a role in the trapping and untrapping of particles and in the distribution of impurities in an azimuthal angle.

3. *Effect of Inhibition of Free Motion along Field Lines*

We have seen that if the field lines are ergodic so that the motion of charges along field lines can freely short out charge fluctuations, then convective transport is greatly reduced. It is clear that if this shorting were somehow inhibited the convective motion would reassert itself. A number of effects can cause such inhibition. First, collisions will inhibit the shorting. An investigation of this effect was carried out by Okuda et al. (1972), and the results appear to explain the experimental results of Gurnee et al. (1972). A second effect which inhibits the free flow of charge is magnetic mirroring (Kamimura and Dawson, 1975). Excess charge can become trapped in a mirror. This cannot be neutralized by untrapped charges for the following reason. For the sake of argument, let us assume the excess trapped charges are ions. Then an excess of electrons is needed to neutralize them. However, untrapped electrons are accelerated as they enter this region, and so they spend less time in this region and give up less than their average charge density to this region. On the other hand, untrapped ions are deaccelerated upon entering this region and thus spend more time there; hence they contribute more than their average density to this region and, as a result, increase the excess of ion density. The excess charge can only be neutralized on a collisional time-scale by the collisional trapping of electrons or untrapping of ions.

Since the magnetic fields in all fusion devices are nonuniform, such magnetic mirroring always exists. Thus the influence of mirroring on convective diffusion would appear to be one of considerable importance. We undertook to investigate this effect, using the two and one-half dimensional model described in Section II. Figure 17 shows the mean square displacement

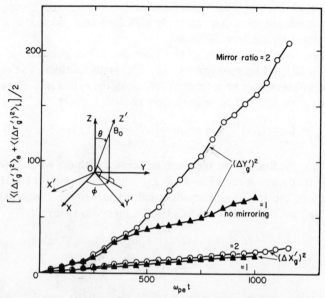

FIG. 17. $\langle \Delta r^2 \rangle$ versus t for the two and one-half dimensional model with mirroring. Guiding center diffusion 64×64 grid, 2^{14} electrons and ions, $m_i/m_e = 1.25$, $n\lambda_{De}^2 = 36$, $\rho_e = 3$, $\omega_{ce}/\omega_{pe} = 1$, $\theta = 10°$, $\phi = 60°$.

of the guiding centers of a set of test particles perpendicular to a magnetic field for various mirror ratios. The expected enhancement of diffusion by mirroring is clear, the value of Δr^2 being twice as large at $\omega_p t = 1000$ for a mirror ratio of 2 as it is for no mirror. As yet, an extensive parameter study has not been carried out. However, more details of this calculation can be found in (Kamimura and Dawson, 1975).

4. Turbulent Electron Heat Transfer Due to Ion Fluctuations

The diffusion due to convective modes causes both ions and electrons to move together because of the low frequency ($\omega \ll \Omega_i$) and the long wavelength ($k_\perp \rho_i < 1$) nature of the convective mode. Since the electrons, on the other hand, are much more easily subject to the low frequency fluctuations because of their small gyroradius and the fast gyration, it is quite possible that some low frequency fluctuations, other than the convective mode, may cause the anomalous transfer of electrons leading to an enhanced heat transfer. The frequency of the fluctuation can be larger than the ion gyrofrequency Ω_i, and the perpendicular wavelength can be shorter than the ion gyroradius for the collective transport of electrons. All that is required is that $\omega \ll \Omega_e$ and $k_\perp \rho_e \lesssim 1$. The fact that the level of fluctuations associated with the high frequency mode can be appreciably greater than that associated with the convective mode suggests that relatively high frequency fluctuations may be more important for electron diffusion.

Let us consider the motion of test electrons in the presence of low frequency ($\omega \ll \Omega_e$) and long wavelength ($k_\perp \rho_e \lesssim 1$) fluctuations. The cross-field diffusion coefficient may be estimated following the orbits in the presence of such fluctuations. The diffusion coefficient is

$$D_\perp = \lim_{t\to\infty} \langle (\Delta x)^2 \rangle / t = c^2/B^2 \sum_k (E^2)_k (k_\perp^2/k^2) \, \text{Im}(1/\omega + i\gamma), \quad (35)$$

where $(E^2)_k$ is the fluctuation spectrum associated with the wave of frequency ω, and γ is its damping rate. For $k_\perp \gg k_\parallel$, which is important for the cross-field diffusion, the low frequency fluctuations are at the lower hybrid frequency. Then we have (Okuda et al., 1975; Chu et al., 1975)

$$(E^2)_k/8\pi = T/2(1 + \omega_{pe}^2/\Omega_e^2),$$
$$\omega = \omega_{pi}/(1 + \omega_{pe}^2/\Omega_e^2)^{1/2} \quad (36)$$

and

$$\gamma = -(\pi/8)^{1/2} \exp(-3/2) \omega^4/k^3 v_i^3 - 4k_\perp^2 D_\perp/(1 + \Omega_e^2/\omega_{pe}^2). \quad (37)$$

The damping is due to ion Landau damping and to turbulent electron diffusion. For $k_\perp \sim k_\parallel$, the important modes are plasma oscillations with

$$(E^2)_k/8\pi = T/2,$$
$$\omega = \omega_{pe} k_\parallel/k, \quad (38)$$

and

$$\gamma = -(\pi/8)^{1/2} \omega_{pe} \cos\theta \exp(-3/2) \exp(-1/2k^2\lambda_D^2)/k^3\lambda_D^3, \quad (39)$$

where we assumed $\Omega_e \gtrsim \omega_{pe}$ throughout this section.

We first estimate the diffusion due to the lower hybrid mode ($k_\parallel = 0$). Assuming the turbulent damping dominates over the Landau damping for $k\lambda_D < 1$, we find (Canosa et al., 1975)

$$D_\perp = (T/2L)^{1/2} (c/B) \{\ln[(\omega^2 + k_{max}^4 D_\perp^2)/\omega^2]\}^{1/2}/(1 + \omega_{pe}^2/\Omega_e^2)^{1/2}, \quad (40)$$

where L is the length of the system along the B field, k_{max} is the minimum of λ_D^{-1} or ρ_e^{-1}, and we assumed $\gamma = k_\perp^2 D_\perp$ along the particle orbit.

The diffusion found above is appreciably greater than the classical value. In fact, this diffusion, in general, exceeds the convective diffusion discussed in the previous papers by roughly $(m_i/m_e)^{1/2}$ for $\omega_{pe} \sim \Omega_e$ because the fluctuation level associated with the lower hybrid oscillation is m_i/m_e times greater than that associated with the convective mode for $\omega_{pi} > \Omega_i$ (Okuda and Dawson, 1973a).

One can similarly estimate the diffusion due to finite $k_{\parallel}$ plasma oscillations. This gives a diffusion which is very similar to the classical diffusion, namely,

$$D_\perp \approx \pi^{3/2} \rho_e^2 \omega_{pe}/(n\lambda_D^3) \ln(m_i/m_e) \ln(k_{max}/k_{min}), \qquad (41)$$

where $k_{max} \approx \lambda_D^{-1}$ and k_{min} is the wavenumber at which the collisional damping takes over the Landau damping ($k_{min} \lambda_D \approx 0.1$ for many cases).

Several three-dimensional simulations were carried out to study the heat transfer for both a uniform field and a sheared field. The magnetic field is in the z-direction, and the electron temperature varies with x as $T_e(x) \sim \sin \pi x/L$, whereas the ion temperature as well as the plasma density were taken to be uniform. The heat conductivity was measured from the decay of $T_e(x)$ with time. $32 \times 32 \times 32$ grids and particles were used with $m_i/m_e = 400$.

Figure 18 shows the results with a uniform field for both closed and open

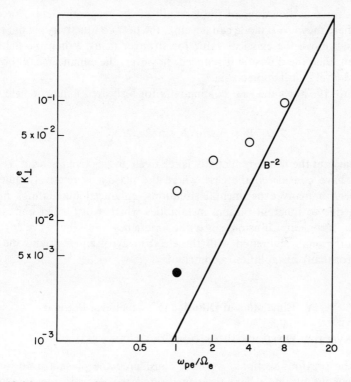

FIG. 18. Measurement of the electron heat conductivity for closed and open field lines of force in three-dimensions. ○, Closed lines; ●, nonclosed lines.

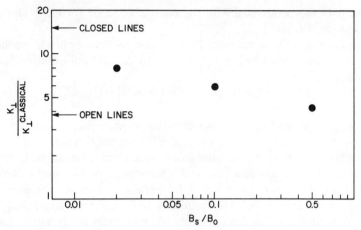

FIG. 19. Electron heat conductivity in a sheared magnetic field showing a weak dependence on shear.

lines of force as functions of the strength of the magnetic field. For closed lines where the $k_\parallel = 0$ mode can develop, the heat conducitvity is appreciably enhanced above the classical value for stronger fields. When the field lines are open, then the diffusion is reduced; however, the enhancement above the classical level is still appreciable.

Figure 19 shows the heat conductivity for a sheared magnetic field of the form

$$\mathbf{B} = b_0 \hat{z} + B_s \sin(2\pi x/L) \tilde{y}. \qquad (42)$$

It is clear that the heat conductivity is relatively insensitive to shear.

We have considered the case where the plasma is relatively quiescent. However, in many experimental situations, enhanced fluctuations may be present due to inherent plasma instabilities which will further enhance the electron turbulent diffusion since the frequency and the wavelength are sufficiently small compared with the electron gyrofrequency and the gyroradius for many instabilities to contribute.

V. Simulation of Diffusion in Nonuniform Plasmas

A. Convective Diffusion in an Inhomogeneous Plasma

In the previous sections, we have considered the plasma diffusion in a homogeneous plasma where the density and the temperature are uniform across the magnetic field. Most laboratory plasmas, however, are inhomogeneous because of the presence of the vacuum vessel which absorbs plasma particles.

In this section, we consider the effects of density and temperature inhomogeneities on the convective cells. It turns out that the convective mode is considerably modified due to the presence of density and temperature gradients. This is because of the fact that the diamagnetic drift frequency of the ions due to the density gradient is considerably greater than the damping rate of the convective mode, so that the drift wave becomes predominant over the convective mode. As we see below, the coupling of the convective mode with the drift mode destabilizes either the drift wave or the convective mode for flute-type perturbations.

To see exactly what happens to the convective mode due to the presence of density gradients, we will make use of the two fluid equations. We consider an inhomogeneous plasma in the x-direction with the magnetic field in the z-direction. Assuming a flute-type perturbation across the magnetic field, the linear dielectric constant can be obtained straightforwardly, and one finds (Rukhadze and Silin, 1968)

$$\varepsilon_\perp(k_\perp, \omega) = 1 + (\omega_{pi}^2/\Omega_i^2)(\omega + ik_\perp^2 \mu_i/\omega)(\omega - \omega_i^*/\omega), \qquad (43)$$

where

$$\omega^* = -(k_\perp Tc/eB)(\partial \ln N/\partial x)$$

is the ion diamagnetic drift frequency and $k_\perp = k_y$. In deriving Eq. (43), we have assumed that the density gradient is sufficiently large, so that $\omega_i^* \gg k_\perp^2 \mu_i$ is satisfied for $k_\perp \rho_i < 1$. This condition is trivially satisfied for most laboratory plasmas except perhaps at the center of the plasma column where the density gradient vanishes.

The dispersion relation can be obtained from $\varepsilon_\perp(k_\perp, \omega) = 0$ and is found to be

$$\omega = \omega_i^*/(1 + \Omega_i^2/\omega_{pi}^2) + ik_\perp^2 \mu_i/(1 + \Omega_i^2/\omega_{pi}^2) \qquad (44)$$

or

$$\omega = -ik_\perp^2 \mu_i/(1 + \Omega_i^2/\omega_{pi}^2), \qquad (45)$$

corresponding, respectively, to the ion flute mode and the convective mode which takes the same form as for the homogeneous plasma.

We find that the convective mode is unchanged and the ion flute mode is destabilized by the presence of ion viscosity. We can see immediately that the convective diffusion in an inhomogeneous plasma may be considerably different from that for the homogeneous case because the mode structure is quite different for the two cases. In fact, the drift mode will cause more diffusion than the convective mode since the former makes the plasma unstable, causing turbulent diffusion which is not stabilized until the original density gradient is completely wiped out, as we shall see below.

Another reason that the diffusion due to the convective mode may not be as large as that for a homogeneous plasma is that the fluctuation field energy associated with the convective mode for an inhomogeneous plasma is much smaller than that for the homogeneous case. This is because the drift frequency is much greater than the damping of the convective mode, and the partition of the low frequency fluctuation field energy between the drift and convective modes is proportional to their frequency ratio. Strictly speaking, one cannot use the fluctuation theorem for an unstable plasma. However, one can imagine, for example, a case where a small shear is introduced in the system so that the drift mode is stabilized.

In order to check the above argument, a few numerical simulations were carried out using an inhomogeneous, two-dimensional model. The parameters for the simulation were 64×64 grid, 128×128 particles, $\Omega_e = 2\omega_{pe}$, and $m_i/m_e = 4$.

Figure 20 shows the growth of the most unstable mode propagating in the y-direction. The corresponding particle diffusion is shown in Fig. 21, where the diffusion for a homogeneous plasma and the diffusion due to collisions are both included. Collisional diffusion is obtained by using a two and one-half dimensional model with finite $k_\parallel$.

One observes that the diffusion in an inhomogeneous plasma is three times greater than that for a homogeneous plasma, which itself is several times enhanced above the collisional level. The field energy does not saturate until the density gradient is completely wiped out. One can see that the growth rate of the flute mode, which is proportional to $k_\perp^2 \mu_i \approx k_\perp^2 D_\perp$, may be en-

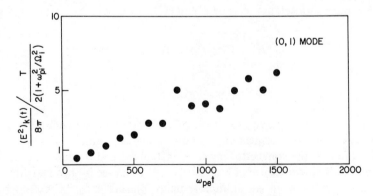

FIG. 20. Growth of the electrostatic field energy for $(0, 1)$ mode associated with the ion flute mode propagating in the y-direction.

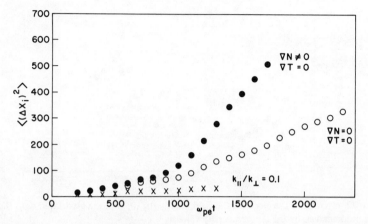

FIG. 21. Test particle diffusion due to classical collisions ($\times$), due to thermal convective cells (O), and due to flute mode (●).

hanced as the instability grows since the viscosity is the destabilizing factor for this case. This will further enhance the turbulent diffusion and renders incorrect approximate turbulent diffusion treatments which equate D to γ/k^2.

So far, we have assumed that the temperature is uniform across the magnetic field. When we consider temperature gradients, assuming the density is uniform, we find that the ion flute mode becomes stable while the convective mode is unstable with the dispersion relation given by (Rukhadze and Silin, 1968)

$$\omega = \omega_i^*/(1 + \Omega_i^2/\omega_{pi}^2) - ik_\perp^2 \mu_i \qquad (46)$$

or

$$\omega = ik_\perp^2 \mu_i,$$

where

$$\omega_i^* = -(k_\perp T_i c/eB)(\partial \ln T_i/\partial x). \qquad (47)$$

The results of simulation for this case are shown in Figs. 22 and 23 where the parameters of simulation are 64×64 grid, 128×128 particles, $\Omega_e/\omega_{pe} = 1$, and $m_i/m_e = 25$. The fluctuation level and the associated diffusion is smaller than the case of inhomogeneous density, because the convective mode has a much smaller amplitude compared with the drift mode. However, here again the growth is proportional to the viscosity and may be enhanced as the wave grows.

We have shown that the convective diffusion in a homogeneous plasma is considerably modified by the presence of density and temperature gradients

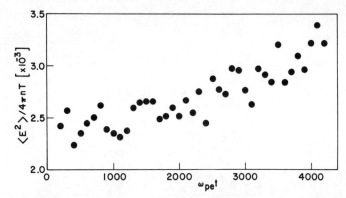

FIG. 22. Growth of the electrostatic field energy for (0, 1) mode associated with the convective mode driven by the temperature gradient.

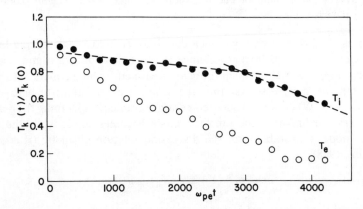

FIG. 23. Relaxation of the ion (●) and the electron (○) temperatures. The fast decay of electron temperature is due to ion lower-hybrid wave.

which couple the ion flute mode and the convective mode to drive them unstably through the ion viscosity. Although the quantitative estimate of convective diffusion is not easy, one expects that the convective diffusion in an inhomogeneous plasma will dominate that for a homogeneous case because of the presence of dissipative instabilities as described here.

B. Energy Transport across Magnetic Field by Plasma Waves

In a high temperature, collisionless plasma in a strong magnetic field, classical diffusion is quite small and the convective diffusion studied in the previous articles of this volume becomes the dominant mechanism for plasma transport. Introduction of strong shear significantly reduces the convective enhancement in the long mean free path region (Okuda and Dawson, 1973b).

While the particles are not able to move across a strong magnetic field, the electrostatic plasma waves propagate across the magnetic field carrying the energy associated with the wave. Since the mean free path for absorption for long wavelength waves is long, the energy transfer due to waves becomes significant.

Rosenbluth and Liu (1972) calculated explicitly the energy flux across the magnetic field due to electrostatic waves near thermal equilibrium for an infinitely strong magnetic field. Consider a plasma slab extending from $x = -L$ to L; the magnetic field is in the z-direction. We now assume that the equilibrium temperature distribution is given by

$$T(x) = T_0(1 - x^2/L^2). \tag{48}$$

The fluctuation field energy associated with the plasma waves is given by $(E^2)_k/8\pi = T_e/2(1 + k^2\lambda_D^2)$. These plasma waves are spontaneously excited due to the Cerenkov radiation and are damped as they propagate across the magnetic field. Since the amplitude of the plasma waves excited at the center of the plasma column is larger than that of the waves excited near the boundary, there is a net energy flux from the plasma center toward the edge which flattens the equilibrium distribution.

In order to calculate the net energy flux due to emission and absorption of plasma waves, let us make use of the wave kinetic equation for a weakly nonuniform plasma. Assuming a steady state, the kinetic equation for the energy density per mode in local thermal equilibrium is (Shafranov, 1967)

$$\partial E(\mathbf{k}, x)/\partial x + \lambda(\mathbf{k}, x) E(\mathbf{k}, x) = \lambda(\mathbf{k}, x) T(x) \tag{49}$$

with

$$E(-L) = 0.$$

The boundary condition used in (49) is the reflective one for electrostatic waves, as these cannot escape the boundary. Here, we consider a two-dimensional slab plasma in order to compare the theoretical results with the simulation results obtained from a two and one-half dimensional code. Thus, the magnetic field is $\mathbf{B} = B\mathbf{y}$, and the temperature variation is in the x-direction. This geometry eliminates the convective modes and so they do not obscure the wave heat transport. The absorption probability per unit length for a particular k mode is given by

$$\lambda(x, \mathbf{k}) = 2\gamma_L/v_x \tag{50}$$

where γ_L is the damping rate and v_x is the group velocity of the wave in the x-direction.

For strong magnetic fields ($\Omega_e > \omega_{pe}$), the dispersion relation takes the form of (Canosa and Okuda, 1975)

$$\omega = (\omega_{pe} \cos\theta/a^{1/2})(1 + \tfrac{3}{2}K^2 a) - i(\pi/8)(\omega_{pe} \cos\theta/a^{1/2})(1/K^3 a^{3/2})$$
$$\cdot \exp(-1/2aK^2) \cdot \exp(-3/2), \tag{51}$$

$$K = k\lambda_D \quad \text{and} \quad a = 1 + (\omega_{pe}^2/\Omega_e^2)\sin^2\theta.$$

The absorption probability is then

$$\lambda(\mathbf{k},x) = (\pi/2)^{1/2}(1/4.5)(1/\lambda_D)(1/k_x k\lambda_D^2)(1/a^{3/2})(1 + \varepsilon\cos^2\theta - 4.5k^2\lambda_D^2)$$
$$\cdot \exp[-T_0/2ak^2\lambda_D^2 T(x)] \tag{52}$$

with

$$\varepsilon = \omega_{pe}^2/\Omega_e^2.$$

For a weak magnetic field ($\omega_{pe} \gtrsim \Omega_e$), the dispersion relation is

$$\omega = \omega_{pe}[1 + \sin^2\theta\,\Omega_e^2/2\omega_{pe}^2 + (3/2)K^2\cos^4\theta]$$
$$- i(\pi/8)^{1/2}(1/n!)(\omega_{pe}/K^3\cos\theta)(\lambda/2)^n \tag{53}$$

with

$$\lambda = k_\perp^2 v_e^2/\Omega_e^2,$$

where for simplicity we have assumed that the ratio ω_{pe}/Ω_e is sufficiently close to an integer n so that the cyclotron damping at the nth cyclotron harmonic dominates the Landau damping.

The absorption probability is then

$$\lambda(\mathbf{k},x) = (\pi/2)^{1/2}(1/n!)(1/2^n)[\sin^{2n-1}\theta(\omega_{pe}/\Omega_e)^{2n}/\lambda_D\cos^3\theta(\varepsilon + 3K_y^2\lambda_D^2)]$$
$$\cdot (k\lambda_D)^{2n-2}(1 - x^2/L^2)^{n-1}. \tag{54}$$

Since the wave transport is nonlocal in nature, we will calculate the net energy flux at the $x = 0$ plane. This is obtained by integrating over all the waves giving

$$A(x) = [1/(2\pi)^2] \int d^2k\, v_x\, \partial E/\partial x$$
$$= [1/(2\pi)^2] \int d^2k\, 2\gamma_L(x)[T(x) - E(k,x)]. \tag{55}$$

After mathematical manipulation, one obtains

$$A(0) = 3.34 \times 10^{-2} \omega_{pe} T_0 (1 - \omega_{pe}^2/\Omega_e^2)/[L\lambda_D \ln^2(L/\lambda_D)]$$
$$\cdot [1 + 2.25 \ln^{-1}(L/\lambda_D) + O(\ln^{-2} L/\lambda_D)] \tag{56}$$

for a strong magnetic field ($\Omega_e > \omega_{pe}$) and

$$A(0) = 3.2 \times 10^{-3} \omega_{pe} T_0/(L\lambda_D)(1 + 1.67 \Omega_e^2/\omega_{pe}^2) \tag{57}$$

for $\omega_{pe} > \Omega_e$.

Numerical simulations were carried out to study the wave transport, in particular, and to check the theoretical calculations. The model employed is a two and one-half dimensional code as shown in Fig. 24. The magnetic field is in the y-direction, while the temperature gradient is taken as

$$T(x, 0) = T_b[1 + 3 \sin(\pi x/2L)] \tag{58}$$

which is parabolic at the plasma center as assumed in the theory. All the modes lie in the x–y plane in this model.

One of the advantages of the models shown in Fig. 24 is that it can completely eliminate the energy transfer due to electron convective cells because the $c\mathbf{E} \times \mathbf{B}/B^2$ drift is in the z-direction. On the other hand, because of the discreteness of the model, it is not possible to eliminate the energy transfer due to binary collisions completely. However, the collisional diffusion can be measured independently from the temperature diffusion, and as we show below there is an unambiguous way of separating the energy transfer by plasma waves from the collisional transfer.

In order to measure the energy transfer, the relaxation of temperature is

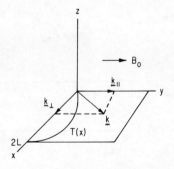

FIG. 24. Sketch of the two and one-half dimensional particle model used for the wave energy transport.

measured instead of directly measuring the energy flux. To do this, we use the energy conservation law given by

$$\tfrac{3}{2}\partial n T(x,t)/\partial t = (\partial/\partial x) D_\perp (\partial n T/\partial x) + [1/(2\pi)^2]\int d^2k\, 2\gamma_L (E-T), \quad (59)$$

where the first term on the right-hand side is the classical heat conduction due to diffusion. For the strong magnetic field case, the diffusion term is negligible compared with the wave transport as will be shown in a moment. This is confirmed since the decay of the temperature takes place only in the "parallel" temperature $T_\parallel$ and no appreciable relaxation is observed for the "perpendicular" temperature $T_\perp$. This also ensures that the collisional coupling between parallel and perpendicular temperatures is negligible for the time-scale considered here. Therefore, one can conclude that the decay of $T_\parallel$ is due to wave transport and collisions, while the decay of $T_\perp$ is solely due to classical collisions. This can be checked independently from the measurement of cross-field particle diffusion.

Figure 25 shows the result of simulation for a strong field ($\Omega_e/\omega_{pe} = 4$). Decay of the parallel, perpendicular, and average temperatures are shown. It is clearly seen that the decay of $T_\parallel$ is nearly twenty times as large as that of $T_\perp$. The decay rate for $T_\parallel$ is observed to be 5.0×10^{-5} which is also the value predicted by theory.

Figure 26 illustrates the case of $\Omega_e/\omega_{pe} = 2$ where $T_\perp$ decays to a noticeable degree, although it is still one-third of the parallel decay rate. For weaker

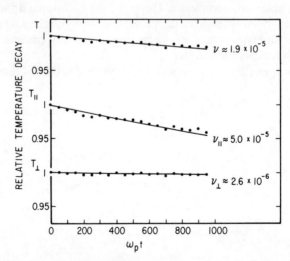

FIG. 25. Relative temperature relaxation at the plasma center for $\Omega_e/\omega_{pe} = 4$ showing the decay of $T_\parallel$ much greater than that of $T_\perp$. 64×64 grid and 256×256 particles.

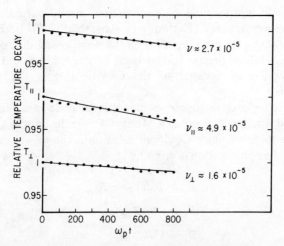

FIG. 26. Relative temperature relaxation at the plasma center for $\Omega_e/\omega_{pe} = 2$. Collisional relaxation increased while the wave relaxation stays the same level.

fields, it was observed that the decay of $T_\parallel$ and $T_\perp$ approaches equality since the collisional decay increases as B^{-2} and comes to dominate the wave transport.

The agreement between the theoretical predictions and the simulation results confirm the validity of the present model for the study of wave transport. The model can also be used, for example, for wave transport in a weakly turbulent plasma where the fluctuations are enhanced above the thermal level.

C. Neoclassical Diffusion in a Toroidal Magnetic Field

So far, we have discussed particle diffusion and heat transfer in relatively simple magnetic field configurations. However, most of the experimental devices for nuclear fusion research such as Stellarators, Tokamaks, and mirror machines have, in general, much more complicated magnetic field structures. Many of the characteristic transport properties in such magnetic field configurations may be simplified and modeled by an idealized geometry, as shown in the previous chapters.

On the other hand, there are certain kinds of problems where the structure of the magnetic field configuration determines the plasma transport. This is particularly true for a high temperature plasma in which the mean free paths of the ions and the electrons are much longer than the size of the machine. The transport coefficients are geometry-dependent for such situations.

In this section, we will study the neoclassical diffusion of plasma in a model toroidal magnetic field configuration (Tsang *et al.*, 1975). The study of plasma diffusion in a toroidal system has been carried out quite extensively in the past several years (Hinton and Rosenbluth, 1973). Experimental observations have also been reported on the neoclassical diffusion (Ohkawa et al., 1972), although precise comparisons between theory and experiments are not always easy because of the uncertainties in the experimental measurements.

We have developed a simple, computational code in a model toroidal magnetic field which makes use of a Monte Carlo collision operator to scatter particles due to small angle Coulomb collisions. The model magnetic field employed in the computation is given by

$$B_z = B_0/(1 + x/R_0),$$
$$B_x = -\Theta B_0 \, y/a, \qquad (60)$$
$$B_y = \Theta B_0 \, x/a,$$

where the x and y coordinates are in the poloidal plane, the z coordinate is along the local toroidal direction, R_0 is the distance from the magnetic axis to the major axis of the torus, Θ is a small parameter characterizing the size of the poloidal field compared with the major toroidal field, and a is the size of the plasma column. The model corresponds to straightening a torus into a cylinder, keeping the toroidal $1/R$ variation in the strength of the toroidal field unchanged. The poloidal field in (60) is generated by a uniform current profile in the z-direction.

The model magnetic field lacks the curvature of the toroidal field and therefore, in the model magnetic field, particle drifts are mainly due to ∇B drift; drifts due to motion along field lines are not included. It is well known that in the banana regime, the dominant contribution to diffusion is from the trapped particles whose parallel velocities are small. Therefore, our model will be essentially the same as a toroidal system. In the plateau regime, the particles contributing primarily to diffusion also have small parallel velocities. In the Pfirsch–Schlüter regime, the bulk of the particles contribute to diffusion and the curvature drift is as important as the gradient drift. Then, in our model, the Pfirsch–Schlüter enhancement is a factor of two smaller than the diffusion in a torus.

The collision operator employed in this work was described by Shanny et al. (1967). It gives accurate, small angle Coulomb collisions for the electrons scattered by the fixed ion background (Lorentz gas collision model), while neglecting the electron self-collisions. Since the ambipolar diffusion arises from the electron–ion collisions, this model collision operator provides a good approximation for the study of neoclassical diffusion.

For each time-step of integration, the velocity of each particle is scattered by the angle given by

$$\Delta\theta = [-2\langle\theta\rangle^2 \Delta t \ln(1 - \alpha)]^{1/2}, \tag{61}$$

where α is a random number between 0 and 1, Δt is the time-step of integration, and $\langle\theta\rangle^2 = 2v_{ei}$ with v_{ei} being the electron–ion collision frequency (Shanny et al., 1967).

In a toroidal field given by (60), the unperturbed guiding center orbit is a helix, and the drift surface in the poloidal plane is a circle. The diffusion coefficient is determined by measuring the radial displacement of a particle from its original flux surface, r, and then averaging over the particles, i.e.,

$$D_\perp = \lim_{t\to\infty} \langle(\Delta r)^2\rangle/2t.$$

Initially, 1024 electrons were loaded uniformly on a circle at a fixed radius with a Maxwell velocity distribution.

Figure 27 shows typical orbits for a trapped and a circulating particle. The magnetic axis is on the left of the figure. The banana orbit is clearly seen with the small fluctuations corresponding to the fast gyromotion.

Figure 28 gives two examples of the variation of $\langle(\Delta r)^2\rangle$ with time for two different collision frequencies corresponding to the Pfirsch–Schlüter and the banana regimes. For the Pfirsch–Schlüter regime, $\langle(\Delta r)^2\rangle$ increases linearly with time almost from the beginning of the computation, indicating the high rate of collisions compared with the drift time, whereas in the banana regime $\langle(\Delta r)^2\rangle$ increases as t^2 for a while and then linearly with t later. This t^2 behavior is due to ∇B drift and is given by

$$\langle(\Delta r)^2\rangle = \langle(v_t^2/2R_0\Omega_e)^2 \sin^2\phi\rangle t^2, \tag{62}$$

where Ω_e is the gyrofrequency, v_t is the thermal velocity, and ϕ is the azimuthal angle in the poloidal cross section. Using $v_t = 2$, $R_0 = 100$, and $\Omega_e = 5$, which were used in the computation, gives

$$\langle(\Delta r)^2\rangle = 8 \times 10^{-6} t^2 \tag{63}$$

which agrees fairly well with the numerical result. The diffusion coefficient is determined from the gradient of the $\langle(\Delta r)^2\rangle$ curve at late times when $\langle(\Delta r)^2\rangle$ increases linearly with time.

Figure 29 shows the results of diffusion measurements varying with the collision frequency over a wide range. The parameters used for the computation are $R_0 = 100$, $a = 20$, $\Theta = 0.2$, $\rho_e = 0.4$, and $\Omega_e = 5$. Here, the timescale and length unit are in terms of electron plasma frequency and the spatial

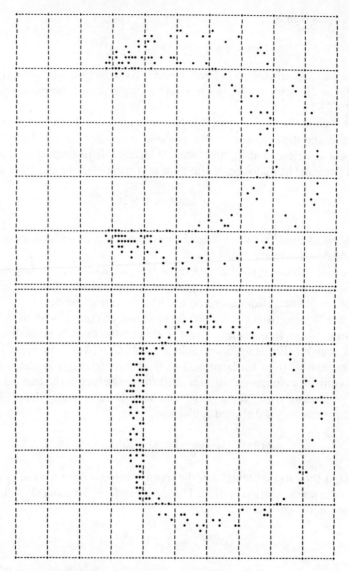

FIG. 27. Sketch of typical particle orbits for both trapped and circulating particles in the model toroidal field given in the text.

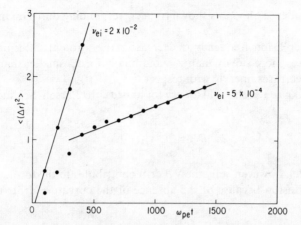

FIG. 28. $\langle (\Delta r)^2 \rangle$ versus t for two collision frequencies. The ∇B drift of the guiding centers becomes clear for small-collision frequency.

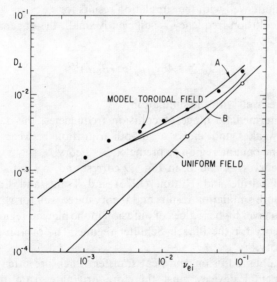

FIG. 29. Diffusion coefficient for a model toroidal field, $B_T/B_p = 5$ and $a/R_0 = 1/5$, changing the collision frequency from 10^{-1} to 5×10^{-4}. Curve A corresponds to the theory of Hazeltine and Hinton (1973), and B corresponds to that of Tsang and Callen (1975).

grid length. The inverse aspect ratio and the safety factor at the particle location are $\varepsilon = 0.2$ and $q = rB_T/R_0 B_p = 1$. The trapped particle bounce frequency $\omega_b = v_{te}(2\varepsilon)^{1/2}/qR_0 = 0.018$ and the trapped particle banana width $\Delta\gamma_b = 2v_t \varepsilon^{1/2}/B_p = 2.5$.

From Fig. 29, we observe that for ν_{ei} greater than ω_b, $D_\perp$ approaches a linear dependence on v_{ei}, as shown for a corresponding diffusion in a uniform field.

As the collision frequency is decreased, the toroidal diffusion levels off and decreases more slowly than v_{ei}, and finally for v_{ei} smaller than 5×10^{-3}, $D_\perp$ again decreases linearly with respect to v_{ei}.

According to neoclassical theory for a torus, the Pfirsch–Schlüter diffusion is given by

$$D_\perp = \rho_e^2 v_{ei}(1 + q^2). \tag{64}$$

In our model, however, only the ∇B drift contributes to the toroidal enhancement of diffusion because of the absence of the curvature drift; then (64) is modified to

$$D_\perp = \rho_e^2 v_{ei}(1 + \tfrac{1}{2}q^2) \tag{64}$$

which agrees quite well with the simulation results for $v_{ei} > \omega_b$.

When the collision frequency is sufficiently small, the banana diffusion is given by

$$D_\perp \cong 1.46 v_{ei} \rho_e^2 q^2 (R_0/a)^{3/2} \tag{65}$$

which compares well with the simulation.

For the intermediate range of collision frequencies, the dependence of $D_\perp$ on v_{ei} is weaker and shows a smooth transition between the Pfirsch–Schlüter and the banana region as predicted by theory.

The solid curves, A and B, in Fig. 29 correspond to the theoretical predictions of Hazeltine and Hinton (1973) and Tsang and Callen (1975), respectively. Both simulation results and theory agree well for $v_{ei} \lesssim 5 \times 10^{-4}$. The simulation gives higher values of diffusion in the plateau region and agrees again with theory for the Pfirsch–Schlüter regime. The overall agreement is quite satisfactory.

The anomalous diffusion and heat transfer which are often observed in many experimental devices must therefore originate from the collective behavior of plasmas. Besides the collective transport processes described in the preceding sections, trapped particle instabilities (Kadomtsev and Pogutse, 1971), disruption of magnetic surfaces, enhanced convective cells, and the energy transport due to various electrostatic waves may be responsible for the enhancement. All these collisionless processes may be simulated using the model toroidal field described here, including the self-consistent electric and magnetic fields. It is also straightforward to mock-up the curvature drift by adding the centrifugal force $m v_\parallel^2 / R_0$ for each particle.

References

Armentrout, C. J., Barter, J. D., Bruin, R. A., Cavallo, A. J., Drake, J. K., Etzweiler, J. T., Greenwood, J. R., Guss, W. C., Kerst, D. W., Post, R. S., Rudmin, J. W., Schmidt, G. L., Sprott, J. C., and Wong, K. L. (1974). *Proc. Conf. Plasma Phys. Controlled Nucl. Fusion Res. 5th, 1974* Paper B 4–2.
Bohm, D. (1949). In "The Characteristics of Electrical Discharges in Magnetic Fields" (A. Guthrie and R. K. Wakerling, eds.), Chapter 2, Sect. S. McGraw-Hill, New York.
Buneman, O. (1967). *J. Comput. Phys.* **1**, 517.
Canosa, J., and Okuda, H. (1975). *Phys. Fluids* **18**, 335.
Canosa, J., Krommes, J. A., Oberman, C., Okuda, H., Tsang, K., Dawson, J. M., and Kamimura, T. (1975). "Plasma Physics and Controlled Nuclear Fusion Research," Vol. II, p. 177. IAEA, Vienna.
Chen, L., and Okuda, H. (1975) *J. Comput. Phys.* **19**, 339.
Chu, C., Dawson, J. M., and Okuda, H. (1975). *Phys. Fluids* **18**, 1762.
Dawson, J. M., Okuda, H., and Carlile, R. M. (1971). *Phys. Rev. Lett.* **27**, 491.
Dupree, T. H. (1974). *Phys. Fluids* **17**, 100.
Grieger, G., Ohlendorf, W., Pacher, H. P., Wabig, H., and Wolf, G. H. (1971). "Plasma Physics and Controlled Nuclear Fusion Research," Vol. III, p. 37. IAEA, Vienna.
Gurnee, M. N., Hooke, W. M., Goldsmith, G. J., and Brennan, M. H. (1972), *Phys. Rev. A* **5**, 158.
Harries, W. L. (1970). *Phys. Fluids* **13**, 140.
Hazeltine, R. D., and Hinton, F. L. (1973). *Phys. Fluids* **16**, 1883.
Hinton, F. L., and Rosenbluth, M. N. (1973). *Phys. Fluids* **16**, 836.
Hockney, R. W. (1966a). *Stanford Univ. Tech. Rep.* No. 53.
Hockney, R. W. (1966). *Phys. Fluids* **9**, 1826.
Kadomtsev, B. B., and Pogutse, O. P. (1971). *Nucl. Fusion* **11**, 67.
Kamimura, T., and Dawson, J. M. (1975). UCLA Physics Department Report PPG-228.
Kruer, W. L., Dawson, J. M., and Rosen, B. (1973). *J. Comput. Phys.* **13**, 114.
Langdon, A. B. (1970). *J. Comput. Phys.* **6**, 247.
Ohkawa, T., Gilleland, J. R., and Tamano, T. (1972). *Phys. Rev. Lett.* **28**, 1107.
Okuda, H., and Dawson, J. M. (1972). *Phys. Rev. Lett.* **28**, 1625.
Okuda, H., and Dawson, J. M. (1973a). *Phys Fluids* **16**, 408.
Okuda, H., and Dawson, J. M. (1973b). *Phys. Fluids* **16**, 1456.
Okuda, H., and Dawson, J. M. (1973c). *Phys. Fluids* **16**, 2336.
Okuda, H., Dawson, J. M., and Hooke, W. M. (1972). *Phys. Rev. Lett.* **29**, 1658.
Okuda, H., Chu, C., and Dawson, J. M. (1974). *Phys. Fluids* **18**, 243.
Okuda, H., Chu, C., and Dawson, J. M. (1975). *Phys. Fluids* **18**, 243.
Rosen, B., Okuda, H., and Di Massa, L. G. (1972). Princeton Plasma Physics Laboratory Report MATT-890.
Rosenbluth, M. N., and Liu, C. S. (1972). *Eur. Conf. Controlled Fusion Plasma Phys., 5th, 1972* Vol. I, p. 12.
Ruhkadze, R. R., and Silin, V. P. (1968). *Usp. Fiz. Nauk* **96**, 87 *Sov. Phys.—Usp.* **11**, 659 (1969).
Shafranov, V. D. (1967). *Rev. Plasma Phys.* **3**, 144.
Shanny, R., Dawson, J. M., and Greene, J. M. (1967). *Phys. Fluids* **10**, 1281.
Tamano, T., Prater, R., and Ohkawa, T. (1973). *Phys. Rev. Lett.* **30**, 431.
Tsang, K. T., and Callen, J. D. (1975). "Neoclassical Diffusion from the Banana to Pfirsch–Schlüter Regimes," ORNL-TM 4848.
Tsang, K. T., Matsuda, Y., and Okuda, H. (1975). *Phys. Fluids* **18**, 1282.

Electromagnetic and Relativistic Plasma Simulation Models

A. BRUCE LANGDON AND BARBARA F. LASINSKI

LAWRENCE LIVERMORE LABORATORY
UNIVERSITY OF CALIFORNIA
LIVERMORE, CALIFORNIA

I. Introduction 327
II. Simulation of Collisionless Plasmas 328
III. Electromagnetic Codes Working Directly with E and B 330
    A. Time Integration of the Fields 331
    B. Time Integration of the Particles 333
    C. Coupling Particles and Fields 335
    D. Boundary Conditions 338
    E. Diagnostics 346
    F. Applications 348
    G. LTSS 6600-7600 Implementation: ZOHAR 356
IV. Algorithms with Special Stability Properties 361
    A. A One-Dimensional Algorithm 361
    B. Two-Dimensional Fourier Transform Codes 362
V. SUPERLAYER 363
    References 364

I. Introduction

AT THE TIME VOLUME 9 of this series was written, on applications to plasma physics, large-scale plasma simulations with only Coulomb interactions had been underway for some time, but codes including the full electromagnetic field were only starting to come into heavy use. By now the evolution of these codes has converged to such an extent that it is possible to describe the numerical and coding principles held in common by several successful codes in different problem areas. Most of the development to date has been guided by heuristic and pragmatic considerations. Only recently has the mathematical analysis of the algorithms approached the stage where there is hope for guidance in constructing and improving practical codes; that analysis is beyond the scope of this article, but is referred to where appropriate.

In collisionless plasmas, each particle is interacting to a comparable degree with a large number of particles. At the same time, nearest-neighbor interactions are relatively unimportant, as are quantum effects in many collective phenomena. The successful simulation algorithms exploit these properties. In Section II we discuss this subject as it pertains to the development of present-day codes.

In Section III we discuss one type of electromagnetic plasma simulation code and our implementation, ZOHAR. This code has been used to model the interaction of intense laser light with plasmas, and for some electron beam–plasma studies. Several applications illustrate the uses and limitations of such codes.

A differently motivated algorithm is described in Section IV. It is useful because of its special stability properties, but it may also be of interest as the first fully electromagnetic algorithm, developed ten years ago before present applications of such codes were under study.

In some problems field retardation is important only to transient behavior. In Section V we discuss a code that combines qualities of the fully electromagnetic and the Darwin models. This code is also unusual in that it had great influence on the direction of a CTR experimental program.

II. Simulation of Collisionless Plasmas

There are two main classes of plasma simulation algorithms. "Particle" models represent the plasma by a large number of computational particles which move according to classical mechanics in the self-consistent electromagnetic fields. "Vlasov" codes begin with the kinetic equation governing the distribution function $f(x, v, t)$. This equation is then integrated forward in time as is, or f may be expanded as a series (see, e.g., *Methods in Computational Physics*, Vol. 9). While particle codes have proven their versatility in multidimensional problems, the Vlasov codes have not yet proven their utility in two-space dimensions, although they have had some successes in one dimension. While this situation may change in the future, we shall confine our discussion to the dominant simulation tool—particle codes.

We can present the basic ideas underlying present codes in the simple context of a plasma model including only Coulomb interactions. Early simulations found the acceleration of each particle simply by summing over all the other particles. In one dimension this sum becomes trivial if the particle coordinates are kept ordered. Special care with particle crossings must be taken in the time integration.[1] In two dimensions the force sums for N particles

---

[1] For details of this approach and plasma applications, see Dawson (1970, and references therein). For a complete bibliography including many papers on simulation of electronic devices (where the subject was born), see Van Duzer and Birdsall (1971).

involve $N(N-1)$ interaction terms which may be quite complicated in form, depending on the boundary conditions. Furthermore, near encounters require small time-steps or other special handling, and also exaggerate the importance of large-angle scattering. Very little plasma work has been done in this way, although such models are used to study stellar cluster evolution.

By 1963 a completely different approach was in use at Stanford by Buneman and his co-workers. This permitted much more rapid computation of the particle accelerations, and a fast simple time integration of their equations of motion. By now, all large-scale particle simulations are done by elaborations of their method, even in one dimension. Basically, this approach introduces a fixed spatial grid on which charge densities, potentials, and fields are defined. To form the charge density, the charge of each particle is associated with the nearest grid point (NGP). A new, fast algorithm was devised to solve a difference form of Poisson's equation, and this potential was then differenced to give the electric field:

$$\rho_{i,j} = (\phi_{i+1,j} - 2\phi_{i,j} + \phi_{i-1,j})\Delta x^{-2} + (\phi_{i,j+1} - 2\phi_{i,j} + \phi_{i,j-1})\Delta y^{-2}, \quad (1)$$

$$E_{x,i,j} = -(\phi_{i+1,j} - \phi_{i-1,j})/(2\Delta x),$$

where $\phi_{i,j} = \phi(i\Delta x, j\Delta y)$. Particle accelerations were given by the field at the nearest grid point. The particle velocities and then positions were advanced in time using a simple leapfrog scheme:

$$v_x^{n+1/2} = v_x^{n-1/2} + a_x^n \Delta t, \qquad x^{n+1} = x^n + v_x^{n+1/2}\Delta t, \quad (2)$$

where superscript n denotes time $n\Delta t$ (Hockney, 1965; Yu et al., 1965; Burger et al., 1965; Buneman, 1967). Superficially, it seems a wonder that such a crude scheme can give useful results. It can succeed at all only because of the properties of collisionless plasma. This force calculation suppresses the divergence of the Coulomb force at separations less than the grid size, reducing the unwanted large-angle scattering which was exaggerated due to the small number of simulation particles used (Hockney, 1966).

In addition to being fast, this algorithm is second-order accurate in space and time (the relative errors are of order Δx^2 and Δt^2) and conserves momentum exactly. The time reversibility of Eq. (2) ensures that there is no numerical damping or growth of plasma oscillations.

A straightforward modification greatly improving the performance was made by Birdsall and his co-workers (Birdsall and Kamimura, 1966; Birdsall and Fuss, 1969). This was to use bilinear interpolation to obtain the particle force, and to collect the charge density using the same weights as for the interpolation. This can be regarded as associating the charge with the nearest four grid points ("charge sharing"). Bilinear interpolation was also adopted in plasma simulation codes developed at Los Alamos (Morse and Nielson, 1969), in analogy to the "particle-in-cell" (PIC) hydrodynamic codes (Harlow, 1956–1957, 1964). The descriptive names "PIC" for these codes, and "area-weighting" for the charge sharing scheme, are now commonly used.[2]

Much insight was obtained into the meaning of the modified force law by Hockney (1966), Birdsall and Fuss (1969), and others, who viewed it as Coulomb interaction between particles with extended tenuous charge ("clouds"). This viewpoint led to a theoretical description of the effects on plasma waves, fluctuations, and collisions (e.g., Langdon and Birdsall, 1970) which eventually included an exact treatment of the effects of finite cell and time-step sizes (Langdon, 1970a,b).

III. Electromagnetic Codes Working Directly with E and B

The main emphasis of this article is on algorithms for electromagnetic plasma simulation for application to the interaction of intense laser light with hot plasmas, and for some electron beam–plasma studies. We choose for discussion the algorithms we consider best for these applications; for the same reason, we chose these algorithms for our own code ZOHAR.[3] The ideas are not necessarily unique to ZOHAR; some have been thought of independently elsewhere, and we have borrowed freely from the work of others. We have not sought originality as such, but a good implementation of a selection of the best algorithms. (A formal approach to algorithm synthesis has been developed by Lewis, 1972.)

Some electromagnetic codes integrate A and ϕ forward in time; others like ZOHAR advance E and B directly. In choosing the latter, we were influenced by Occam's razor and by the smaller field array storage requirement. The relation between the two methods has been discussed by Langdon (1972). In terms of the numerical and physical properties of the algorithms, there seem to be few grounds for preference either way, unless the particle mover makes use of the canonical momentum (Godfrey, 1975).

ZOHAR can be run in four configurations, differing in boundary conditions and number of vector components of velocity. There are two space dimensions and either two or three velocity components (two is faster and requires less storage, but some physics requires three). The boundary conditions are periodic in the y-direction and either periodic in x or "open-sided." The

[2] The next refinement would be parabolic splines (see, e.g., Langdon, 1973).

[3] Zohar means brilliance or radiance in Hebrew. It is also the name of a book central to the caballah (de Leon, ~1280). The qualities of brilliance and mysticism seemed appropriate to the applications of the code and to certain features of simulation methodology itself.

former is useful for simulating phenomena occurring on a scale length small enough that the plasma may be considered uniform. The latter version is used to model the interaction of incoming light with a segment of the inhomogeneous plasma corona.

A. Time Integration of the Fields

The fields are integrated forward using their time derivatives as given by Faraday's law and the Ampere–Maxwell law:

$$\partial \mathbf{B}/\partial t = -c\mathbf{V} \times \mathbf{E}, \tag{3}$$

$$\partial \mathbf{E}/\partial t = c\mathbf{V} \times \mathbf{B} - \mathbf{J}. \tag{4}$$

These equations have been written in rationalized cgs (or Heaviside–Lorentz) form which eliminates almost all occurrences of factors of 4π during problem design and physical interpretation of the results (Panofsky and Phillips, 1962; Jackson, 1962).

By defining the fields at the positions shown in Fig. 1, it is possible to difference Maxwell's equations in a very simple way whose accuracy is second order in space and time (Buneman, 1968; Boris, 1970; Sinz, 1970; Morse and Nielson, 1971). All derivatives become central differences; for example

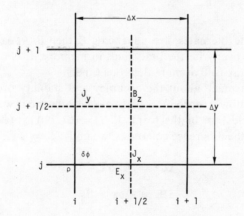

FIG. 1. Spatial layout of the two-dimensional fields and their source terms as they appear during their integration forward in time via the differenced Maxwell equations. In the text this is referred to as the "field grid." The "particle grid" is the same as for the charge density ρ. This grid is used during the time integration of the particle coordinates via the difference Newton-Lorentz equations. In the two and one-half dimensional case, B_x and B_y are collocated with E_x and E_y, while E_z and J_z are collocated with B_z. Subscripts i and j range from 0 to N_x and 0 to N_y, respectively, to cover the rectangular region $0 \leq x \leq L_x$, $0 \leq y \leq L_y$.

we write the time derivative as

$$(\partial_t E_x)_{i+1/2,j}^{n+1/2} \equiv (E_{x,i+1/2,j}^{n+1} - E_{x,i+1/2,j}^n)/\Delta t, \tag{5}$$

where $E_{x,i+1/2,j}^n \equiv E_x([i+\tfrac{1}{2}]\Delta x, j\Delta y, n\Delta t)$, etc. Define ∂_x and ∂_y analogously. The gradient $\mathbf{V}$ becomes ∂_x. What makes this notation helpful later is that these operators, applied to fields defined on our space–time grid, commute. Therefore the difference equations can be manipulated in the same ways as the similar appearing differential equations. The differenced Maxwell equations become

$$(\partial_t B_z)_{i+1/2,j+1/2}^n = -c(\partial_x E_y - \partial_y E_x)_{i+1/2,j+1/2}^n, \tag{6}$$

$$(\partial_t E_x)_{i+1/2,j}^{n+1/2} = (c\,\partial_y B_z - J_x)_{i+1/2,j}^{n+1/2}, \tag{7}$$

$$(\partial_t E_y)_{i,j+1/2}^{n+1/2} = (-c\,\partial_x B_z - J_y)_{i,j+1/2}^{n+1/2}. \tag{8}$$

When $B_z^{n-1/2}$ and E^n are known, Eq. (6) determines $B_z^{n+1/2}$. The electric field is then advanced similarly. For example, Eq. (7) expands to

$$\frac{E_{x,i+1/2,j}^{n+1} - E_{x,i+1/2,j}^n}{\Delta t} = c\frac{B_{z,i+1/2,j+1/2}^{n+1/2} - B_{z,i+1/2,j-1/2}^{n+1/2}}{\Delta y} - J_{x,i+1/2,j}^{n+1/2}. \tag{7a}$$

In the code one alternates, first advancing E, then B. At each step the new values for a field overwrite the old values in memory. It is not necessary to retain values for any field at more than one time.

One can learn much about the accuracy and stability properties of this scheme by seeing how it reproduces plasma electromagnetic waves in vacuum. Assuming the fields are of the form $(\mathbf{E}, \mathbf{B}) = (\mathbf{E}_0, \mathbf{B}_0)\exp(i\mathbf{k}\cdot\mathbf{x} - i\omega t)$ and substituting into the difference equations, we find

$$\Omega \mathbf{B} = c\boldsymbol{\kappa} \times \mathbf{E},$$

$$\Omega \mathbf{E} = -c\boldsymbol{\kappa} \times \mathbf{B},$$

where $\Omega = (\sin\omega\Delta t/2)/(\Delta t/2)$, $\kappa_x = (\sin k_x \Delta x/2)/(\Delta x/2)$. In the continuum limit Ω and $\boldsymbol{\kappa}$ reduce to ω and $\mathbf{k}$. Eliminating E and B yields

$$\Omega^2 = c^2\kappa^2. \tag{9}$$

One can readily verify that there are no phase or magnitude errors between E

and B, and that there is no damping or growth (ω is real) provided that the Courant condition is satisfied:

$$\Delta t^{-2} > c^2(\Delta x^{-2} + \Delta y^{-2}). \tag{10}$$

The errors in the magnitude of ω, and the relative directions of the fields and $\mathbf{k}$, are second order in Δx, Δy, and Δt. All these properties are a direct result of the centered differencing in space and time. Similar accuracy with noncentered differencing schemes would require a much more complicated algorithm.

When the condition Eq. (10) is violated, $\sin^2 \omega \Delta t/2$ exceeds unity for $k_x \Delta x$, $k_y \Delta y$ near π. The ω roots are now complex, with one root giving growth which can be very rapid.

In two space dimensions the fields E_z, B_x, and B_y are not coupled through Maxwell's equations to B_z, E_x, and E_y. Therefore the grids for the two sets of fields can be given any desired relative location. We have chosen to locate E_z and J_z with B_z, B_x with E_x, and B_y with E_y. This makes the code's indexing and boundary conditions analogous.

There are two other Maxwell equations. We will now show that the difference equations have the property, as do the continuum equations, that if the divergences of $\mathbf{E}$ and $\mathbf{B}$ are correct initially, they will remain correct. That is,

$$\partial_t(\partial_x \cdot \mathbf{B}) = \partial_x \cdot (\partial_t \mathbf{B}) = -c\, \partial_x \cdot \partial_x \times \mathbf{E} \equiv 0.$$

Similarly,

$$\partial_t(\partial_x \cdot \mathbf{E} - \rho) = \partial_x \cdot \mathbf{J} - \partial_t \rho.$$

Therefore, if $\mathbf{J}$ and ρ satisfy the continuity equation, Gauss' law will remain satisfied if it holds initially. We will have more to say on this point in Section III, C.

B. Time Integration of the Particles

Initially, let us assume that E and B are somehow interpolated from the grid fields to the particles at time $t^n = n\Delta t$. We desire a centered-difference form of the Newton–Lorentz equations of motion.

The simplest case is nonrelativistic motion, with only one magnetic field component, B_z. There is a large class of physical problems in which v_z, E_z, B_x and B_y are unimportant. If these quantities are zero initially, the equations of motion show they will remain zero. Following Buneman (1967), we write

$$\frac{\mathbf{v}^{n+1/2} - \mathbf{v}^{n-1/2}}{\Delta t} = \frac{q}{m}\left(\mathbf{E}^n + \frac{1}{c}\frac{\mathbf{v}^{n+1/2} + \mathbf{v}^{n-1/2}}{2} \times \mathbf{B}^n\right). \tag{11}$$

The extension over Eq. (2) is the magnetic field force. Its centering is achieved simply by time averaging; this is also how B^n is obtained [see Eq. (16)]. In obtaining the solution of Eq. (11) for $v^{n+1/2}$ it has been common to subtract $c\mathbf{E} \times \mathbf{B}/B^2$ from $\mathbf{v}$. The difference equation for the remainder just expresses a cyclotron rotation. It turns out that the relativistic generalization is easier if we instead substitute

$$\mathbf{v}^{n-1/2} = \mathbf{v}^- - q\mathbf{E}^n \Delta t/2m$$
$$\mathbf{v}^{n+1/2} = \mathbf{v}^+ + q\mathbf{E}^n \Delta t/2m \qquad (12)$$

into Eq. (11) to obtain

$$(\mathbf{v}^+ - \mathbf{v}^-)/\Delta t = (q/2mc)(\mathbf{v}^+ + \mathbf{v}^-) \times \mathbf{B}^n.$$

By taking the inner product of this equation with $(\mathbf{v}^+ + \mathbf{v}^-)$ to obtain $(v^+)^2 = (v^-)^2$, one sees that it expresses a rotation; the angle can be shown to be $\theta = -2 \arctan(qB\Delta t/2mc)$. This plane rotation can be done very concisely by an algorithm due to Buneman (1973):

$$v_x' = v_x^- + v_y^- t,$$
$$v_y^+ = v_y^- - v_x' s, \qquad (13)$$
$$v_x^+ = v_x' + v_y^+ t,$$

where $t = -\tan\theta/2 = qB_z \Delta t/2mc$, $s = -\sin\theta = 2t/(1+t^2)$.

In the more general case in which $\mathbf{E}$, $\mathbf{B}$, and $\mathbf{v}$ all can have three nonzero components we use an algorithm requiring little arithmetic and as few as possible arithmetic registers in an optimized particle "mover" code. Following Boris (1970), who gives a geometrical motivation for the steps:

$$\mathbf{v}' = \mathbf{v}^- + \mathbf{v}^- \times \mathbf{t},$$
$$\mathbf{v}^+ = \mathbf{v}^- + \mathbf{v}' \times \mathbf{s}, \qquad (14)$$

with $\mathbf{t} = q\mathbf{B}\Delta t/2mc$, $\mathbf{s} = 2\mathbf{t}/(1+t^2)$. The error in the angle between $\mathbf{v}^-$ and $\mathbf{v}^+$ is $\simeq t^3/3$ and does not seem worth correcting.

For the relativistic generalization we use $\mathbf{u} \equiv \gamma\mathbf{v}$ rather than $\mathbf{v}$. The addition of electric impulses carries over with no formal change,

$$\mathbf{u}^- = \mathbf{u}^{n-1/2} + q\mathbf{E}^n \Delta t/2m,$$
$$\mathbf{u}^{n+1/2} = \mathbf{u}^+ + q\mathbf{E}^n \Delta t/2m. \qquad (12a)$$

In the rotation the angle is reduced by the factor γ. Therefore $\mathbf{t} = q\mathbf{B}\Delta t/2\gamma^n mc$ with $(\gamma^n)^2 = 1 + (u^-)^2$. Since $(\gamma^n)^2 = 1 + (u^+)^2$ also, this scheme is time reversible[4] and the overall momentum integration is second-order accurate (Boris, 1970).

In all cases the position is advanced according to

$$\mathbf{x}^{n+1} = \mathbf{x}^n + \mathbf{v}^{n+1/2}\Delta t = \mathbf{x}^n + \mathbf{u}^{n+1/2}\Delta t/\gamma^{n+1/2} \tag{15}$$

with $(\gamma^{n+1/2})^2 = 1 + (u^{n+1/2})^2$. This step also is reversible and produces a second-order error in the particle orbit.

Algorithms have been used elsewhere which are second order but not time-centered or reversible (Nielson and Lindman, 1973b). While no physical or computational advantages have been established for them, they perform adequately.

In many laser–plasma applications, the magnetic field does not significantly affect the ion motion and the ion transverse current is negligible. In such cases some time may be saved by using a mover for the ions in which only the electric field accelerates the ions, relativity is ignored, and only the ion charge density, not the current, is collected. The manner in which the longitudinal current is taken into account is discussed in the next section.

C. Coupling Particles and Fields

In coupling the particle and field integrations, we have to relate quantities given at different locations and times. The simplest case is the magnetic field, given at half-integer times in the field equations, and needed at integer times for the particles. Since Faraday's law can be used to advance B to a time ahead of E, one simply time averages B:

$$\mathbf{B}^n = \tfrac{1}{2}(\mathbf{B}^{n-1/2} + \mathbf{B}^{n+1/2}). \tag{16}$$

This is to be used in the particle mover and also in certain diagnostics. In practice the average does not appear explicitly. To avoid using additional computer storage, the B integration is split into two steps (Boris, 1970). As the last step of the field integration we advance B only halfway, obtaining

$$\mathbf{B}^n = \mathbf{B}^{n-1/2} - c\Delta t\, \partial_x \times \mathbf{E}^n/2 \tag{17}$$

---

[4] This γ^n may be used to calculate a time-centered kinetic energy. On short word-length computers it is best to use the identity $(\gamma - 1)mc^2 = mu^2(\gamma + 1)^{-1}$; the second form of the kinetic energy is far less susceptible than the first to roundoff error.

which replaces $\mathbf{B}^{n-1/2}$ in memory. The particles are integrated, and then B is advanced in the same way to $\mathbf{B}^{n+1/2}$ as the first step in the following field integration.

To compute the current density from the particle velocities and positions, we use an average of the weights for the two positions $\mathbf{x}^n$ and $\mathbf{x}^{n+1}$, together with $\mathbf{v}^{n+1/2}$, to obtain a time-centered $\mathbf{J}^{n+1/2}$ (Boris, 1970). Another way is to use weights for the position $\mathbf{x}^{n+1/2} = \mathbf{x}^n + \mathbf{v}^{n+1/2}\Delta t/2$ (Morse and Nielson, 1971). The two are similar in computational expense. We chose the former in the hope that it would have better noise properties, but comparisons have been inconclusive (Nielson and Lindman, 1973b). However, recent work (Godfrey and Langdon, 1975) has shown that our choice is less susceptible to numerical instabilities.

The reader may easily verify that neither current density satisfies the continuity equation with ρ calculated by any method which depends only on the present particle locations. This may be seen even in the $\Delta t \to 0$ limit, e.g., by considering a particle which moves in a small circle inside a quarter-cell. After each turn, $\mathbf{V} \cdot \mathbf{J}$ has a nonzero average, yet ρ is the same.

Methods for calculating a charge conserving $\mathbf{J}$ have been discussed for ρ obtained by NGP-weighting (Buneman, 1968) and by area-weighting (Morse and Nielson, 1971). The latter authors found that the noise level in the electromagnetic fields rose in time at an inconveniently rapid rate. They then developed a Coulomb gauge $\mathbf{A}, \phi$ model in which, as for Boris (1970), the longitudinal $\mathbf{E}$ field is determined from ρ by a Poisson solution. It was later shown that the noise properties observed by Morse and Nielson were related to the methods of forming $\mathbf{J}$, not to the use of potentials. An equivalence was established between these methods, and reasons suggested for the slower buildup of noise (Langdon, 1972).

Boris' alternative method may be motivated as follows: The current density is computed as described earlier. A correction to $\mathbf{J}$ is then computed which is purely longitudinal and such that the corrected $\mathbf{J}$ satisfies the continuity equation. Using this current to advance $\mathbf{E}$ ensures that the Gauss law, $\mathbf{V} \cdot \mathbf{E} = \rho$, remains satisfied. This correction to $\mathbf{J}$ is $-\nabla \xi$, where

$$\nabla^2 \xi = \mathbf{V} \cdot \mathbf{J} + \partial \rho / \partial t.$$

It is easy to see that this algorithm is time-centered and therefore second-order accurate. Note that the difference form for the Laplacian, consistent with the gradient and divergence operators, is just the simple five-point operator $\partial_x{}^2$.

Computationally it is more convenient to advance $\mathbf{E}$ using the uncorrected $\mathbf{J}$, then apply a correction $-\nabla \delta\phi$ to $\mathbf{E}$ afterward, where $\partial_t \delta\phi = \xi$ or

$$-\nabla^2 \delta\phi = \rho - \mathbf{V} \cdot \mathbf{E}. \tag{18}$$

This asymmetric-looking procedure produces exactly the same final fields as does the previous, more obviously time-centered, algorithm. The NRL codes and ZOHAR all use the correction to $\mathbf{E}$.[5]

We have already discussed in Section II the manner in which the electric field at the particle position may be obtained by interpolation from the field grid. The most obvious thing to do is to interpolate separately from each of the three sets of grid locations shown in Fig. 1, as was done in several early codes (Boris, 1970; Sinz, 1970; Morse and Nielson, 1971). However, this is an inconvenience to the particle mover. Since the mover accounts for the majority of the computer time, more recent codes redefine the fields beforehand to a single set of grid locations. This may be done simply by a spatial average to the ρ grid positions. There are other advantages: the longitudinal part of $\mathbf{E}$ now is the same as for the momentum-conserving electrostatic codes (Section II; Boris and Lee, 1973), and the additional smoothing decreases short wavelength noise. Diagnostics are also simplified. The same benefits have been realized in an $\mathbf{A}, \phi$ code by changing the differencing used to derive $\mathbf{E}$ and $\mathbf{B}$ from $\mathbf{A}$ and ϕ (Nielson and Lindman, 1973a, b).

After the particle integration the fields might be restored to the field grid by a further spatial averaging. However, this would produce damping of electromagnetic waves that is unacceptably rapid in our applications. Boris has observed that the *original* E_x, for example, can be easily reconstructed from the averaged E_x if the values of E_x at one side are saved before averaging. For B_z the simplest procedure is to "unaverage" in x first, along with E_x, then in y along with E_y. In this way we can redefine the fields to a common grid and restore them without adding appreciably to computer storage requirements.

As with the redefinition of field grids, it is advantageous to collect $\mathbf{J}$ by area-weighting to a single set of grid points, collocated with ρ in our case, and then spatial average to obtain the currents at the locations shown in Fig. 1, where they are needed for the field equations. No restoration of the unaveraged $\mathbf{J}$ is needed, of course.

As a final complication, we discuss a method by which one can use smaller time-steps for the fields than for the particles. If we consider a reasonable set of parameters for some problems in which $c\Delta t \simeq \Delta x/2$, Debye length $\lambda_D = \Delta x/2$, and thermal velocity $v_t = c/20$, then we find $\omega_p \Delta t = 0.05$. For many applications, this time-step is smaller than needed for the particles. Since the

---

[5] It may be worthwhile to spatial filter ρ before subtracting $\mathbf{V} \cdot \mathbf{E}$ in Eq. (18), to boost the medium wavelengths and perhaps suppress short wavelength noise. The former is chosen to correct the finite-difference errors throughout the field and force calculations which decrease or even reverse the Bohm-Gross dispersion of plasma oscillations when grid spacings larger than a few Debye lengths are used. Quantitative results needed to derive the form of this "pre-emphasis" appear in Section 5 of Langdon (1970a).

particle integration is expensive, it is helpful to advance the particles less often than the fields.[6] In order to illustrate this extension and to summarize this section, we outline the operations taken in one particle time-step for the case where the fields are integrated twice as often. Superscript n will represent the *particle* step number.

Start with $\mathbf{E}^n$, $\mathbf{B}^n$, $\mathbf{x}^n$, and $\mathbf{u}^{n-1/2}$
0. Average fields to the particle grid.
1. Advance $\mathbf{u}^{n-1/2}$ to $\mathbf{u}^{n+1/2}$, $\mathbf{x}^n$ to $\mathbf{x}^{n+1}$; form $\mathbf{J}^{n+1/2}$ and ρ^{n+1}.
2. Average $\mathbf{J}$ to the field grid. Reform E and B on the field grid.
3. Advance $\mathbf{B}^n$ to $\mathbf{B}^{n+1/4}$.
4. Advance $\mathbf{E}^n$ to $\mathbf{E}^{n+1/2}$ using $\mathbf{J}^{n+1/2}$.
5. Advance $\mathbf{B}^{n+1/4}$ to $\mathbf{B}^{n+3/4}$.
6. Advance $\mathbf{E}^{n+1/2}$ to $\mathbf{E}^{n+1}$, using $\mathbf{J}^{n+1/2}$.
7. Advance $\mathbf{B}^{n+3/4}$ to $\mathbf{B}^{n+1}$.
8. Correct $\nabla \cdot \mathbf{E}^{n+1}$ using ρ^{n+1}.

Checking for time-centering, we note that steps 3 and 7, and steps 4 and 6 are symmetric. The longitudinal part of $\mathbf{E}^{n+1}$ is affected only by $\mathbf{J}^{n+1/2}$, and is the same no matter how many time-steps are used for the fields in reaching time $n + 1$. Therefore the argument used earlier still holds, that the divergence correction does not in fact affect the time centering.

We have recently discovered that this splitting of the time-step can lead to numerical instability. While this has so far affected only one physics problem we have run, the instability is capable of reaching very high amplitudes over a few thousand time-steps. A theoretical analysis has been done which shows the combinations of circumstances which lead to this instability. This analysis is being checked with test problems and will be reported elsewhere.

D. Boundary Conditions

1. *Periodic*

As with electrostatic codes, much interesting work can be done simulating a system which is periodic in x and y. In this case, boundary conditions offer no conceptual problems, with one possible exception: As formulated so far, the $\mathbf{k} = 0$ component of the electric field is given by

$$0 = \mathbf{J}_{\text{total}} = (\mathbf{J} + \partial \mathbf{E}/\partial t)_{k=0}$$

and is not zero, in general [case (a)]. Contrast this to the usual electrostatic

---

[6] Another approach is to use an algorithm for the fields which is not subject to the Courant condition. See Nielson and Lindman (1973a, b) and Section IV.

code in which $\mathbf{E}(\mathbf{k} = 0) = 0$ or a specified function of time, as usually coded [case (b)]. ZOHAR has a switch to select between these options.

One difference is that (a) provides the restoring force for $\mathbf{k} = 0$ plasma oscillations. This means there is no "hole" in the spectrum of modes available for nonlinear processes such as parametric instabilities. The authors found this made a drastic difference in a study of high frequency decays occurring at one-quarter the critical density (Langdon et al., 1973).

Of course, this choice arises in electrostatic codes as well. Orens et al. (1970) observed an important qualitative change in plasma heating in counter-streaming plasmas when this boundary condition choice was changed. The point is that the electrostatic and electromagnetic codes arrive naturally at different choices.

2. *Open-Sided Boundary Conditions*

The work-horse PIC codes in laser–plasma interaction studies are periodic in y and open-sided, in some sense, on the left and right. One wants to illuminate the plasma from one side, and allow scattered light emerging from the plasma to leave the system. We will discuss ways to match the fields to vacuum on the outside of the system. Some procedures will be discussed for dealing with particles which reach a boundary. Also, many more diagnostics are needed.

a. *The Longitudinal Field.* For both ϕ and the correction $\delta\phi$ we wish to obtain a potential solution in the system that will be the same as if the grid had been extended to infinity on the left and right, with no charge outside the simulated portion of space. This does not appear to be possible with cyclic reduction methods for the Poisson solution (Hockney, 1970). Following Buneman (1973), we Fourier analyze in y and then use the known properties of Laplace's equation to fit boundary conditions to ϕ on the sides. Applying the discrete transform

$$\rho_{i,m} = \Delta y \sum_{j=0}^{N_y-1} \rho_{i,j} \exp(-ik_{y,m} y_j)$$

with $k_{y,m} = 2\pi m/L_y$, $y_j = j\Delta y$ to the five-point Poisson equation yields

$$\phi_{i-1,m} - 2\alpha_m \phi_{i,m} + \phi_{i+1,m} = -\Delta x^2 \rho_{i,m} \quad (19)$$

for $i = 0$ to N_x, where

$$\alpha_m = 1 + 2[(\Delta x/\Delta y) \sin(k_{y,m} \Delta y/2)]^2. \quad (20)$$

Subscript m is suppressed hereafter. Outside the grid, where $\rho = 0$, ϕ_i for

$m \neq 0$ has the form

$$\phi_i = ar^i + br^{-i},$$

where r is the larger root of the quadratic

$$r + r^{-1} = 2\alpha \tag{21}$$

(the other root is r^{-1}). Choosing vacuum solutions that do not diverge at $i = \pm \infty$, we must have

$$\begin{aligned} r\phi_{-1} &= \phi_0, \\ r\phi_{N_x+2} &= \phi_{N_x+1}. \end{aligned} \tag{22}$$

These boundary conditions and Eq. (19) constitute a tridiagonal system of simultaneous equations for ϕ_i. Applying the usual Gaussian elimination of the subdiagonal, one finds that all but the last of the new diagonal elements are now equal to r. The reason the solution of this set of equations is simpler is due to the existence of the factorization pointed out by Buneman (1973):

$$-\phi_{i+1} + 2\alpha\phi_i - \phi_{i-1} = (1 - r^{-1}E^{-1})(r - E)\phi_i,$$

where $E^{\pm 1}\phi_i \equiv \phi_{i\pm 1}$. Thus one can write

$$\begin{aligned} \rho_i \Delta x^2 &= (1 - r^{-1}E^{-1})\psi_i = \psi_i - r^{-1}\psi_{i-1}, \\ \psi_i &= (r - E)\phi_i = r\phi_i - \phi_{i+1}. \end{aligned} \tag{23}$$

The boundary conditions on ϕ mean that $\psi_{-1} 0 =$ and $(r - r^{-1})\phi_{N_x+1} = \psi_{N_x+1}$. We solve for the ψ's from left to right, and the ϕ's from right to left. These are the same operations as in the simplified Gaussian elimination.

For $k_y = 0$ the equations are underdetermined. A solution of Eq. (19) which is symmetric and is independent of the location of the boundaries of the grid is

$$\phi_i = -\tfrac{1}{2}\Delta x^2 \sum_{i'} |i' - i| \rho_{i'}.$$

This may be used to determine ϕ_{-1} and ϕ_0, which provides the needed boundary conditions for the solution of Eq. (19).

Finally we perform the inverse

$$\phi_{i,j} = (1/L_y) \sum_{m=0}^{N_y-1} \phi_{i,m} \exp(ik_{y,m} y_j)$$

to obtain the Poisson solution exact to within roundoff error.

The fast Fourier routine we use transforms a complex sequence, but we use it to transform two columns of real data at a time in the standard manner (Cooley et al., 1970). No additional storage is needed; the real parts are stored in rows 0 through $N_y/2$, and the imaginary parts are most easily placed in reverse order in rows $N_y - 1$ through $N_y/2 + 1$.

b. Absorbing Outgoing Electromagnetic Waves in a Dissipative Region. A common way to prevent light leaving the plasma from being reflected at the sides and reentering the plasma, is to place a dissipative region at the sides. The most obvious way to implement this is to introduce a resistive current into the Maxwell–Ampere law, Eq. (4). A disadvantage of this approach is that the resistive region must be quite thick in order to avoid both penetration by and reflection of the longest wavelength light.

Some improvement may be obtained by *also* introducing a "current" into the Faraday law, which becomes

$$-c\mathbf{V} \times \mathbf{E} = \partial \mathbf{B}/\partial t + \mathbf{J}_m,$$

with $\mathbf{J}_m = \sigma_m(x)\mathbf{B}$. This corresponds to a flux of magnetic monopoles. A method equivalent to this is to multiply $\mathbf{B}$ by a factor somewhat less than unity after advancing it in time (J. P. Boris, private communication, 1972). With σ_m equal to the electric conductivity, one finds that a normally incident wave ($k_y = 0$) is not reflected even if σ becomes large in a short distance. However this is not true for oblique incidence, and we will see that normal incidence is trivial to handle by other means.

J. P. Boris (private communication, 1972) has an interesting method for making the electric resistivity purely transverse, i.e., the current does not respond to the longitudinal part of the field, and the current does not itself produce any charge separation. The intent is that electrostatic activity in the plasma not be damped due to resistive currents driven by its fringe fields. The method appears to be equivalent to adding a loop current around each cell proportional to $\mathbf{V} \times \mathbf{E}$ in the cell. This is like a magnetization current, and is equivalent to adding a current

$$\mathbf{J} = c\mathbf{V} \times \mathbf{M} = -\mathbf{V} \times [\alpha(x)\mathbf{V} \times \mathbf{E}]$$

which has zero divergence and to which $-\mathbf{V}\phi$ does not contribute.

The disadvantage of a dissipative region is that it occupies a lot of memory space. To give good absorption, rather than reflection, for a range of frequencies and angles of incidence, the region must be about a wavelength thick, through which all fields must be defined.

c. A Simple Closure of the Maxwell Equations at the Open Boundaries. A simple boundary condition which solves the problem of closing the differenced Maxwell equations has been used by K. Sinz (private communication) and is an option in ZOHAR. It works well in some applications and illustrates several points applicable to more complicated boundary conditions to be discussed later.

By closure we mean the following: Consider the left side, and assume E_x, E_y, and B_z for $x \geq 0$ are to be advanced using the Maxwell equations. When we come to advance $E_{y,0,j+1/2}$ we need $B_{z,-1/2,j+1/2}$ which must be given by some additional condition in order to make the set of equations self-contained.

For a plane wave incident in the $-x$-direction we have $E_y = -B_z$. Sinz uses a time average of E_y and a space average of B_z to obtain the extra condition needed:

$$B_{z,-1/2,j+1/2}^{n+1/2} + B_{z,1/2,j+1/2}^{n+1/2} + E_{y,0,j+1/2}^{n} + E_{y,0,j+1/2}^{n+1} = 0. \qquad (24)$$

This is solved for $B_{z,-1/2,j+1/2}^{n+1/2}$ simultaneously with

$$(\partial_t E_y + c\, \partial_x B_z)_{0,j+1/2}^{n+1/2} = 0 \qquad (8a)$$

which involves the same quantities. The value of B_z thus obtained is stored in the appropriate column of the B_z array. The interior ($x \geq \Delta x/2$) values of B_z are advanced to time $t^{n+1/2}$ with Faraday's law. Then E_x and E_y for $x \geq -\Delta x/2$ are advanced using the Ampere–Maxwell law. This algorithm is all that is needed for a one-dimensional code and is used in this form in K. G. Estabrook's OREMP code.

At the end of the field integration we use the same idea, but without the time average, to obtain

$$B_{z,-1/2,j+1/2}^{n+1} + B_{z,1/2,j+1/2}^{n+1} + 2E_{y,0,j+1/2}^{n+1} = 0. \qquad (25)$$

The field averaging then gives all fields for $x \geq 0$.

In order to examine errors caused by nonnormal incidence and by the averaging, we consider the reflection of a plane wave incident on the boundary. Take

$$B_z(x,y,t) = B_i \exp[i(-k_x x + k_y y - \omega t)] + B_r \exp[i(+k_x x + k_y y - \omega t)]$$

for $x \geq \Delta x/2$ and eliminate $B_{z,-1/2,j+1/2}$ using Eq. (24), then express E_x and E_y in terms of B_z and solve for the ratio

$$B_r/B_i = (ck_x \cos \omega \Delta t/2 - \Omega \cos k_x \Delta x/2)/(ck_x \cos \omega \Delta t/2 + \Omega \cos k_x \Delta x/2). \qquad (26)$$

At small angles $c\kappa_x \to \Omega$ and the averages are the largest sources of error,

$$B_r/B_i = (\cos \omega \Delta t/2 - \cos k_x \Delta x/2)/(\cos \omega \Delta t/2 + \cos k_x \Delta x/2)$$
$$\approx (1/16)[(k_x \Delta x)^2 - (\omega \Delta t)^2].$$

In practice this ranges up to about 0.5%.

When the angle of incidence θ is not so small the largest error comes from assuming $E_y = -B_z$:

$$B_r/B_i \approx (ck_x - \omega)/(ck_x + \omega) = -\tan^2(\theta/2).$$

At 45° the reflection is 17%, or 3% in terms of energy. For many applications this is good enough. If only one angle of incidence is of interest, Eqs. (24) and (25) can be modified to correspond to that angle.

There is a flaw in all this so far: we have assumed that **E** is purely transverse. Suppose a point charge is held fixed near the boundary. The steady-state fields are $B_z = 0$ everywhere and $E_y = 0$ on the boundary, which is the same as the electrostatic field of the charge with a *conducting* boundary. The point charge is therefore attracted toward the boundary, which can be very troublesome in some cases. The correction to $\mathbf{V} \cdot \mathbf{E}$ has no effect on this.

The cure is to subtract $-\partial \phi/\partial y$ from E_y in Eqs. (24) and (25). Then the steady-state fields equal the electrostatic field as obtained by the Poisson solver with open-sided boundary conditions, and there is no force on the charge.

Incidentally, these boundary conditions are nearly equivalent to those used in an early version of the Los Alamos code WAVE (Nielson and Lindman, 1973). A difference is that the boundary conditions in the Poisson solver in WAVE are either $\phi = 0$ or $\partial \phi/\partial y = 0$ at the boundary. Thus a charge is attracted to, or repelled by, respectively, image charges outside the system.

In laser plasma interaction studies one often requires a wave $B_{z0} \cos(k_0 x - \omega_0 t)$ to propagate in from the left side (say). The simplest way to do this is to add $4B_{z0} \cos \omega_0 t$, evaluated at the appropriate times, to the right-hand sides of Eqs. (24) and (25). The fields of the incoming wave clearly satisfy the modified equations, as do outgoing waves. This procedure generalizes to more complicated input waveforms much more easily than does the common current sheet antenna method.

The boundary conditions on the fields B_x, B_y, and E_z are directly analogous, because of their similar spatial locations.

Lastly, we point out that these boundary conditions do not cause $\mathbf{V} \cdot \mathbf{E}$ or $\mathbf{V} \cdot \mathbf{B}$ to be altered at the sides. To see this, consider that the boundary conditions are used to determine only B_z and E_z at the sides. The fields E_x, E_y, B_x, and B_y are then advanced everywhere by the differenced Maxwell equations, and our remarks in Section III,C, regarding the preservation of $\nabla \cdot \mathbf{E}$ and $\nabla \cdot \mathbf{B}$, continue to apply.

d. Boundary Conditions for Waves Incident at "Any" Angle. In many problems light is scattered from the plasma in more than one direction or one has incoming light that is not a simple plane wave. In such cases the simple boundary condition of the last section is not adequate. Here we describe the boundary conditions normally used in ZOHAR to meet these requirements.

Lindman (1973) decomposes a complex wave propagating out of the system into a superposition of plane waves. For each plane wave a relation

$$c\, \partial A/\partial x = C(ck_y/\omega)\, \partial A/\partial t$$

holds, where A is the y- or z-component of the vector potential in the WAVE code, and $C \approx \cos\theta = (1 - c^2 k_y^2/\omega^2)^{1/2}$. The Coulomb gauge condition $\nabla \cdot \mathbf{A} = 0$ is used to determine $\partial A_x/\partial x$. He then regards C as a linear operator involving $\partial/\partial y$ and $\partial/\partial t$, which can be evaluated at the boundary without extrapolation. The form he found to be both stable and accurate was a partial fraction expansion. This concept was adapted to apply directly to the fields by E. Valeo, and our further discussion will mainly concern this form as incorporated into ZOHAR. After including incoming light, we use at the left side

$$B_z(0, y, t) + C^{-1}(-c\, \partial_y/\partial_t) E_y(0, y, t) = 2B_{z0}(y, t), \tag{27}$$

where B_{z0} is the desired incoming wave at $x = 0$, and

$$\begin{aligned} C^{-1} &\approx (1 - c^2 \partial_y^2/\partial_t^2)^{-1/2} \\ &= 1 + \sum_n \alpha_n/(\partial_t^2/c^2 \partial_y^2 - \beta_n). \end{aligned} \tag{28}$$

In terms of the computer code this means that

$$-B_z + 2B_{z0} = C^{-1} E_y = E_y + \sum_n E_n, \tag{29}$$

where E_n is the solution of

$$(\partial_t^2/c^2 - \beta_n \partial_y^2) E_n = \alpha_n \partial_y^2 E_y. \tag{30}$$

As in the preceding section, $-\partial_y \phi$ must be subtracted from E_y in these relations, and we must do time and space averages in order to make Eq. (29) second-order accurate. The only difference between Eqs. (29) and (24) is the E_n terms; these are known at the same times and positions as E_y, and therefore are time-averaged in the boundary condition.

Considering once again the reflection of a plane wave incident on the boundary, we find

$$B_r/B_i = (c\kappa_x \cos \omega \Delta t/2 - \Omega C \cos k_x \Delta x/2)/(c\kappa_x \cos \omega \Delta t/2 + \Omega C \cos k_x \Delta x/2).$$

Neglecting finite-difference errors, this becomes

$$B_r/B_i = (\cos \theta - c)/(\cos \theta + c).$$

This shows that the reflection in steady state is half the relative error in the expansion $C(ck_y/\omega)$. This is to be kept in mind when choosing the α and β coefficients.

Useful diagnostic information may be obtained from $B_z - B_{z0} \approx (B_z - C^{-1}E_y)/2$, which is the field of the outgoing waves only.

We have achieved good results with only two terms in the expansion. However, coefficients for a more accurate, three-term expansion have recently been given by Lindman (1975). His coefficients are $\alpha = (0.3264, 0.1272, 0.0309)$, $\beta = (0.7375, 0.98384, 0.9996472)$. With so few terms, the amount of computer memory and computation required is much less than with a dissipative region. Presumably because the β's are less than unity, Eqs. (30) do not require for stability any reduction in time-step below what is required by Maxwell's equations.

Lindman also discusses a difficulty with the transient response of the boundary conditions. If the prescribed incoming wave is turned on too suddenly, it takes a long time for the fields to settle into a steady state. If the angle of entry θ is close to 90°, then "too suddenly" may be an inconveniently long time. He describes a more complicated expansion which improves the transient response. We had experimented with boundary conditions which were other linear combinations of E_x, E_y, and B_z, different from Eq. (27), and we also found poor transient response, to varying degrees. Equation (27) is better in this regard, and has been very satisfactory so far. The reasons for the differences in transient response are not fully understood as yet. An analysis has shown how differences can arise among our forms and those of Lindman, but the analysis has not been carried to the point of quantitative comparison with observed performance.

As Lindman (1975) points out, the transient behavior is due in large part to the fact that a general disturbance contains Fourier components with $\omega < ck_y$. These do not propagate away from the boundary and the expansion

Eq. (28) cannot approximate the analytic continuation of $(1 - c^2 k_y^2/\omega^2)^{-1/2}$ for $\operatorname{Re} \omega < ck_y$ and $\operatorname{Im} \omega > 0$. Lindman's newer expansion differs in that it does approximate the analytic continuation.

We have observed another problem also caused by behavior of the expansion for $\omega < ck_y$. An instability in the fields E_x, E_y, and B_z occurs when there is a density jump parallel to and near the boundary. The jump supports a surface wave with $\omega < ck$. For some frequency intervals in that range, the expansion C is negative. Thus the direction of the Poynting flux is reversed, and energy flows into the system to drive up the surface wave. We have had no difficulty except when the distance from the density jump to the boundary is less than about $4c/\omega_p$. Perhaps Lindman's newer expansion would avoid the problem. However, all that is necessary to suppress the instability is to have an expansion which remains positive.

e. Particle Boundary Conditions. With periodic boundary conditions, when the particle coordinate y exceeds L_y after it is advanced in the mover, we simply subtract L_y before collecting this particle's charge density and the second half of its current density. If the new y is less than 0, then we add L_y.

In the open-sided version, particles which approach the left or right boundaries cannot continue to be advanced by the Newton–Lorentz equation, but must be disposed of in some manner. In ZOHAR we specify for each species an interval in x. When a particle moves outside this interval, the particle mover turns it over to a separate routine. We presently have three options:

1. The particle is deleted. A record is kept of the contribution to the charge density due to "lost" particles at their last positions.
2. The particle is reintroduced with velocity drawn from a half-Maxwellian (weighted by v_x) plus the oscillatory velocity as obtained from the time-integrated electric field at that location.
3. The particle is deleted as in option 1. At the same time, new particles are introduced as in 2 but at an average rate corresponding to a specified density and temperature rather than at the rate at which they arrive.

In each case we keep a tally of particle energies and any other property of interest. Options 2 and 3 allow the modeling of a neighboring reservoir of plasma, and/or prevent expansion of the plasma slab. For examples of how this may be used, see Section III, F.

E. Diagnostics

We describe a minimal set of diagnostics for understanding the results of laser–plasma interaction simulations. Many are obvious; the need for others is appreciated after experience both with and without them on specific problems. It is these parts of the code which are most frequently changed and should be kept flexible.

1. Particles

The most familiar particle diagnostic is the phase space scatter plot. Dots are plotted at positions given by two of the particle's coordinates, e.g., u_x and u_y. Often these plots will make direct connection to theoretical descriptions; particle trapping in waves is a well-known example.

While conceptually simple, implementation is complicated by not having all the particles in memory at once (see Section III, G, 2). If a phase space plot is to be made on the next time-step, ZOHAR scans the particles for minima and maxima, to see if plot limits must be expanded. The other problem concerns making more than one phase space plot at the same time-step. As an alternative to plotting several on one frame or running several plot channels, we specify an offset as well as a plotting interval. Thus one plot is made at steps 0, 100, 200, ...; another may be made at steps 49, 99, ..., etc. Usually the slight difference in time of plotting does not hinder comparisons.

An elaboration is to be able to plot one linear combination of particle coordinates versus another, skipping particles not satisfying two linear constraints. For example, plotting $u_x - u_y$ versus $x - y$ for particles with $a < x < b$ will display trapping in waves propagating at 45° in a slice of the system. Morse and Nielson (1971) present a useful application.

Also handled by the same subroutine are plots of $f(q)$ and contours of $f(q_1, q_2)$, where q, q_1, and q_2 are any particle phase space coordinates, and q may be u^2 (these are projections over the other coordinates). This includes, for example, $f(u_x)$, $f(u^2)$, and $\rho(x, y)$. In the open-sided version, the transform $\rho(x, k_y)$ is often used, and we also plot statistics on particles leaving the system.

2. Fields

Obvious diagnostics are contour plots for B_z, E_z, and ϕ. We also have plots of $E_x - E_y$, $B_x - B_y$, and $-\nabla\phi$ consisting of an array of little arrows whose directions and lengths indicate the vector value at that point in space.

Fourier mode energies for B_z, E_z, and ϕ are indicated by an array of vertical lines whose lengths are proportional to the logarithms of the energies of the modes, over a range of 10^4. This is used to point out modes which should be watched more closely by other means.

In the open-sided version, plots of $B_z(x, k_y)$, etc., for specified y modes, aid both in interpretation of results and in separating a signal from other effects or from noise.

3. Histories

Many quantities are saved each time-step to provide "history" plots. Most of these are energies and momenta for fields and particle species. Also saved are specified field probes, mode energies, and mode amplitudes, e.g., for $E_z(x, k_y)$. The latter include E_z and B_z at $x = 0$ for the outgoing waves only, computed as in Section III, D, 2, d.

4. Remarks

Usually it is not possible to foresee exactly which plots will be decisive in understanding a computer run. Compared to an electrostatic code, there are many times more quantities to monitor. In order to get what is needed the first time, much thought is given to a judicious choice of diagnostics, but we prefer to err on the side of inclusion rather than omission. The result is a little like a telephone book: a lot of data, much of which will not be useful, but you want to have it all because you do not know in advance what you will need.

For such reasons, we believe this type of computing cannot be seriously pursued without access to a high resolution and high volume graphical output device, such as a cathode-ray tube and camera. Mechanical plotters are too slow, especially for contours and phase space plots. Impact "printer plots" are too coarse, too often hiding or distorting valuable detail. Microfiche is the most compact and easily handled format.

Sometimes one must resort to numerical printouts. Again the telephone book analogy holds, and microfiche is the best available format.

F. APPLICATIONS

Representative results obtained with doubly-periodic boundary conditions have been discussed, e.g., by Morse and Nielson (1971) on the Weibel instability, Lee and Lampe (1973) on electron beam filamentation, and Langdon and Lasinski (1975) on laser light filamentation. Here it is more interesting to illustrate problem definition and diagnostics with problems run on the open-sided version of ZOHAR.

The examples all concern instabilities in underdense plasma irradiated with intense coherent light, proceeding from a physically simple case to more involved problems. We simulate a section of the plasma corona encompassing quarter-critical density (i.e., where the plasma frequency ω_{pe} is half the laser frequency ω_0). Two high frequency instabilities are possible in this region: (1) $2\omega_p$—the decay of the incident wave into two electron plasma waves near quarter-critical density; (2) "Raman" scattering—the decay into an electron plasma wave and an electromagnetic wave, at or below quarter-critical density. In addition, stimulated Brillouin scattering—the decay into a sound wave and an electromagnetic wave—is possible throughout the underdense region. Other processes such as filamentation are not discussed here.

First, a digression on units: for laser–plasma interaction problems we set c, the laser frequency ω_0, and $-q_e/m_e$ equal to unity in the code. This prescribes a set of units in which numbers in the code are easily related to ratios of physical interest. Since now $\omega_{pe}^2 = -\rho_e$ in Heaviside–Lorentz units, density is measured in units of the critical density. For the potential, $\phi = 1$ means $-q_e \phi = m_e c^2$, i.e., 511 keV. The cyclotron frequency $\omega_{ce} = |q_e| B_z/(m_e c) = B_z$, so that unit B_z means $\omega_{ce} = \omega_0$, which occurs near 100 MG for a laser wavelength of 1.06 μm.

1. The "$2\omega_p$" Decay Instability with Fixed Ions

In our first example, with the two-dimensional configuration of ZOHAR, we use a fixed ion background to eliminate all decay processes except $2\omega_p$ and Raman. The density profile is shown in Fig. 2. Laser light propagates as a plane wave along the x-direction from the left, with $E_y = B_z = 0.1$ (peak) which gives an electron oscillation amplitude of $0.1c$, corresponding to $\approx 10^{16}$ W/cm^2 for 1.06 μm wavelength. Sinz-type boundary conditions are used for the transverse fields since we will not look for light waves emerging from the slab at angles, and we wish to save the space required to stabilize our version of Lindman's boundary conditions.

The nonuniform electron density is created by varying the electron spacing. This is more efficient than loading a uniform number density and varying the "weight" (charge and mass) of the particles, as some have done. During the run, the electrons in a region of interest will have come from a large area and

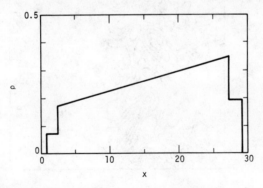

Fig. 2. Density profile of the plasma slab for the example of $2\omega_p$ decay with stationary ions. The linear portion rises from $\rho = 0.1725$ to 0.35 between $x = 2.4$ and 27.3. The short sections at each end are quarter-wavelength matching sections with densities chosen to minimize reflection of the laser light.

have various weights. For a given total number of particles, statistics and thermal noise are then somewhat poorer than if the electrons had equal weights. Note also that in two dimensions the number of particles per Debye square, $n_e \lambda_{De}^2$, is independent of n_e for electrons of equal weight. The electron charge is chosen to give the prescribed charge density, and the mass is obtained from q/m. The ion charge density is obtained from a function $\rho_i(x)$ constant in time.

For this problem the system size is 30 by 25.6 and we use a grid with 160 by 128 cells. This gives $\lambda_D/\Delta x = 0.5$ at density $\rho = 0.25$, more than enough to avoid the grid instability (Langdon, 1970a,b), and some smoothing is obtained without interfering much with Bohm–Gross dispersion. We use $N_e = 10^5$ simulation electrons, for which $n_e \lambda_D^2 = 1.4$ and $n_e \Delta x \Delta y = 5.2$ at density 0.25. This is a very low particle density compared to what is required in some plasma problems; we will have more to say on this point in Section III, F, 4. The particle velocities are taken from a random number generator, with a Maxwellian distribution at temperature 1.28 keV ($v_t = 0.05c$). ("Quiet starts" have not seemed promising in these applications.) Particles hitting either end of the slab are reemitted at this temperature.

We have found it desirable to "turn on" the laser gradually, bringing the amplitude from 0 to 0.1 smoothly over a time 20 in this case. If this is not done the sudden arrival of the light gives a y-independent perturbation to the plasma, which artificially enhances Raman backscatter relative to the $2\omega_p$ decay.

By time 180, plasma waves from $2\omega_p$ decay are visible in the noise. As can be seen from the ϕ contours in Fig. 3, the waves initially show localized growth,

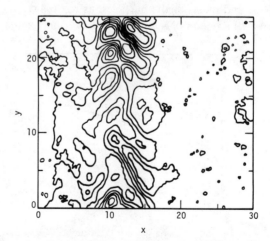

FIG. 3. Potential contours at time 280 showing the localized initial growth of the plasma waves. Plasma waves from Raman backscatter would be y-independent. The interval in ϕ between adjacent contours is 0.01. The wider-spaced contours elsewhere are due to thermal noise.

suggesting that $2\omega_p$ decay is absolute in x and y. The details of spatial location depend on the random loading and vary from run to run. Later, the waves spread out in y, with modes 0–4 the strongest. Finally, mode 2 dominates.

For this intensity and gradient length $[L = (d\rho/dx)^{-1} = 140]$, we found Raman decay to saturate at a level where little electron trapping occurred, in one-dimensional simulations (Fig. 4). This effect has been noted independently by Biskamp and Welter (1975), who call this the "nondissipative" regime. However, we have never seen this behavior in two-dimensional simulations; as in this example, we find instead that $2\omega_p$ decay and heating dominate.

With particle time-step $\Delta t = 0.4\omega_0^{-1}$ (and $0.1\omega_0^{-1}$ for the fields), this run took 2500 steps at 5–6 sec/step on the CDC 7600, including diagnostics.

How much smaller could this problem have been made and still produce interesting information? We have run the same problem with half as many particles. The thermal electric field noise is doubled, as expected, and the $2\omega_p$ growth is about 50 time units earlier, but the mode numbers which appear, the growth rate, and initial saturation levels are very close. General features seem satisfactory with 50,000 particles.

In a number of runs with smaller y dimensions, $L_y = 8$–13, we were able to reach conclusions (e.g., on the competition with Raman, absorption percentages versus gradient length and intensity) which still seem valid. Large L_y, which gives close spacing of k_y modes, is necessary more for questions of turbulent spectra and the absolute versus convective nature of the instability. Incidentally, we divide system energies, etc. by L_y before plotting, so that with our

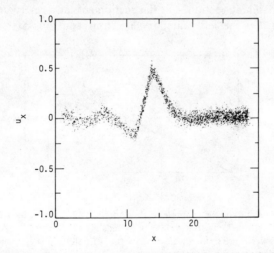

FIG. 4. Phase space for a one-dimensional (x only) simulation, which therefore includes only Raman scattering. For the same light intensity, density gradient, etc., as in the two-dimensional example, the plasma wave saturates at a level just below where wave breaking or appreciable particle trapping would occur; as seen here, only a few electrons "break away" from the instability as heated particles.

other normalizations the diagnostics do not vary with changes in N_y, L_y, N_e, etc., unless the physics has changed.

To shorten the system in x one could probably trim off some of the slab at the right. We have simulated $2\omega_p$ decay with gradient lengths as small as $L = 63$; this instability will operate with even smaller L, but the character of the light absorption already has changed relative to larger L cases. The largest L we have used is 660.

2. *The $2\omega_p$ Decay Instability with Mobile Ions*

In the above example the decay field magnitudes reach ~ 0.2 at frequency ≈ 0.5. This implies oscillation velocities of ~ 0.4, compared to the thermal velocity $v_t = 0.05$. Therefore ponderomotive forces greatly exceed thermal forces, and ion motion must be included in long runs.

The next example has a mass ratio $m_i/m_e = 100$, sufficiently large to separate electron and ion time-scales. In addition to ion motion driven by the decay waves, Brillouin scatter now occurs, and the slab expands hydrodynamically. The density profile shown in Fig. 5 minimizes interference from these effects. A matching section cuts reflection from the right end of the slab, which otherwise is amplified by Brillouin scatter to high levels. Electrons are reemitted thermally from the end of the matching section. Ions are reemitted thermally from a point in the flat section; a fixed ion density is used to the right of that point. This prevents expansion of the slab to the right, and maintains the matching section. The flat section acts as a reservoir for expansion to the left. On the left boundary electrons and ions are "absorbed."

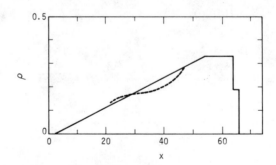

FIG. 5. Density profile for the example of Section III, F, 2. The gradient scale length is $L = 160$. The system size is 75 by 16 with 300 by 64 cells giving $\lambda_D/\Delta x = 0.4$. Again the laser is normally incident from the left, with $E_y = v_{osc} = 0.1$. Initially, $\lambda_{Di} = \lambda_{De}$; one could argue for an even higher ion temperature on the grounds that Brillouin scattering will heat the ions, as does occur at this intensity ($v_{osc}/v_t = 2$). The dashed line shows density steepening at time 1280 due to the ponderomotive force of the decay waves which first depresses the density locally, and then the left side of the depression blows off to the left.

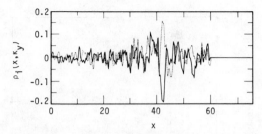

FIG. 6. Real (solid line) and imaginary (dotted line) parts of $\rho_i(x, k_y)$ for mode 3 at time 640, showing ion perturbations driven by the decay waves. Quarter-critical is at $x = 21.7$. Modes 1–4 show similar behavior.

Lindman's boundary conditions are used for the fields because now we wish to observe the spectrum of harmonics of $\omega_0/2$ emitted in various directions.

We had expected the intense decay wave fields to depress the local density. It had also been suggested that the decay waves would themselves decay via the oscillating two-stream (OTS) instability. In the simulation, the ponderomotive force from the superposition of several decay waves drives ion perturbations, as seen in Fig. 6, without need for the OTS instability feedback mechanism. This later subsides; the lasting feature is a steep density jump through quarter-critical (Fig. 5) which stabilizes $2\omega_p$ at a modest level.

3. Raman and Brillouin Sidescatter

Next we rotate the polarization of the laser light by 90°, so its fields are E_z and B_y (in the two and one-half dimensional configuration of ZOHAR). This lets us study the processes occurring in the other plane—Brillouin and Raman sidescatter—but eliminates $2\omega_p$. In this geometry the scattered light is produced preferentially in the $\pm y$-directions. It then diffracts and refracts toward the left and exits at an angle. Therefore Lindman's boundary conditions are used for E_z and B_y, while Sinz's are used for E_y and B_z as in the example of Section III, F, 1. The system size is 60 by 90; the large y dimension minimizes spurious effects due to periodicity in y. The density profile is similar to that of the example of Section III, F, 2.

We monitor the first 15 Fourier modes $E_z(x, k_y)$ at various x positions, so that the spectrum, direction, and point of origin of scattered light components can be determined. Figure 7 shows the spectrum of scattered light (mainly by Raman) emerging at the left boundary in mode 7. The spectra of modes 1–6 are weaker but otherwise similar. Most modes show a time dependence of the scattering amplitude which is associated with density profile changes. Brillouin shows most strongly as sidescatter in mode 12; it too shows time dependence. Its spatial structure is shown in Fig. 8.

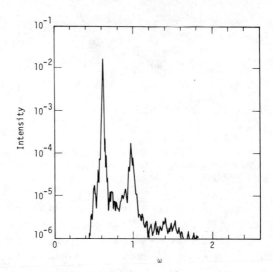

FIG. 7. The spectrum of scattered light in mode 7 at the left boundary shows mainly Raman-scattered light (frequency $\omega \approx 0.6$, exit angle 52°) and a smaller amount of Brillouin scattering.

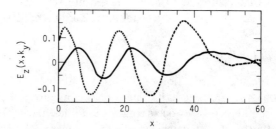

FIG. 8. Spatial structure of Brillouin sidescattered light for mode 12. The point of origin is near its "turning point" ($k_x = 0$); at lower densities leftward propagation is more rapid (k_x increases), and the light emerges at an angle of 57°. Solid line: ReE_z; dashed line: ImE_z.

Late in the run an interesting density depression forms (Fig. 9). This sort of structure has developed in other runs in this polarization; the factors determining their formation are under study.

4. Discussion

The progression above, studying the nonlinear evolution and competition among the instabilities occurring in underdense plasma in successively more complicated plasma conditions, has been carried further. Extensions have included the use of a spatially-shaped laser beam (a "finite spot") in place of

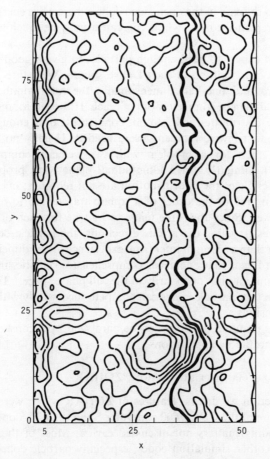

FIG. 9. Ion density in the sidescatter run (example of Section III, F, 3), showing the density depression which forms near the quarter-critical surface (heavy line). The minimum density at this time ($t = 1120$) is 0.04 critical.

the simple plane wave, and initial plasma densities which are functions of y as well as x. Details on the physical results will be given elsewhere.

We conclude by discussing two matters strongly affecting the practicality of this type of simulation. The first concerns collision times, while the second is the density of particles required to represent the velocity distribution in multidimensions.

It has often been stated that to obtain "collisionless" behavior, the collision times must be longer than the length of the run. Fortunately this is overly pessimistic. What appears to be closer to the truth is that collision times should exceed, e.g., instability exponentiation times and trapping times.

Particularly in the case of the open-sided code, where wave energy and thermal particles are replenished, the whole run can usually be much longer than the latter times.

One might expect to need enough particles to give a good representation of the velocity distribution in each Debye square. With three velocity components ("two and one-half dimensional") the total number of particles required would then be completely out of reach. However, for our applications, details of the distribution in v_z do not concern us; only enough particles to represent the first few v_z moments were needed. Usually no more particles were needed than for a problem with two velocity components ("two-dimensional"). Similarly, for a longitudinal plane wave propagating in the x-direction, details of the v_y distribution are not required; one needs enough particles in an area of the order of a square wavelength to represent the v_x distribution well. Since the same tends to be true for each of several superimposed waves traveling in various directions, the particle density required is much lower than one might infer from experience in one dimension. Another way of saying this is that one needs comprehensive statistics in *projections* of the distribution, and not necessarily in the full phase space.[7] This economy is a main reason that particle codes can compete successfully with Vlasov codes in multidimensions.

Caution is required, but one can be paralyzed by a conservative attitude into missing profitable applications.

G. LTSS 6600-7600 Implementation: ZOHAR

In this section we discuss features of ZOHAR which were influenced by the architecture of the CDC-7600 computer [8] and the LTSS operating system, and other coding matters not discussed earlier. Most of these features are applicable to other simulation codes, especially particle codes, and may be of wider interest since this machine and operating system have been selected for the National CTR Computer Center. Many similar considerations will apply for machines of different architecture.

1. *Field Storage*

In an electromagnetic code there are more fields to be stored than in an electrostatic code. Processing the particles involves six fields in two dimen-

---

[7] Of course, for some phenomena in CTR, such as cyclotron harmonic waves, the dimensionality of the relevant projection is not as low as in the example here, and many times more particles may be needed for such problems.

[8] Reference to a company or product name does not imply approval or recommendation of the product by the University of California or the U.S. Energy Research and Development Administration to the exclusion of others that may be suitable.

sions $(E_x, E_y, B_z, J_x, J_y, \rho)$ and ten in two and one-half dimensions (add B_x, B_y, E_z, J_z). The fast memory on the 7600 ("small core," or SCM) is only 2^{16} words, of which about $57,000_{10}$ are available to the user under LTSS. For many applications of current interest, one of these fields at most can be fit into SCM together with the code itself. While the particles are being integrated we have chosen to store all these fields in "large core" (LCM) and use the available SCM for collection of diagnostic information and other purposes. There are $400,000_{10}$ words available in LCM, of which some are used to copy the portion of SCM used.[9]

LCM may be accessed in two ways. The more common way is a block copy of up to 1023 contiguous words from LCM to SCM or vice versa. The transfer takes a long time to start, but then goes at the very rapid rate of one word per minor cycle (27.5 ns). During this time the CPU is not otherwise usable. One can also transmit single words between LCM and CPU registers. A single word read from LCM takes longer than from SCM, but much less than the time to start a block copy, and the CPU can be used during the wait. Furthermore, LCM is organized into groups of eight words; reading one word causes all eight to be copied to a "bank register," from which any of the eight can be accessed very quickly until that bank register is overwritten.

We have chosen to exploit this feature in the particle mover. Values for several fields at the same grid point are stored contiguously, followed by the values for the next grid point to the right. Careful grouping of fields and choice of subscript limits greatly decrease the time lost waiting for LCM.

ZOHAR has one array in SCM, slightly larger than the number of cells. This limits the problem size to about 30,000 cells even with the use of overlays for the code. However, in two and one-half dimensions the size of LCM imposes a similar limit on problem size. The availability of the extra field array greatly facilitates many diagnostics and other operations. Having it in SCM improves the efficiency, especially of Fourier transforms and Poisson solutions.

The spatial Fourier transforms rely on a Compass assembly-coded "Fast Fourier Transform" routine (FFT2) written by R. Singleton of Stanford Research Institute for the 6600 and "tuned" for the 7600 by L. Sloan. This routine was selected for speed and simplicity of use from a large collection of FFT routines. Its speed on long sequences is good (e.g., 6.8 ms for length 1024) but it performs especially well on the short sequences arising in two-dimensional problems, due to its ingeniously simple indexing. It is not necessary

---

[9] When two or more codes fit in LCM, this feature enables the system to make SCM available very quickly to another code in LCM when the currently executing code can no longer make use of the central processor (CPU); for example, when it must wait for input/output to complete. However, for a large code needing all available memory, this represents a loss of up to 57,000 words of LCM storage.

to copy the rows or columns out of the two-dimensional array into separate working storage, nor to set up tables of sines or "bit-reversed" indices. Most readers are advised not to attempt writing their own FFT for production use; it takes a lot of thought and familiarity with the algorithm just to appreciate what Singleton has done, without having improved on it.

Most efficient FFT routines, including FFT2, suffer from the well-known limitation that the sequence length must be a power of 2. We found all too often that we were forced to double the grid size when a 50% increase would have been adequate for the physics. The excess grid size requirement meant extra expense or sometimes made the problem too large for memory. To alleviate this without losing much speed, ZOHAR now allows the grid size to include a single factor of 3, i.e., to be $3 \cdot 2^n$ as well as 2^n. For lengths of the form $3 \cdot 2^n$ we use the FFT factoring principle in a routine which calls FFT2 three times with length 2^n and does the necessary other manipulations. The cost per cell is little higher than for powers of 2. (In the open-sided version the x-dimension is arbitrary.)

2. *Particle Processing, I/O, and Data Structure*

For most two-dimensional problems of interest, there will not be enough LCM space to contain all the particle coordinates, so that disk storage must be used. On the other end of the size scale, we can only have a few particles at a time in SCM where they can be most efficiently processed. In between, we would like to transfer at least 10,000 words at a time between disk and LCM in order to decrease the fraction of time spent waiting for disk arm positioning. In this section we discuss measures to decrease the I/O and handle it efficiently without unnecessary complexity, and a data structure which aids bookkeeping operations, such as deletion and creation of particles.

While being advanced and diagnosed, the particle coordinates are in SCM a few hundred at a time, stored as x, y, u_x, u_y for each particle in consecutive words in floating-point format. At the beginning of the "block" of particles is a 24-word "header" containing such information as species identity, which mover to use, how many particles are currently in the block, time-step, etc.

The particle integrators ("movers") were written in Compass assembly code (by R. Judd and R. Williams with a fast square-root routine by H. Nelson), except for the portion dealing with boundary conditions on the left and right sides in the aperiodic version. That is handled by a subroutine called as needed by the mover and written in FORTRAN to facilitate changes.

While the block is in LCM or on disk, the coordinates are "packed," e.g., x and u_x are stored in parts of one 60-bit word. The low order bits of the floating-point velocity are replaced by the position in fixed-point, normalized to the lowest power of 2 not less than the dimension. A checksum for the block is stored in the header. Packing roughly halves the I/O time and one's cross section for disk errors. Unpacking the coordinates in SCM simplifies diagnostics. The conversions are done by a Compass assembly-coded routine (written by M. DeGraff) and take very little time.

The resulting ratio of I/O to CPU time is low enough (with CDC type 817 or 819 disks) that a simpler alternative to the usual triple-buffering (see, e.g., Boris and Roberts, 1969) is preferable. With two buffers in LCM, while the particles in one are processed the other buffer (previously processed) is written to disk. When the write ends, a read is begun of the particles to be processed next. The read is to be finished by the time the processing of the first buffer is complete. This sequence is repeated with the roles of the buffers reversed. Double buffering has the advantage over triple that only one working file or set of files is needed, instead of two which must be on separate disks. Also, record lengths are larger for a given amount of total LCM buffer space, saving on disk seek time. When the ratio of I/O to CPU time is small enough, we are aware of no disadvantages.

Following D. Fuss, at the end of the particle integration, the last two buffers have not been written back to disk. On the next time-step, processing begins with these buffers and continues in reverse order. This saves, for example, half the I/O when half the particles will fit in LCM.

3. *Histories*

There is far too much information to be left in memory, so we use a procedure to write the information to disk efficiently. The various parts of the code, which calculate the quantities to be recorded at each time-step, call a subroutine entry which stores an item in a buffer. The buffer length is $512 + N$, where N is the total number of different items to be recorded. If the items are being stored past word 512, at the end of the time-step the first 512 words are written in disk file HISTORY and the remaining items are copied to the beginning of the buffer. The reason for this procedure is that the type 817 disk is most efficiently written in full sectors, which begin every 512 words. The number of quantities saved is typically 15–40, so no disk write occurs for most time-steps.

From time to time during a run, the code plots the saved quantities, and others derivable from them, over an interval of time overlapping with the last interval plotted. The particle buffers provide a convenient large block of storage into which the necessary portion of the HISTORY file can be read and manipulated. These plots are sufficient for a survey of the results of a run. For more careful work, we have an interactive postprocessor, ZED, which reads the HISTORY file, performs operations on the data (such as calculating power spectra), and makes plots, under teletype control.

The HISTORY file format facilitates interpretation by ZED. Word 0 is a pointer to where the modes begin. Words 1 through N are ASCII labels identifying the N quantities to be recorded, which are not the same from problem to problem. The following records of $N + 1$ words each are the step number n ($n = -1, 0, ...$) followed by the quantities at time $n\Delta t$, except for the particle momenta which are at time $(n + 0.5)\Delta t$.

The modes part of the file is structured similarly. The first word contains N, the number of modes saved. A number identifying the quantity and the two subscripts are packed into each of the next N words. The following records of $N + 1$ words each are the step number $(0, 1, ...)$ and the N quantities. Included with the modes are other doubly indexed quantities, such as fields at specified positions.

4. Operational Features

a. Restarts. Only the smallest problems are run to completion at one time. One needs a means to save the state of the problem while its progress is considered before continuation. Under LTSS the simplest way to do this is for the code to close all output files and terminate leaving its core image on disk (the "drop file"). We chose instead to have the code write the field arrays and other necessary data into a disk file, STATE, which is then saved along with HISTORY and the particle files. This procedure has several advantages: One may resume the run with a new version of the code; for example, with a new diagnostic. Particularly complicated initial conditions may be generated by a separate code. Or the problem may be altered before continuation by manipulating STATE.

b. Run-Time Interaction. Often the user chooses to run part of a problem himself, for example, at the start of a run, when an error in problem generation can often be caught before much time is wasted. It is useful to be able to interact to some degree with the executing code through the teletype. At a convenient point during the time-step, ZOHAR checks for a waiting message from the teletype. This is most commonly a change in diagnostic plotting parameters or a request to empty output buffers so that plots can be examined while the code continues running.

c. Error Detection and Recovery. Any code whose problems run for many hours will encounter hardware errors and system software errors. For a particle code on LTSS these are usually disk errors on the particle files, or memory parity errors. It is prudent to save the state of the problem every ten minutes or so in order to be able to recover after such an error. In ZOHAR this is done by emptying the particle and HISTORY buffers and writing the STATE file as for a normal termination, then copying these files. The copies are reread and checksums compared before continuing. If the disks are not in top condition, there is an unacceptable possibility that a disk error will occur during this large volume of I/O. Avoiding this requires two sets of backup files; our restart files are copied to alternate sets so that one set is always good. At longer intervals everything is written onto tape.

In order to detect failures not reported by the system to the code, each particle block header contains the step number and a checksum for the block, which are checked and updated each step. This measure is essential; it is in this way that frequent errors caused by deadstarts are detected.

When an error occurs, the "controller" program runs a utility routine which disposes of the plot and printer files, and copies the latest good set of backup files into the working files. The controller then restarts ZOHAR in the usual way.

IV. Algorithms with Special Stability Properties

For problems with many relativistic electrons, the fact that the wave phase velocity [given by Eq. (9)] drops well below c at short wavelengths produces unphysical disturbances. J. P. Boris (private communication, 1972), Boris and Lee (1973), and Haber *et al.* (1973) have mentioned noise produced by Cerenkov emission from particles whose velocities exceed the minimum phase velocity. Godfrey (1974) has examined collective instabilities involving interaction between relativistic electron beams and these slow light waves. This has created interest in codes in which vacuum light waves travel at the correct velocity and other features also improve stability. The best examples seem to be an old one-dimensional algorithm and its two-dimensional generalization.

A. A One-Dimensional Algorithm

Apparently the first electromagnetic algorithm was devised by J. M. Dawson in 1965. He separated the transverse fields into left- and right-going components. By choosing $\Delta x = c\Delta t$, these components are advanced in time simply by shifting the values over one cell and adding current contributions from particles whose paths intersect the light rays (Langdon and Dawson, 1967). We will discuss a variant of this algorithm as formulated and implemented by Langdon and revived for laser–plasma studies by Cohen *et al.* (1975).

By adding and subtracting the Maxwell equations governing E_y and B_z, we obtain the two equations

$$(\pm c\, \partial/\partial x + \partial/\partial t)^{\pm}F = -\tfrac{1}{2}J_y \tag{31}$$

for the left- and right-going fields

$$^{\pm}F = \tfrac{1}{2}(E_y \pm B_z),\tag{32}$$

in terms of which $E_y = {}^+F + {}^-F$, $B_z = {}^+F - {}^-F$, the Poynting flux is $c({}^+F^2 - {}^-F^2)$ and the energy density is ${}^+F^2 + {}^-F^2$. When defined on a grid with $\Delta x = c\Delta t$, the fields are advanced in time according to

$$^{\pm}F_i^{n+1} = {}^{\pm}F_{i\mp 1}^n - \tfrac{1}{4}(J_{y,i\mp 1}^{n+1/4} + J_{y,i}^{n+3/4}),\tag{33}$$

where $J_y^{n+1/4}$ and $J_y^{n+3/4}$ denote current densities computed from velocities $v_y^{n+1/2}$ assigned to the grid by linear weighting according to the positions x^n and x^{n+1}, respectively. This corresponds to integrating (31) along its characteristics. Linear interpolation is used to obtain the fields at the particle positions, as in electrostatic codes.

This scheme, Eq. (33), was devised to be time-centered (reversible and second-order accurate) and to render the radiation self-force accurately.[10] The stability against nonphysical beam modes, noted by B. Cohen and analyzed by Godfrey and Langdon (1976), is an unforeseen bonus.

There are other advantages to this algorithm: The fields are known at integral positions i and times n, for the convenience of particle movers and diagnostics, without the averaging and unaveraging procedures of Section III,C. Also, outgoing wave boundary conditions for ${}^{\pm}F$ are trivial. (In fact the original application involved a bounded plasma radiating into a vacuum.)

B. Two-Dimensional Fourier Transform Codes

In order to achieve dispersionless vacuum wave propagation as in this case, but without the restrictions to one dimension and $\Delta x = c\Delta t$, codes have been developed in which the fields are Fourier transformed in space and advanced in time with the correct phase change for each mode (Haber et al., 1973; Lin et al., 1974). So far, none has generalized the above one-dimensional algorithm in such a way as to retain its special stability properties. However, this can be done by replacing Eqs. (32) and (33) by

$$^{\pm}\mathbf{F} = \tfrac{1}{2}(\mathbf{E} \mp \hat{\mathbf{k}} \times \mathbf{B}), \quad \hat{\mathbf{k}} \equiv \mathbf{k}/k,\tag{32a}$$

$$^{\pm}\mathbf{F}^{n+1} = e^{\mp ick\Delta t}({}^{\pm}\mathbf{F}^n - \tfrac{1}{4}\mathbf{J}^{n+1/4}\Delta t) - \tfrac{1}{4}\mathbf{J}^{n+3/4}\Delta t.\tag{33a}$$

[10] The radiation drag $(\dot{v}_y)_{\text{rad}} = -v_y q^2/2mc$ is proportional to velocity in one dimension, a surprising result which can be reconciled with the point particle three-dimensional case by considering the radiation from a moving spherical charge shell in the limits of large and small ratio (radius/wavelength). The fact that a plasma slab does not quickly radiate away its thermal energy can be understood in terms of the radiative normal modes of the slab (Langdon, 1969).

Expressing (33a) in terms of **E** and **B** gives

$$\mathbf{E}^{n+1} = \mathbf{E}^n \cos ck\Delta t + i\hat{\mathbf{k}} \times \mathbf{B}^n \sin ck\Delta t - \tfrac{1}{2}\Delta t(\mathbf{J}^{n+1/4} \cos ck\Delta t + \mathbf{J}^{n+3/4}), \quad (34a)$$

$$\mathbf{B}^{n+1} = \mathbf{B}^n \cos ck\Delta t - i\hat{\mathbf{k}} \times \mathbf{E}^n \sin ck\Delta t + i\tfrac{1}{2}\Delta t \hat{\mathbf{k}} \times \mathbf{J}^{n+1/4} \sin ck\Delta t. \quad (34b)$$

It seems simplest to correct $\mathbf{V} \cdot \mathbf{E}^{n+1}$; then $\mathbf{E}^n$, $\mathbf{J}^{n+1/4}$, and $\mathbf{J}^{n+3/4}$ can include both transverse and longitudinal contributions.

Note that there is no Courant limitation on Δt for stability in vacuum.

In two dimensions it may be troublesome to collect two current densities. The current density $\mathbf{J}^{n+1/2} = \tfrac{1}{2}(\mathbf{J}^{n+1/4} + \mathbf{J}^{n+3/4})$ of Section III,C could be substituted for both currents in Eq. (34), or could be added halfway through the rotation of $^\pm F$:

$$^\pm\mathbf{F}^{n+1} = e^{\mp ick\Delta t/2}(^\pm\mathbf{F}^n e^{\mp ick\Delta t/2} - \mathbf{J}^{n+1/2}\Delta t/2)$$

which changes the current contributions in Eqs. (34a) and (34b) to $-\mathbf{J}^{n+1/2}\Delta t \cos \tfrac{1}{2} ck\Delta t$ and $i\hat{\mathbf{k}} \times \mathbf{J}^{n+1/2}\Delta t \sin \tfrac{1}{2} ck\Delta t$, respectively. These and other choices (e.g., Haber *et al.*, 1973) agree to order Δt^2 but differ in stability properties for $ck\Delta t \gtrsim \pi/2$.

V. SUPERLAYER

SUPERLAYER was the most elaborate of a series of codes developed to simulate the formation of cylindrical layers of relativistic electrons in the Astron experiment at Livermore. As envisioned, a number of pulses from an electron accelerator were to be "stacked" into one layer until sufficient current density was built up to reverse the direction of the axial magnetic field inside the "E-layer." This configuration provides stable confinement and heating of plasma inside the E-layer.

This particle simulation code is of unique interest as a design tool rather than for the study of small-scale basic plasma phenomena. SUPERLAYER was used to optimize experimental parameters and try new ideas. The degree of influence of the experiment and computation on each other is probably unprecedented in CTR. The code work also stimulated radical revision of some key theoretical impressions of the stacking process. The SUPERLAYER code is discussed by Brettschneider *et al.* (1973); later code work and detailed Astron studies are reported by Byers *et al.* (1974).

Some features of the code itself are of interest in the context of this article.

The fields are found with the aid of the potentials A and ϕ, normally in the Lorentz gauge. The wave equation is integrated with an alternating direction implicit (ADI) scheme. An implicit scheme removes a limitation on the time-step otherwise imposed by the small charge relaxation times of resistive layers in the machine. It also damps the solutions of the wave equations toward the potentials corresponding to the Darwin model (Nielson and Lewis, this volume), while retaining retardation for short time-scale effects. This bridge between the fully electromagnetic and magnetoinductive-electrostatic limits matched the physical requirements of Astron, which featured rapid transients but no oscillatory fields (cavity modes) in the axisymmetric behavior accessible to SUPERLAYER.

This code is now used to study plasma buildup by neutral injection in mirror experiments.

Acknowledgments

The ideas of many people have contributed to the design of ZOHAR: C. K. Birdsall, J. P. Boris, O. Buneman, J. M. Dawson, K. Estabrook, D. Fuss, E. L. Lindman, B. McNamara, and E. Valeo. We have benefited from K. Estabrook's experience in extending and applying a version of ZOHAR. R. Judd and R. W. Williams wrote the assembly-coded particle integrators so necessary for efficient running. We are also grateful to J. P. Boris for encouraging the initiation of the project and participating in the early planning, and to W. L. Kruer for his support and advice then and since.

This work was performed under the auspices of the U. S. Energy Research and Development Administration under Contract No. W-7405-ENG-48.

References

Birdsall, C. K., and Fuss, D. (1969). *J. Comput. Phys.* **3**, 494.
Birdsall, C. K., and Kamimura, T. (1966). "Plasma Computer Experiments in Three Dimensions", Report IPPJ-55. Institute of Plasma Physics, Nagoya University, Nagoya, Japan.
Biskamp, D., and Welter, H. (1975). *Phys. Rev. Lett.* **34**, 312.
Boris, J. P. (1970). *In* "Proceedings of the Fourth Conference on the Numerical Simulation of Plasmas" (J. P. Boris and R. A. Shanny, eds.), Stock 0851 00059, p. 3. U.S. Govt. Printing Office, Washington, D.C.
Boris, J. P., and Lee, R. (1973). *J. Comput. Phys.* **12**, 131.
Boris, J. P., and Roberts, K. V. (1969). *J. Comput. Phys.* **4**, 552.
Brettschneider, M., Killeen, J., and Mirin, A. A. (1973). *J. Comput. Phys.* **11**, 360.
Buneman, O. (1967). *J. Comput. Phys.* **1**, 517.
Buneman, O. (1968). *In* "Relativistic Plasmas" (O. Buneman and W. Pardo, eds.), p. 205. Benjamin, New York.
Buneman, O. (1973). *J. Comput. Phys.* **12**, 124.

Burger, P., Dunn, D. A., and Halstead, A. S. (1965). *Phys. Fluids* **8**, 2263.
Byers, J. A., Holdren, J. P., Killeen, J., Langdon, A. B., Mirin, A. A., Rensink, M. E., and Tull, C. G. (1974). *Phys. Fluids* **17**, 2061.
Cohen, B. I., Mostrom, M. A., Nicholson, D. R., Kaufman, A. N., Max, C. E., and Langdon, A. B. (1975). *Phys. Fluids* **18**, 470.
Cooley, J. W., Lewis, P. A. W., and Welch, P. D. (1970). *J. Sound Vib.* **12**, 315.
Dawson, J. M. (1970). *Methods Comput. Phys.* **9**, 1.
Dawson, J., and Langdon, B. (1967). *Bull. Amer. Phys. Soc.* **12**, 806.
de Leon, Moses (~ 1280). Sefer ha-Zohar.
Godfrey, B. B. (1974). *J. Comput. Phys.* **15**, 504.
Godfrey, B. B. (1975). *J. Comput. Phys.* **19**, 58.
Godfrey, B. B., and Langdon, A. B. (1976). *J. Comput. Phys.* **20**, 251.
Haber, I., Lee, R., Klein, H. H., and Boris, J. P. (1973). *In* "Prodeedings of the Sixth Conference on Numerical Simulation of Plasmas," CONF-730804, Paper B3. Lawrence Livermore Laboratory, Livermore, California.
Harlow, F. H. (1956–1957). *J. Ass. Comput. Mach.* **3–4**, 137.
Harlow, F. H. (1964). *Methods Comput. Phys.* **3**, 319.
Hockney, R. W. (1965). *J. Ass. Comput. Mach.* **12**, 95.
Hockney, R. W. (1966). *Phys. Fluids* **9**, 1826.
Hockney, R. W. (1970). *Methods in Comput. Phys.* **9**, 135.
Jackson, J. D. (1962). "Classical Electrodynamics," pp. 616–618. Wiley, New York.
Langdon, A. B. (1969). Ph.D. Thesis, Princeton University, Princeton, New Jersey.
Langdon, A. B. (1970a). *J. Comput. Phys.* **6**, 247.
Langdon, A. B. (1970b). *In* "Proceedings of the Fourth Conference on the Numerical Simulation of Plasmas" (J. P. Boris and R. A. Shanny, eds.), Stock 0851 00059, p. 467. US Govt. Printing Office, Washington, D.C.
Langdon, A. B. (1972). *Phys. Fluids* **15**, 1149.
Langdon, A. B. (1973). *J. Comput. Phys.* **12**, 247.
Langdon, A. B., and Birdsall, C. K. (1970). *Phys. Fluids* **13**, 2115.
Langdon, A. B., and Lasinski, B. F. (1975). *Phys. Rev. Lett.* **34**, 934.
Langdon, A. B., Lasinski, B. F., and Kruer, W. L. (1973). "Plasma Heating at One-Quarter Critical Density," Report UCRL-75018. Lawrence Livermore Laboratory, Livermore, California.
Lee, R., and Lampe, M. (1973). *Phys. Rev. Lett.* **31**, 1390.
Lewis, H. R. (1972). *J. Comput. Phys.* **10**, 400.
Lin, A. T., Dawson, J. M., and Okuda, H. (1974). *Phys. Fluids* **17**, 1995.
Lindman, E. L. (1973). *In* "Proceedings of the Sixth Conference on Numerical Simulation of Plasmas," CONF-730804. Paper B2. Lawrence Livermore Laboratory, Livermore, California.
Lindman, E. L. (1975). *J. Comput. Phys.* **18**, 66.
Morse, R. L., and Nielson, C. W. (1969). *Phys Fluids* **12**, 2418.
Morse, R. L., and Nielson, C. W. (1971). *Phys. Fluids* **14**, 830.
Nielson, C. W., and Lindman, E. L. (1973a). *In* "Abstracts of the Sherwood Theory Meeting," Paper D9. University of Texas, Austin.
Nielson, C. W., and Lindman, E. L. (1973b). *In* "Proceedings of the 6th Conference on Numerical Simulation of Plasmas," CONF-730804, Paper E3. Lawrence Livermore Laboratory, Livermore, California.
Orens, J. H., Boris, J. P., and Haber, I. (1970). *In* "Proceedings of the Fourth Conference on the Numerical Simulation of Plasmas" (J. P. Boris and R. A. Shanny, eds.), Stock 0851 00059, p. 526. US Govt. Printing Office, Washington, D.C.

Panofsky, W. K. H., and Phillips, M. (1962). "Classical Electricity and Magnetism," p. 461. Addison-Wesley, Reading, Massachusetts.

Sinz, K. (1970). In "Proceedings of the Fourth Conference on the Numerical Simulation of Plasmas" (J. P. Boris and R. A. Shanny, eds.), Stock 0851 00059, p. 153. US Govt. Printing Office, Washington, D. C.

Van Duzer, J., and Birdsall, C. K. (1971). "Plasma Computer Simulation: A Bibliography for 1950–1970." Department of Electrical Engineering and Computer Sciences, University of California, Berkeley.

Yu, S. P., Kooyers, G. P., and Buneman, O. (1965). *J. Appl. Phys.* **36**, 2550.

Particle-Code Models in the Nonradiative Limit

CLAIR W. NIELSON AND H. RALPH LEWIS

LOS ALAMOS SCIENTIFIC LABORATORY
UNIVERSITY OF CALIFORNIA
LOS ALAMOS, NEW MEXICO

I.	Introduction	367
II.	The Darwin Model	368
III.	Hamiltonian Formulations	371
IV.	Lagrangian Formulation	374
V.	Solution of the Field Equations	379
VI.	One-Dimensional Comparisons	381
VII.	A Two-Dimensional Example	384
VIII.	Summary	386
	References	387

I. Introduction

MOST OF THE COLLECTIVE effects that are encountered in plasmas of interest for controlled fusion with magnetic confinement are nonrelativistic and nonradiative in character. Even when radiation is important, separate study of the remaining phenomena may be useful. Such separation is routine in much analytic plasma theory, where approximations valid in a particular parameter range are made and the consequences explored in detail. In particular, the radiative terms in a dispersion relation may be identified and discarded, or the displacement current may be neglected at the outset.

In the case of particle simulation, a compelling additional reason exists for excluding radiation when it is not of crucial physical importance. Because of the limited number of simulation particles that can be used in practical cases, the fluctuation levels for charge and current densities are usually much higher than in the real plasma being simulated. Smoothing techniques improve the situation substantially, but do not completely eliminate the problem. In a closed simulation system, the normal modes of the electromagnetic field will eventually be excited to an artificially high level which may distort or obscure important physical effects. In such cases, nonradiative plasma simulation codes are of considerable utility.

The preponderance of work in the field of particle simulation has indeed been nonradiative—it has been completely electrostatic. On the other hand, many recent developments in electromagnetic simulation have necessarily been fully electromagnetic since the major application has been to study the laser–plasma interaction (Haber et al., 1973; Lindman, 1973; Godfrey, 1974). The literature of nonradiative, electromagnetic simulation is somewhat limited. Auer et al. (1961) introduced a one-species, one-dimensional, nonradiative sheet model to study collisionless shock waves propagating perpendicular to a magnetic field. In their model the magnetic field and current are self-consistent, but charge separation effects are treated in the quasi-neutral approximation. Hasegawa and Birdsall (1964) devised a one-dimensional nonradiative sheet model for the study of ion–cyclotron waves with self-consistent magnetic and transverse electric fields. Electrostatic effects were not considered. Hasegawa and Okuda (1968) described in more general terms a one-dimensional nonradiative sheet model including electrostatics. They require that there be no magnetic field in the direction of problem variation. Dickman et al. (1969) implemented a two-dimensional code using a particle and cell formulation which includes those self-consistent electromagnetic effects due to currents normal to the plane of computation. This code was used to study the nonlinear evolution of mirror instabilities and tearing modes. Haber et al. (1970) described a completely self-consistent, one-dimensional, nonradiative algorithm, using a particle and cell computational method. Their code and codes similar to it elsewhere have been used extensively to study nonlinear microturbulence effects in one dimension. One such code, the EMI program of Forslund et al. (1972), has been distributed in a simplified, relatively machine-independent form.

It is somewhat surprising that these codes, even in one dimension, are considerably more complex than their fully electromagnetic counterparts. Moreover, it is found that the generalization of the one-dimensional algorithms to two dimensions, or of the special two-dimensional algorithm to full generality, is by no means straightforward. It is the purpose of this article to place previous work in a general framework and to formulate a completely general radiation-free plasma simulation algorithm.

II. The Darwin Model

The proper small v/c approximation of the equations of the Maxwell–Lorentz system was first given by Darwin (1920), who used it to generalize the Bohr–Sommerfeld atom. Readable accounts of this derivation are given by Landau and Lifshitz (1951) and Jackson (1962). Additional features of the model are deduced by Kaufman and Rostler (1971). In the present section we shall develop only those properties of the model which we subsequently use.

The interaction Lagrangian for the motion of a charged particle in specified external electric and magnetic fields is well known,

$$L_{\text{int}} = -q\phi + (q/c)\mathbf{v} \cdot \mathbf{A}, \tag{1}$$

and the many-particle Lagrangian is obtained from this by summation (Goldstein, 1953). If the fields are not specified *a priori*, but rather are at least partly the self-consistent fields due to the particles themselves, then the complete Lagrangian formulation of the field–particle system requires that the modes of the electromagnetic field be included as additional degrees of freedom (Goldstein, 1953, p. 366; Low, 1958, Sturrock, 1958; Lewis, 1970).

The objective of the Darwin derivation is to find the most accurate possible Lagrangian for the field–particle system that is expressible directly in terms of nonretarded particle positions and velocities. Such a Lagrangian must necessarily exclude radiation since its structure implies instantaneous action-at-a-distance. Offhand one might expect that no more than electrostatics and magnetostatics could be described in this way. The surprise in the Darwin result is that the Lagrangian is correct through order $(v/c)^2$, and includes those electromagnetic effects sometimes referred to as inductive, that is, the effects associated with Faraday's law.

For the present purpose, the explicit form of the Lagrangian is not required. The efficiency of current particle and cell simulation techniques, as compared to the sheet models that preceded them, is largely a consequence of performing computations in terms of space-dependent fields rather than in terms of binary interactions, thus reducing the number of operations from being proportional to N^2 to being proportional to N, where N is the number of simulation particles. Therefore, we wish to express the field quantities of of the Darwin approximation in terms of particle positions and velocities, and then obtain the force on each particle from these fields in subsequent operations. The fact that the forces could have been deduced directly from a multiparticle Lagrangian is a valuable reminder that the equations represent an instantaneous model without retardation; however, the multiparticle Lagrangian itself is not of direct computational utility.

By working in the Coulomb gauge, no approximation to the scalar potential need be made; it is already free of retardation and satisfies Poisson's equation,

$$\nabla^2 \phi = -4\pi\rho. \tag{2}$$

The exact vector potential satisfies the wave equation

$$\nabla^2 \mathbf{A} - (1/c^2)(\partial^2 \mathbf{A}/\partial t^2) = -(4\pi/c)\mathbf{J}_t, \tag{3}$$

where $\mathbf{J}_t$ is the divergence-free part of the current. Using the continuity equation it is straightforward to show (Jackson, 1962, p. 182) that

$$\mathbf{J}_t = \mathbf{J} - (1/4\pi)\nabla(\partial\phi/\partial t). \tag{4}$$

The solution of Eq. (3) may be written (Jackson, 1962, p. 185), using the retarded potential method, as

$$\mathbf{A} = (1/c)\int [\mathbf{J}_t(\text{retarded})/R]\, dV, \tag{5}$$

where $\mathbf{J}_t(\text{retarded})$ is the retarded value of the transverse current at the source point. Since the only difference between this solution and the solution of Poisson's equation *is* the retardation, neglecting retardation is equivalent to neglecting the time derivative term in Eq. (3). A less trivial result is the fact that this approximation preserves accuracy in the interaction Lagrangian up to order $(v/c)^2$. This is true for three reasons: first, the vector potential $\mathbf{A}$ is multiplied by (v/c) in the interaction Lagrangian; second, the lowest-order contribution to $\mathbf{A}$ in Eq. (5) is already of order (v/c); and third, the first retardation correction to $\mathbf{A}$, being due to motion of particles in time (R/c), is proportional to $(v/c)^2$, so that the retardation correction to the interaction Lagrangian is of order $(v/c)^3$.

Combining this approximation to Eq. (3) with Eq. (4), we finally obtain the equation for the Darwin approximation to the vector potential,

$$\nabla^2 \mathbf{A} = -(4\pi/c)\mathbf{J} + (1/c)\nabla(\partial\phi/\partial t), \tag{6}$$

with $\nabla \cdot \mathbf{A} = 0$. This differs from the more naive notion of "neglecting the displacement current" in that the longitudinal part of the displacement current, $-\nabla(\partial\phi/\partial t)$, is retained. This term is required in order to guarantee a charge continuity equation.

It is informative to express the equations directly in terms of the electric and magnetic fields, which are gauge invariant and easier to interpret. To do this, we split the electric field into longitudinal and transverse parts, denoted by $\mathbf{E}_l$ and $\mathbf{E}_t$, respectively, defined by the requirements $\nabla \times \mathbf{E}_l = 0$ and $\nabla \cdot \mathbf{E}_t = 0$. The field equations for the Darwin model may now be expressed as

$$\nabla \cdot \mathbf{E}_l = 4\pi\rho, \quad \nabla \times \mathbf{E}_l = 0,$$
$$\nabla \times \mathbf{B} = 4\pi\mathbf{J}/c + (1/c)(\partial\mathbf{E}_l/\partial t), \quad \nabla \cdot \mathbf{B} = 0, \tag{7}$$
$$\nabla \times \mathbf{E}_t = -(1/c)(\partial\mathbf{B}/\partial t), \quad \nabla \cdot \mathbf{E}_t = 0.$$

The field equations thus written differ by only one term from the exact Maxwell equations—specifically, in the omission of the time derivative of the transverse electric field. However, this simple alteration profoundly changes their character: the approximate Maxwell equations, when coupled by the particle equations through the source terms, are elliptic in type rather than hyperbolic. A detailed demonstration of this will be given later, but we note here that the ellipticity is of considerable computational significance. When solving the full Maxwell equations the time derivatives may be used to advance the fields, while the divergence equations serve as initial conditions which are preserved in time because of the charge continuity equation. Indeed, some fully electromagnetic simulation algorithms proceed in precisely this fashion (Buneman, 1968; Boris, 1970; Morse and Nielson, 1971). However, any attempt to use the time derivatives in the Darwin field equations in a similar fashion will lead to violent numerical instabilities. These instabilities reflect the fact that the equations are now elliptic and represent instantaneous action-at-a-distance. Any attempt to propagate information across the mesh at a finite velocity is inconsistent with the nonretardation approximation. A transformation of the Darwin model field equations that displays their elliptic character explicitly will be given in Section IV.

III. Hamiltonian Formulations

Several Darwin model simulation codes were referred to in Section I. Most of these are expressed in Hamiltonian form, by which we mean that the particle motion is described in terms of the Hamiltonian variables, position $\mathbf{x}$, and canonical momentum $\mathbf{p}$. Since none of the previous Darwin-model algorithms can be generalized immediately to more dimensions, we shall restate the Hamiltonian formulation in a way suited to such generalization. It turns out that when our formulation is specialized to one dimension it yields a method similar to but simpler than those previously used.

To obtain the particle Hamiltonian from the Lagrangian one follows the standard procedure (Goldstein, 1953, p. 222) to obtain

$$H = \sum_i \{(1/2m_i)[\mathbf{p}_i - (q_i/c)\mathbf{A}]^2 + q_i\phi\}, \tag{8}$$

where $\mathbf{p}_i$ is the canonical momentum of the ith particle, $\mathbf{A}$ and ϕ are evaluated at the positions of the particles, and q_i and m_i are the charge and mass of the ith particle. Hamilton's equations of motion then follow as

$$d\mathbf{p}_i/dt = (q_i/m_ic)(\nabla_{\mathbf{x}_i}\mathbf{A}) \cdot [\mathbf{p}_i - (q_i/c)\mathbf{A}] - q_i\nabla_{\mathbf{x}_i}\phi, \tag{9a}$$

$$d\mathbf{x}_i/dt = (1/m_i)[\mathbf{p}_i - (q_i/c)\mathbf{A}], \tag{9b}$$

where the potentials are evaluated at the position of the ith particle.

The charge and current densities, ρ and $\mathbf{J}$, as functions of position $\mathbf{x}$ and time t, can be expressed in terms of particle variables as

$$\rho = \sum_s q_s n_s, \tag{10a}$$

$$\mathbf{J} = \sum_s q_s n_s \langle \mathbf{v}_s \rangle$$

$$= \left(\sum_s (q_s/m_s) n_s \langle \mathbf{p}_s \rangle \right) - \left(\sum_s (q_s^2/m_s c) n_s \right) \mathbf{A}, \tag{10b}$$

where s denotes particle species, n_s is the spatial density of species s, and $\langle \mathbf{v}_s \rangle$ and $\langle \mathbf{p}_s \rangle$ are the average velocity and canonical momentum for species s as functions of position and time.

Inserting the charge and current densities into the equations for the Darwin potentials, Eq. (2) and Eq. (6), we obtain

$$\nabla^2 \phi = -4\pi \left(\sum_s q_s n_s \right), \tag{11a}$$

$$\nabla^2 \mathbf{A} = -\sum_s (4\pi q_s/m_s c) n_s \langle \mathbf{p}_s \rangle + \left[\sum_s (4\pi q_s^2/m_s c^2) n_s \right] \mathbf{A} + (1/c) \nabla (\partial \phi/\partial t). \tag{11b}$$

When spatial variation is maintained in only one direction, these equations simplify: the transverse components of $\mathbf{A}$ are completely decoupled from ϕ, and the component of $\mathbf{A}$ in the direction of variation is determined to be constant by imposing the condition $\nabla \cdot \mathbf{A} = 0$. Previous Darwin model algorithms made essential use of these simplifications. However, one-dimensional spatial variation actually *adds* some complication since a constant magnetic field in the direction of variation cannot be obtained as the curl of the vector potential, but must be included in a special manner.

When all of the terms are included in a two- or three-dimensional algorithm, the $\nabla \phi$ term couples components of the equation for the vector potential. The first essential observation when attempting to solve these equations is that $\dot{\phi}$ should not be computed by time differencing. Because of the instantaneous nature of the forces in this model, there should be no need to do so, and, indeed, the equation for $\mathbf{A}$ and the condition $\nabla \cdot \mathbf{A} = 0$ are four conditions on four unknowns, which are sufficient to determine both $\mathbf{A}$ and ϕ, assuming proper boundary conditions are given. The detailed algorithm required to solve this system of coupled field equations has proved to be the most difficult aspect of implementing the Darwin approximation in two dimensions, and Section V is devoted to describing the method of solution.

Time differencing of the equations of motion for the particles, Eqs. (9a) and (9b), can proceed in a variety of ways. Heretofore, exact reversibility has often been a requirement imposed on simulation algorithms. The reasons given are sometimes more in the nature of philosophy than of numerical analysis, and we feel that reversibility is not essential. While reversible schemes are likely to have good stability properties, they are certainly not the *only* schemes with good stability. Other second-order-accurate schemes have been used in fully electromagnetic codes with results that are practically indistinguishable from results obtained from codes having reversible algorithms (Morse and Nielson, 1971; Nielson and Lindman, 1973).

In the case of the Darwin approximation, most of the algorithms previously mentioned do indeed impose reversibility as a requirement. However, this leads to highly implicit algorithms for the field equations that are completely out of the question in two or three dimensions. By relaxing the requirement for reversibility when differencing the particle equations of motion, we have obtained equations whose solution is reasonably efficient, regardless of the dimensionality of the problem. In one dimension, the simplification saves both storage space and execution time. Detailed comparisons presented in Section VI suggest that in the Darwin approximation, as in the fully electromagnetic case, no significant difference can be detected as being due to exact reversibility in the simulation algorithm.

We now specify the particular time differencing we use for the particle equations of motion. Let the momenta $\mathbf{p}$ be saved at the half time-steps and the positions $\mathbf{x}$ be saved at the whole time-steps. Moreover, assume that the potentials $\mathbf{A}$ and ϕ may be obtained at either whole or half time-steps as needed. Then the momenta and positions are advanced by

$$(\mathbf{p}^{1/2} - \mathbf{p}^{-1/2})/\Delta t = (q/mc)(\nabla \mathbf{A}^0) \cdot [\tfrac{1}{2}(\mathbf{p}^{1/2} + \mathbf{p}^{-1/2}) - (q/c)\mathbf{A}^0] - q\nabla\phi^0, \tag{12a}$$

$$(\mathbf{x}^1 - \mathbf{x}^0)/\Delta t = (1/m)[\mathbf{p}^{1/2} - (q/c)\mathbf{A}^{1/2}], \tag{12b}$$

where the superscripts represent typical time-step indices. Solving the implicit equations for the components of $\mathbf{p}$ is not difficult since they involve only the coordinates of a single particle. These equations would be reversible if the algorithm for finding $\mathbf{A}^0$, $\mathbf{A}^{1/2}$, and ϕ^0 were reversible. Indeed, this is straightforward in a fully electromagnetic code using explicit time advance of the wave equation. However, for the Darwin model, reversibility is not practical in more than one dimension.

It should be noted from Eqs. (12) that the potentials need be correct only to order Δt at any time if the differences for $\mathbf{p}$ and $\mathbf{x}$ are required to be of accuracy $(\Delta t)^2$, as they are if the accumulated errors in $\mathbf{p}$ and $\mathbf{x}$ are to be of order Δt. [The situation here should be distinguished from that in the fully electromagnetic case, where the potentials propagate in time and, therefore, their differencing must be of accuracy $(\Delta t)^2$.] This being the case, we obtain $\mathbf{A}^0$ in the following way. At the end of the previous time-step, before taking the weighted values of momentum to be used in the source term of the field equation for $\mathbf{A}$, we project each $\mathbf{p}$ ahead an additional half time-step to $\tilde{\mathbf{p}}^0$, using the same rate of change that was used in obtaining the saved value $\mathbf{p}^{-1/2}$. This value, $\tilde{\mathbf{p}}^0$, will be only first-order accurate, but since it used only once, first-order accuracy is sufficient. During the field solution, the summed values of $\tilde{\mathbf{p}}^0$ are used to solve for $\mathbf{A}^0$, which will also be correct to order Δt. The value $\mathbf{A}^{1/2}$ required for advancing the positions in more than one dimension is found by simple extrapolation from $\mathbf{A}^0$ and saved values of $\mathbf{A}^{-1}$. The accumulation of density required in both field equations is also made at the end of the time-step, and in this instance is actually accurate to order $(\Delta t)^2$, a fact of no particular consequence. The above sequence of steps amounts to a predictor–corrector method, with the predictor being linear extrapolation from past values of the potential, and the corrector being Ampere's law as expressed in Eq. (11b).

Numerical stability is not guaranteed by the consideration of accuracy discussed above. Indeed, an analogous procedure in a Lagrangian formulation is violently unstable. However, empirical studies of the Hamiltonian algorithm in one dimension have demonstrated stability and accuracy comparable to that of the fully reversible algorithms (see Section VI). Moreover, as will be shown in the next section, the instability in the Lagrangian case is due to explicit time differencing of $\partial \mathbf{A}/\partial t$, a differencing not required in the Hamiltonian form. Therefore, we believe that the foregoing Hamiltonian algorithm will be stable and accurate in two-dimensional and three-dimensional codes.

IV. Lagrangian Formulation

The expression of the Darwin approximation in terms of the Lorentz force, Newton's law, and a set of field equations is appealing because of the direct and intuitive character of such a formulation. Most simulation codes are expressed in such terms, and we refer to the formulation as Lagrangian since the particle motion is expressed in terms of the Lagrangian variables, position $\mathbf{x}$ and velocity $\mathbf{v}$. However, using the Lagrangian formulation in a fashion completely analogous to that used in fully electromagnetic codes, one finds that the computation is destroyed in a few time-steps by a numerical instability. This difficulty has been referred to earlier in this paper and was stated to be due to an inconsistency in the treatment of the fact that the Darwin model represents instantaneous action-at-a-distance. Before defining a satisfactory Lagrangian algorithm, we shall further elucidate the nature of this instability by giving a simple example.

Consider charged fluid motion transverse to the x-direction in the Darwin approximation. Let all quantities vary only in the x-direction, let $\mathbf{A}$, the fluid velocity $\mathbf{v}$, and the current $\mathbf{J}$ be transverse to the x-direction, and take the density n to be constant. Then the equations for the system in terms of the Lagrangian variables are

$$\mathbf{E} = -(1/c)(\partial \mathbf{A}/\partial t), \quad m\, d\mathbf{v}/dt = q\mathbf{E},$$
$$\nabla^2 \mathbf{A} = -4\pi \mathbf{J}/c = -(4\pi/c)qn\mathbf{v}. \tag{13}$$

A typical differencing, including extrapolation of $\mathbf{A}$ to obtain $\mathbf{E}$, is

$$\mathbf{E}^0 = -(\mathbf{A}^{-1/2} - \mathbf{A}^{-3/2})/c\Delta t, \quad m(\mathbf{v}^{1/2} - \mathbf{v}^{-1/2})/\Delta t = q\mathbf{E}^0,$$
$$\nabla^2 \mathbf{A}^{1/2} = -(4\pi/c)qn\mathbf{v}^{1/2}. \tag{14}$$

Combining terms and using $\omega_p^2 = 4\pi n q^2/m$, we have

$$\nabla^2 \mathbf{A}^{1/2} - \nabla^2 \mathbf{A}^{-1/2} = (\omega_p^2/c^2)(\mathbf{A}^{-1/2} - \mathbf{A})^{-3/2}. \tag{15}$$

Assuming $\mathbf{A} = \mathbf{A}^0 e^{i(kx+\omega t)}$ we obtain

$$e^{i\omega \Delta t} = -\omega_p^2/(kc)^2, \tag{16}$$

which has the solution

$$\omega = (2j+1)(\pi/\Delta t) - i\gamma, \tag{17}$$

where

$$e^{\gamma \Delta t} = \omega_p^2/(kc)^2 \tag{18}$$

and j is an integer. This implies that the solution diverges with alternating sign for any value of k less than ω_p/c. For problem lengths of several c/ω_p, this is a violent instability which destroys the calculation in a few time-steps. Extrapolating $\mathbf{A}$ with higher order extrapolation formulas makes quantitative but not qualitative changes in the behavior and is of no practical value.

A fundamentally different approach to the determination of $\mathbf{E}_t$ is required. First, the determination of $\mathbf{E}_l$ and $\mathbf{B}$ is carried out in a relatively straightforward manner from the solution of the set of equations

$$\rho = \sum_s q_s n_s, \quad \nabla^2 \phi = -4\pi\rho, \quad \mathbf{E}_l = -\nabla\phi,$$
$$\mathbf{J} = \sum_s q_s n_s \langle \mathbf{v}_s \rangle, \quad \nabla^2 \mathbf{A} = -(4\pi \mathbf{J}/c) + (1/c)\nabla\dot{\phi}, \tag{19}$$
$$\nabla \cdot \mathbf{A} = 0, \quad \mathbf{B} = \nabla \times \mathbf{A}.$$

As in the Hamiltonian formulation, because of the condition $\nabla \cdot \mathbf{A} = 0$, these equations yield $\dot{\phi}$ directly without any time differencing. To obtain an equation for $\mathbf{E}_t$, we differentiate Ampere's law in Eqs. (19) with respect to time and use $\mathbf{E}_t = -(1/c)(\partial \mathbf{A}/\partial t)$ to obtain

$$\nabla^2 \mathbf{E}_t = (4\pi/c^2)\dot{\mathbf{J}} - (1/c^2)\nabla \dot{\phi}. \tag{20}$$

Attempting to find $\dot{\mathbf{J}}$ at the appropriate time by extrapolating $\mathbf{J}$ would lead to the same type of numerical instability as before. However, if $\dot{\mathbf{J}}$ can be expressed in a way that uses only present values of the particle parameters, the system of equations consisting of Eq. (20) and $\nabla \cdot \mathbf{E}_t = 0$ will become truly elliptic and no time-differencing instabilities will be generated. This can in fact be done by using the Vlasov equations for the various species,

$$\partial f_s/\partial t + \mathbf{v} \cdot \nabla_x f_s + (q_s/m_s)(\mathbf{E} + \mathbf{v} \times \mathbf{B}/c) \cdot \nabla_v f_s = 0. \tag{21}$$

Following steps completely analogous to those used in the derivation of the momentum transfer equation (Krall and Trivelpiece, 1973), we obtain an expression for the time derivative of the current density due to species s,

$$\begin{aligned}
\partial \mathbf{J}_s/\partial t &= \int q_s \mathbf{v}(\partial f_s/\partial t) d^3\mathbf{v} \\
&= -\int q_s \mathbf{v}(\mathbf{v} \cdot \nabla_x f_s) d^3\mathbf{v} - \int (q_s^2/m_s)\mathbf{v}(\mathbf{E} + \mathbf{v} \times \mathbf{B}/c) \cdot \nabla_v f_s d^3\mathbf{v} \\
&= -\int q_s \mathbf{v}(\mathbf{v} \cdot \nabla_x f_s) d^3\mathbf{v} + (q_s/m_s)\rho_s \mathbf{E} + (q_s/m_s c)\mathbf{J}_s \times \mathbf{B},
\end{aligned} \tag{22}$$

where ρ_s is the charge density due to species s. We define

$$\begin{aligned}
\mathbf{D}_s &= -\int q_s \mathbf{v}(\mathbf{v} \cdot \nabla_x f_s) d^3\mathbf{v} \\
&= -\nabla_x \cdot \int q_s \mathbf{v}\mathbf{v} f_s d^3\mathbf{v},
\end{aligned} \tag{23}$$

so that $\partial \mathbf{J}_s/\partial t$ may be written as

$$\partial \mathbf{J}_s/\partial t = \mathbf{D}_s + (q_s/m_s)\rho_s \mathbf{E} + (q_s/m_s c)\mathbf{J}_s \times \mathbf{B}. \tag{24}$$

As is shown by the last term of Eq. (23), $\mathbf{D}_s$ is the divergence of the current transfer tensor for species s. Although the tensor itself could be accumulated and the divergence taken before solving the field equations, such a procedure would require more storage than does accumulating the divergence directly. Therefore, we do accumulate the quantities $\mathbf{D}_s$ and shall have no further occasion to refer to the tensor.

Splitting the electric field into transverse and longitudinal parts, substituting Eq. (24) into Eq. (20), and summing over species yields

$$\nabla^2 \mathbf{E}_t - \eta \mathbf{E}_t = \boldsymbol{\xi} + \eta \mathbf{E}_l + \boldsymbol{\zeta} \times \mathbf{B} - (1/c^2)\nabla\ddot{\phi}, \quad (25)$$

where

$$\eta = \sum_s 4\pi q_s \rho_s/m_s c^2 = \sum_s \omega_{ps}^2/c^2,$$

$$\boldsymbol{\zeta} = \sum_s (4\pi/c^2)(q_s/m_s c)\mathbf{J}_s, \qquad \boldsymbol{\xi} = \sum_s (4\pi/c^2)\mathbf{D}_s,$$

and

$$\nabla \cdot \mathbf{E}_t = 0.$$

The various moments required in an actual simulation must be accumulated on a spatial grid. To reduce noise one almost always uses some kind of charge sharing technique. Using the notation of Langdon (1970a), we associate with the simulation particles a shape function $S(\mathbf{x})$ such that, for example, $qS(\mathbf{x})$ is the charge density of a particle with total charge q whose center is at the origin. Then, if a uniform mesh is introduced into the problem with mesh points at $\mathbf{x}_c$, the charge density at mesh point $\mathbf{x}_c$ for species s is given by

$$\rho_s(\mathbf{x}_c) = q_s \sum_p S(\mathbf{x}_c - \mathbf{x}_p), \quad (26)$$

where the summation is over all particles p of species s and $\mathbf{x}_p$ is the position vector of the center of the pth particle. Similarly

$$\mathbf{J}_s(\mathbf{x}_c) = q_s \sum_p \mathbf{v}_p S(\mathbf{x}_c - \mathbf{x}_p), \quad (27)$$

where $\mathbf{v}_p$ is the velocity of particle p. By straightforward time differentiation we obtain

$$\dot{\mathbf{J}}_s(\mathbf{x}_c) = q_s \sum_p \mathbf{v}_p \mathbf{v}_p \cdot \nabla_{\mathbf{x}_p} S(\mathbf{x}_c - \mathbf{x}_p) + q_s \sum_p \dot{\mathbf{v}}_p S(\mathbf{x}_c - \mathbf{x}_p)$$

$$= -q_s \sum_p \mathbf{v}_p \mathbf{v}_p \cdot \nabla_{\mathbf{x}_c} S(\mathbf{x}_c - \mathbf{x}_p)$$

$$+ (q_s^2/m_s) \sum_p S(\mathbf{x}_c - \mathbf{x}_p)[\mathbf{E}(\mathbf{x}_p) + (\mathbf{v}_p/c) \times \mathbf{B}(\mathbf{x}_p)], \quad (28)$$

where we have used the Lorentz force and Newton's law to eliminate $\dot{\mathbf{v}}_p$. The identification of the terms of this equation with the corresponding Vlasov terms in Eq. (22) is immediate.

As an illustration of this means of finding the quantity $\mathbf{D}_s$, we consider a one-dimensional problem with $S(x)$ chosen to be the commonly used tent function (Langdon, 1970b). If $S(x)$ is the unit tent function, $S'(x)$ will be a square wave with amplitude $1/\Delta x$. The prescription that Eq. (28) gives for evaluating $\mathbf{D}_s$ is as follows. For each particle p of species s, if $x_c - \Delta x \leqslant x_p < x_c$, add $q_s \mathbf{v}_p(v_{px}/\Delta x)$ to $\mathbf{D}_s(x_c)$; or if $x_c \leqslant x_p < x_c + \Delta x$, subtract $q_s \mathbf{v}_p(v_{px}/\Delta x)$ from $\mathbf{D}_s(x_c)$. The generalization to other shape factors and to more dimensions is straightforward.

The most serious source of instability has been removed by expressing $\mathbf{E}_t$ as the solution of a properly posed elliptic equation. Other instabilities, weaker ones due to the use of a finite mesh (Langdon, 1970c), are presumably still present, but their effect can be minimized by a judicious choice of mesh spacing, or be altered somewhat by the details of space differencing. The exact details of time differencing for the particle equations appear not to be critical in this formulation.

In a specific two-dimensional code called DARWIN, which is a Darwin model code in a Lagrangian formulation that has been successfully written and tested, we use the same second-order-accurate but not reversible scheme for the particle equations of motion that has been used previously in a fully electromagnetic code (Nielson and Lindman, 1973). The equations for the particle advance are

$$\tilde{\mathbf{v}}^0 = \mathbf{v}^{-1/2} + (h/2)\mathbf{E}^0,$$
$$\mathbf{v}^{1/2} = f\mathbf{v}^{-1/2} + h(\mathbf{E}^0 + g\mathbf{B}^0 + \tilde{\mathbf{v}}^0 \times \mathbf{B}^0), \tag{29}$$
$$\mathbf{x}^1 = \mathbf{x}^0 + \Delta t \mathbf{v}^{1/2},$$

where $h = \Delta t q/m$, $f = 1 - (h^2/2)(\mathbf{B}^0 \cdot \mathbf{B}^0)$, and $g = (h/2)(\mathbf{v}^{-1/2} \cdot \mathbf{B}^0)$. The values $\mathbf{v}^{1/2}$ and $\mathbf{x}^1$ are of accuracy $(\Delta t)^2$ so that the accumulated error is of order Δt. For the purpose of computing the source terms in the field equations, the velocities are further extrapolated for a one-time only use to

$$\tilde{\mathbf{v}}^1 = \mathbf{v}^{1/2} + \tfrac{1}{2}(\mathbf{v}^{1/2} - \mathbf{v}^{-1/2}). \tag{30}$$

These values are used to accumulate the quantities $\mathbf{J}_s^1$ and $\mathbf{D}_s^1$ corresponding to the particles being at positions $\mathbf{x}^1$. The charge density ρ_s^1 is also accumulated at these positions.

The space derivatives are differenced in the standard centered fashion. For example, $(d\phi/dx)_i = (\phi_{i+1} - \phi_{i-1})/(2\Delta x)$. This kind of differencing avoids the complexity of multiple mesh systems and is equivalent to the averaging procedure advocated by Boris and Lee (1973). An inconvenience evident from the form of the source terms in Eqs. (19) and (25) is that they separately involve the moments

$$\sum_s \rho_s, \quad \sum_s q_s \rho_s, \quad \sum_s \mathbf{J}_s, \quad \text{and} \quad \sum_s q_s \mathbf{J}_s.$$

Because of storage limitations, the present code neglects ion current in order that the latter two moments can be obtained from a single array. Computation time is also saved by treating the ions electrostatically. This approximation is in fact very good for most of the problems amenable to simulation at all, since the ion gyroperiod is very much longer than the electron plasma period.

V. Solution of the Field Equations

In one dimension, the simplifications mentioned in connection with Eqs. (11) cause complete decoupling of the components of the equation satisfied by the vector potential, so that the field equations reduce to simple tridiagonal systems which can be solved by standard elimination methods. The Poisson equation is readily solved, even in two or three dimensions; for solving the Poisson equation we use a version of the odd–even reduction or Buneman algorithm (Buzbee *et al.*, 1970). Fast Fourier transform methods would be equally suitable. However, the solution of Eq. (11b) in the Hamiltonian formulation or Eq. (25) in the Lagrangian formulation is a source of difficulty in two dimensions.

For the purposes of this section we will refer to the general system of equations

$$\nabla^2 \mathbf{A} - \eta \mathbf{A} = \boldsymbol{\xi} - \nabla \chi, \tag{31a}$$

$$\nabla \cdot \mathbf{A} = 0. \tag{31b}$$

Equations (11b) and (25) are examples of Eq. (31a). One approach to this set of equations would be to write $\mathbf{A} = \nabla \times \mathbf{W}$, $\nabla \cdot \mathbf{W} = 0$, and take the curl of Eq. (31a), thereby obtaining a single fourth-order equation, which, in two dimensions, involves only one component of $\mathbf{W}$,

$$\nabla^4 \mathbf{W} + \nabla \times (\eta \nabla \times \mathbf{W}) = -\nabla \times \boldsymbol{\xi}. \tag{32}$$

However, solution of this single equation appears even more difficult than solution of the original system, particularly in view of the complicated boundary conditions that would apply to $\mathbf{W}$.

If the coupling due to $\nabla \chi$ and $\nabla \cdot \mathbf{A} = 0$ were not present, the system of Eqs. (31) would reduce to separate scalar equations of the form

$$\nabla^2 \Psi - \eta \Psi = \rho. \tag{33}$$

An equation of this form was solved by Dickman et al. (1969) by means of successive overrelaxation. The emergence of direct methods has, however, emphasized the poor convergence rate of successive overrelaxation. If η is constant, or a function of x alone, direct methods can be used to solve Eq. (33), for example, by Fourier transformation in the y-direction followed by tridiagonal solution in x. If, however, η is an arbitrary function of both x and y, iteration is required. For such problems a method we call global iteration (Nielson et al., 1971) has given convergence rates far better than do the relatively local iteration schemes such as successive overrelaxation or the alternating direction method. The idea is to do as much as possible of the solution by a direct method and to iterate only on the residual. An extensive analysis of a method of this type was given by Concus and Golub (1973). In addition, they traced the origin of such ideas far back into numerical analysis literature.

Specifically, the method we apply to Eq. (33) is as follows. Assuming that η is relatively constant in the y-direction with arbitrary variations in the x-direction, the form of global iteration is

$$\nabla^2 \Psi^{n+1} - \bar{\eta}^y(x) \Psi^{n+1} = \rho - (\bar{\eta}^y - \eta(x, y)) \Psi^n, \tag{34}$$

where $\bar{\eta}^y$ might be the average over y, for fixed x, of the function η. Concus and Golub proved that the best choice for $\bar{\eta}^y$ is the average of the minimum and maximum values of η over y, for fixed x. Even using the simple average, the method converges with extreme rapidity if the inhomogeneity in y is not too severe. For more general variations in η, the natural generalization would be to alternate the direction in which η is averaged on successive iterations, replacing $\bar{\eta}^y(x)$ with $\bar{\eta}^x(y)$ in Eq. (34).

With the coupling of the components of **A** by the $\nabla \chi$ term and the $\nabla \cdot \mathbf{A} = 0$ requirement, one must complicate the iteration algorithm considerably. Again, if η were constant, one could simply calculate an intermediate value of the potential, $\tilde{\mathbf{A}}$, from

$$\nabla^2 \tilde{\mathbf{A}} - \eta \tilde{\mathbf{A}} = \xi \tag{35}$$

and then generate the final solution in the form

$$\mathbf{A} = \tilde{\mathbf{A}} - \nabla \Psi,$$

where

$$\nabla^2 \Psi = \nabla \cdot \tilde{\mathbf{A}}. \tag{36}$$

Substitution then shows that

$$\chi = \nabla^2 \Psi - \eta \Psi. \tag{37}$$

This method is used to solve for **A** in Eq. (19) of the Lagrangian formulation. However, if η is not constant, multiplication and differentiation do not commute and this procedure fails. However, by combining these notions in an iterative cycle we have been able to find a procedure which empirically seems to converge as well as the simpler algorithm does for the uncoupled system of Eq. (33). In the iterative process we repeat the following steps:

1. solve $\quad \nabla^2 \chi^n = \nabla \cdot \xi + \mathbf{A}^n \cdot \nabla \eta \quad$ for $\quad \chi^n;\quad$ (38a)

2. solve $\quad \nabla^2 \tilde{\mathbf{A}}^{n+1} - \bar{\eta}^y \tilde{\mathbf{A}}^{n+1} = \xi + (\eta - \bar{\eta}^y)\mathbf{A}^n - \nabla \chi^n \quad$ for $\quad \tilde{\mathbf{A}}^{n+1};$ (38b)

3. solve $\quad \nabla^2 \Psi^{n+1} = \nabla \cdot \tilde{\mathbf{A}}^{n+1} \quad$ for $\quad \Psi^{n+1};\quad$ (38c)

4. calculate $\quad \mathbf{A}^{n+1} = \tilde{\mathbf{A}}^{n+1} - \nabla \Psi^{n+1}.\quad$ (38d)

In a simulation context, values of the unknowns from the previous time-step are used to start the process. The solution is exact in one pass if η is a function only of x. The fact that one has a solution if this sequence converges is verified by simple substitution. We do not at present have a rigorous criterion for the convergence of the process, but have had such good results with it in practical cases that we now do the fixed number of five iterations in most problems. Further study of the convergence properties is required.

VI. One-Dimensional Comparisons

We have applied four different one-dimensional Darwin model algorithms to simulate the evolution of a one-species, one-dimensional, Weibel-unstable plasma, the object being to see whether the different numerical algorithms behave at all differently. This example is convenient since the growth is rapid and the general behavior is well understood (Morse and Nielson, 1971). Since the example chosen displays a very powerful instability, this comparison is certainly not complete. In particular, runs on stable cases for very long times may reveal more subtle differences, and it is hoped that such comparisons will be made in the future.

The example chosen is a problem of length $25c/\omega_{pe}$, 128 cells, a background magnetic field of $B_x = 0.2mc\omega_{pe}/e$, and an initially bi-Maxwellian velocity distribution with $(v_{\parallel}/c) = 0.1$ and $(v_{\perp}/c) = 0.5$, where m and e are the electron charge and mass, and ω_{pe} is the electron plasma frequency. The codes compared are the fully reversible code EMI of Forslund et al. (1972), the simplified one-dimensional Hamiltonian algorithm described in Section

III and called MICRO (for microturbulence), the Lagrangian algorithm described in Section IV, with the two-dimensional code DARWIN being run in one dimension as a special case, and a variant of MICRO with a different initialization which will be described later. The results of these simulations are displayed in a single diagnostic, the z-component of the magnetic field. Plots of this quantity are displayed in Fig. 1 for each of the four runs at times

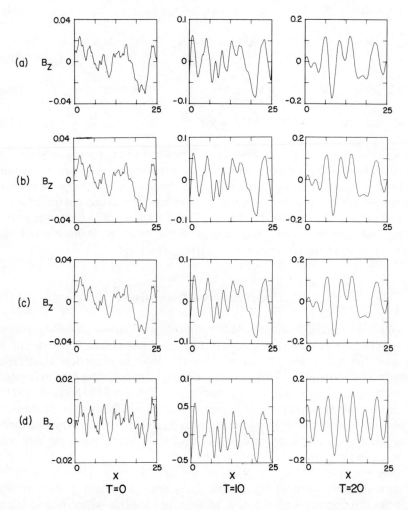

FIG. 1. The z-component of the magnetic field as a function of position at 0, 10, and 20 ω_{pe}^{-1} for (a) the reversible code EMI, (b) the Lagrangian code DARWIN, (c) the simplified Hamiltonian code MICRO, and (d) the code MICRO with a quieter initialization. The position is in units of c/ω_{pe} and the magnetic field in units of $m\omega_{pe}c/e$.

0, 10, and 20 ω_{pe}^{-1}. It is believed that this diagnostic constitutes a more sensitive measure of code behavior than any gross quantity such as the total energy.

Since it is well known that the details in the evolution of a simulation depend on the precise form of the initialization, and since it was expected that the differences between algorithms would be small, it was necessary that the initial conditions be exactly the same in each instance. This required using identical sequences of pseudorandom numbers in generating the velocity distributions. In addition, special care was required in specifying the values of the vector potential which were used to define the initial canonical momenta in the Hamiltonian versions. For the case of the EMI code and the first run of MICRO this was done by first accumulating the currents due to the assigned initial velocities and then finding an initial solution of $\nabla^2 \mathbf{A} = -(4\pi/c) \mathbf{J}$. Close examination of the first three rows of Fig. 1 reveals practically no observable difference between these runs. We therefore conclude that the three algorithms produce identical results to within the precision measured by this test.

A run with MICRO using a different manner of initialization for the canonical momenta is presented in the last row of Fig. 1. Here no special effort was made to obtain a self-consistent $\mathbf{A}$ before initializing the canonical momenta. Instead, the equilibrium limit $\mathbf{A} = 0$ was used. The immediate impression from the diagnostic in Fig. 1 is that of a "quieter" start. The same modes seem to be excited, even with the same phase, but the longer wavelength modes are excited initially to a lesser degree. An understanding of this phenomenon is not difficult. Let $\mathbf{J}^0$ and $\mathbf{A}^0$ represent the current and potential due to the initial self-consistent loading. They satisfy $\nabla^2 \mathbf{A}^0 = -(4\pi/c) \mathbf{J}^0$. In the case of the fourth run, the initial values of the canonical momenta were exactly proportional to the initial velocities, since the potential $\mathbf{A} = 0$ was used. Therefore, the first field solution by this run of the Hamiltonian code was a solution of

$$\nabla^2 \mathbf{A}^0 - (\omega_p^2/c^2) \tilde{\mathbf{A}}^0 = -(4\pi/c) \mathbf{J}^0. \quad (39)$$

Doing a spatial Fourier transform and assuming ω_p constant, as it is initially, we obtain

$$\mathbf{A}_\kappa^0 = -(4\pi/c) \mathbf{J}_\kappa^0 (1/\kappa^2) \quad (40a)$$

and

$$\tilde{\mathbf{A}}_\kappa^0 = -(4\pi/c) \mathbf{J}_\kappa^0 [1/(\kappa^2 + \omega_p^2/c^2)]. \quad (40b)$$

In the case of the fourth run, the first field solution implies a discontinuous change in all of the particle velocities, and therefore it should be regarded, not as part of the first time-step, but rather as a reinitialization. Since κ_{max}

is of the order of ω_p/c, the latter initialization provides a nearly flat spectrum of excitation in **A**, a spectrum generally preferable to the inverse square spectrum of the other cases. We recommend using this technique unless circumstances allow an even quieter initialization, such as one of the type proposed by Byers (1970). Initialization as in the fourth example is almost automatic in Hamiltonian algorithms, whereas in Lagrangian versions it must be specifically provided.

Computer execution times depend on a number of variables in addition to the inherent complexity of the algorithms, and therefore should not be ascribed undue significance. However, rough comparisons may be of interest. These all-FORTRAN programs, run for 100 time-steps with 20,000 particles, required the following execution times on the CDC 7600: the one-dimensional, reversible Hamiltonian code EMI took 3 min and 44 sec; the one-dimensional simplified Hamiltonian code MICRO took 1 min and 5 sec; and the two-dimensional Lagrangian code DARWIN being used on a one-dimensional case took 3 min and 26 sec. Of course, a two-dimensional code used on a one-dimensional case is substantially less efficient than is a similar code designed to apply only in one dimension.

VII. A Two-Dimensional Example

The principal motivation for the development of the two-dimensional code DARWIN was an interest in microturbulence phenomena in the presence of strong gradients. The diffusion of the magnetic field into the surface of a high density theta pinch is an example. It is quite likely that, particularly at early times, the gradients are sufficiently steep so as to make nearly homogeneous approximations inadequate. The hope is that studies with a code of this type will help explain the nature of magnetic field diffusion, cross-field heat transport, ion–electron heat exchange, and other nonlinear phenomena. We have begun to apply the code to such problems and we present here some very preliminary results to illustrate the capability of the code, without any attempt to explain the physical processes involved. This work is being done by Winske and Hewett (1975) as part of their study of magnetic field diffusion in theta pinches.

In the first attempts to apply the simulation code to theta-pinch studies, the code was initialized with a nonequilibrium plasma in order to simulate an aspect of the pinch implosion. However, the complexity of the resulting phenomena made analysis very difficult. No easy differentiation was possible between the effects due to gross implosion dynamics and those due to microinstabilities. Therefore, a means was sought to initialize the code in an inhomogeneous equilibrium. To this end, a new technique for generating Vlasov

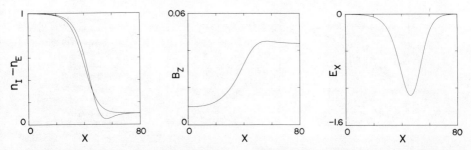

FIG. 2. Properties of the inhomogeneous equilibrium displayed from left to right are the ion and electron densities, the z-component of the magnetic field, and the x-component of the electric field as functions of x over a region 80 c/ω_{pe} in length. The densities are normalized to unity at $x = 0$, the magnetic field is measured in units of $m\omega_{pe}c/e$, and the electric field is in units of $m\omega_{pe}v_{th}/e$, where v_{th} is the thermal speed.

equilibria has been developed (Hewett et al., 1976). The method proceeds from assumed profiles of ion density, ion and electron temperatures, and the assumption that the ions are electrostatically confined. These assumptions are sufficient to determine the electric field, the magnetic field, and the ion and electron distribution functions. Actual generation of the electron distribution function is by means of a Hermite expansion method.

An example of an equilibrium so determined is given in Fig. 2. Here the ion and electron densities drop off to about 10% of their initial values. The small difference in their densities is sufficient to provide the electrostatic confining field for the ions. A mass ratio of 100 is assumed. The fairly complicated electron distribution function, which is not displayed, provides self-consistent currents and pressures.

The output from the equilibrium code is approximated in the initialization of the two-dimensional simulation by matching the first three velocity moments as functions of position. The fact that the resulting ensemble is indeed in approximate equilibrium was verified by collapsing the code to one dimension and observing the persistence of the initial conditions over many tens of plasma periods.

The results of the two-dimensional simulation are shown in Fig. 3. A rectangular region $8c/\omega_{pe}$ wide by $25c/\omega_{pe}$ high is represented by a mesh of 40×128 cells. Ions and electrons of mass ratio 100 are each represented by 64,000 simulation particles. As the simulation proceeds, a powerful microscopic fluting instability is seen to develop. This is illustrated in Fig. 3 by contours of constant ϕ and B_z at 50, 100, and 165 ω_{pe}^{-1}. Note the high degree of correlation between the electric and magnetic contours, and the persistence of the single-mode structure. Whether this persistence is due principally to box quantization or is physically significant is not presently known. Further interpretation of these phenomena awaits further simulation and analysis.

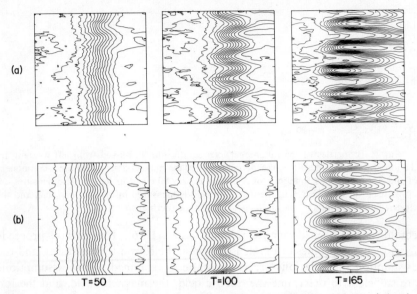

FIG. 3. Results of the two-dimensional simulation at times 50, 100, and 165 ω_{pe}^{-1} showing (a) contours of constant electrostatic potential ϕ, and (b) contours of constant magnetic field component B_z, both displayed in the x–y computational rectangle of dimensions $8\,c/\omega_{pe}$ wide and $25\,c/\omega_{pe}$ high.

VIII. Summary

We have shown that Darwin model codes may be expressed satisfactorily in either Hamiltonian or Lagrangian form. The essential feature of instantaneous action-at-a-distance leads to similar elliptic systems for the field equations in either formulation, and we have devised a means for solving such systems. In the Hamiltonian case, the force due to the transverse electric field is obtained implicitly through the equations of motion for the particles, whereas in the Lagrangian case the transverse electric field is found explictly using the divergence of the current transfer tensor as a source function.

One-dimensional comparisons suggest that exact reversibility is not essential to the accuracy or stability of Darwin-model algorithms, and that non-reversible algorithms are more efficient. The two-dimensional code DARWIN, implementing the Lagrangian algorithm of Section IV, seems completely satisfactory. A two-dimensional Hamiltonian version may offer advantages; superficial estimates suggest that Hamiltonian versions will offer savings in storage, while Lagrangian versions will offer greater computational speed. However, because of the very complex nature of particle simulation codes, a final choice will have to await more theoretical and empirical understanding.

ACKNOWLEDGMENTS

The authors wish to thank D. W. Forslund, J. P. Freidberg, E. L. Lindman, and T. A. Oliphant for many helpful discussions during the course of these investigations, and to thank D. W. Forslund for assistance in the use of the EMI code. This work was performed under the auspices of the USERDA.

REFERENCES

Auer, P. L., Hurwitz, H., and Kilb, R. W. (1961). *Phys. Fluids* **4**, 1105.
Boris, J. P. (1970). *In* "Proceedings of the Fourth Conference on the Numerical Simulation of Plasmas," (J. P. Boris and R. A. Shanny, eds.), Stock 0851 00059, p. 3. US Govt. Printing Office, Washington, D.C.
Boris, J. P., and Lee, R. (1973). *J. Comput. Phys.* **12**, 131.
Buneman, O. (1968). "Relativistic Plasmas," p. 205. Benjamin, New York.
Buzbee, B. L., Golub, G. H., and Nielson, C. W. (1970). *SIAM (Soc. Ind. Appl. Math.) J. Numer. Anal.* **7**, 627.
Byers, J. A. (1970). *In* "Proceedings of the Fourth Conference on the Numerical Simulation of Plasmas," (J. P. Boris and R. A. Shanny, eds.), Stock 0851 00059, p. 496. US Govt. Printing Office, Washington, D.C.
Concus, P., and Golub, G. H. (1973). *SIAM (Soc. Ind. Appl. Math.) J. Numer. Anal.* **10**, 1103.
Darwin, C. G. (1920). *Phil. Mag.* **39**, 537.
Dickman, D., Morse, R. L., and Nielson, C. W. (1969). *Phys. Fluids* **12**, 1708.
Forslund, D. W., Lindman, E. L., Mitchell, R. W., and Morse, R. L. (1972). Los Alamos Report LA-DC-72-721.
Godfrey, B. B. (1974). *J. Comput. Phys.* **15**, 504.
Goldstein, H. (1953). "Classical Mechanics," p. 205. Addison-Wesley, Reading, Massachussetts.
Haber, I., Wagner, C. E., Boris, J. P., and Dawson, J. M. (1970). *In* "Proceedings of the Fourth Conference on the Numerical Simulation of Plasmas" (J. P. Boris and R. A. Shanny, eds.), Stock 0851 0059, p. 126. US Govt. Printing Office, Washington, D.C.
Haber, I., Lee, R., Klein, H. H., and Boris, J. P. (1973). *In* "Proceedings of the Sixth Conference on Numerical Simulation of Plasmas," p. 46. Lawrence Livermore Laboratory, Livermore, California.
Hasegawa, A., and Birdsall, C. K. (1964). *Phys. Fluids* **7**, 1590.
Hasegawa, A., and Okuda, H. (1968). *Phys. Fluids* **11**, 1995.
Hewett, D. W., Nielson, C. W., and Winske, D. (1976). *Phys. Fluids* **19**, 443.
Jackson, J. D. (1962). "Classical Electrodynamics," p. 409. Wiley, New York.
Kaufman, A. N., and Rostler, P. S. (1971). *Phys. Fluids* **14**, 446.
Krall, N. A., and Trivelpiece, A. W. (1973). "Principles of Plasma Physics," p. 86. McGraw-Hill, New York.
Landau, L., and Lifshitz, E. (1951). "The Classical Theory of Fields," p. 180. Addison-Wesley, Reading, Massachusetts.
Langdon, A. B. (1970a). *J. Comput. Phys.* **6**, 252.
Langdon, A. B. (1970b). *J. Comput. Phys.* **6**, 255.
Langdon, A. B. (1970c). *J. Comput. Phys.* **6**, 261.
Lewis, H. R. (1970). *Methods Comput. Phys.* **9**, 307.
Lindman, E. L. (1973). *In* "Proceedings of the Sixth Conference on the Numerical Simulation of Plasmas," p. 42. Lawrence Livermore Laboratory, Livermore, California.

Low, F. E. (1958). *Proc. Roy. Soc., Ser. A* **248**, 282.
Morse, R. L., and Nielson, C. W. (1971). *Phys. Fluids* **14**, 830.
Nielson, C. W., and Lindman, E. L. (1973). *In* "Proceedings of the Sixth Conference on the Numerical Simulation of Plasmas," p. 148. Lawrence Livermore Laboratory, Livermore, California.
Nielson, C. W., Buzbee, B. L., and Rudsinski, L. I. (1971). *In* "Proceedings of the Fifth Conference on the Numerical Simulation of Plasmas," Paper D1.
Sturrock, P. A. (1958). *Ann. Phys. (Leipzig)* [7] **4**, 306.
Winske, D., and Hewett, D. W. (1975). *Phys. Rev. Lett.* **35**, 937.

The Solution of the Kinetic Equations for a Multispecies Plasma*

JOHN KILLEEN, ARTHUR A. MIRIN, AND
MARVIN E. RENSINK

LAWRENCE LIVERMORE LABORATORY
UNIVERSITY OF CALIFORNIA
LIVERMORE, CALIFORNIA

I. Introduction	389
II. Mathematical Model	392
A. Fokker–Planck Equations	392
B. Time-Varying Forces	395
C. Boundary Conditions	397
D. Source and Loss Terms	399
III. Solutions Using Angular Eigenfunctions	401
A. Numerical Methods	401
B. Applications	404
IV. Finite-Difference Solution in a Two-Dimensional Velocity Space	411
A. Numerical Methods	411
B. Applications	412
References	430

I. Introduction

IN THE SIMULATION OF magnetically confined plasmas where the ions are not Maxwellian and where a knowledge of the distribution functions is important, kinetic equations must be solved. The proposition that a stable mirror plasma will yield net thermonuclear power depends on the rate at which particles are lost out of the ends of the device. At number densities and energies typical of mirror machines, the end losses are due primarily to the scattering of charged particles into the loss cones in velocity space by classical Coulomb collisions. The kinetic equation describing this process is the Boltzmann equation with Fokker–Planck collision terms (Rosenbluth et al., 1957).

The use of this equation is not restricted to mirror systems. The heating of plasmas by energetic neutral beams, the thermalization of alpha particles in DT plasmas, the study of runaway electrons in Tokamaks, and the performance of two-energy component fusion reactors are other examples where the solution of the Fokker–Planck equation is required.

*Work supported by the U.S. Energy Research and Development Administration.

The problem is to solve a nonlinear partial differential equation for the distribution function of each charged species in the plasma in terms of seven independent variables (three spatial coordinates, three velocity coordinates, and time). Such an equation, even for a single species, exceeds the capability of any present computer, so several simplifying assumptions are therefore required to treat the problem. Typical approximations that are made in present-day codes are to neglect spatial dependence and to assume that the distribution functions are azimuthally invariant in velocity space (about the direction of the magnetic field). These assumptions reduce the number of independent variables to three—two velocity-space coordinates, v and θ, the speed and pitch angle, and the time, t. Even with these basic assumptions there has been an evolution of numerical Fokker–Planck calculations over the past fifteen years.

The work of Roberts and Carr (1960) and Bing and Roberts (1961) consisted of a solution of the complete Fokker–Planck equation for ions only, ignoring the effects of the electrons, electrostatic ambipolar potential, and spatial inhomogeneities except for the existence of a loss cone. They also investigated the adequacy of an approximation to the Fokker–Planck equation in which the solution was assumed to be approximately separable.

BenDaniel and Allis (1962) extended the work of Roberts and Carr, particularly in the area of approximating the solution by a separated solution. They also made some progress toward approximate solutions in the case where spatial inhomogeneities exist.

Killeen and Futch (1968) and Fowler and Rankin (1966) solved the Fokker–Planck equations for both ions and electrons, assuming that the evolution of the distribution functions can be described by the equations for isotropic distributions, with certain factors included to take the presence of the loss cone into account. This assumption and the ensuing approximations are directly related to the assumption of separability. Fowler and Rankin have included the effect of the ambipolar potential. Killeen and Futch also do this, and they include the effect of charge exchange as well in a time-dependent calculation of a plasma formed by neutral injection.

Killeen and Marx (1970) developed a code which solved the unseparated Fokker–Planck equation in v and θ for a single ion species, under the assumption that the electrons can be represented by a Maxwellian distribution function with loss cone removed. The ion equation is solved using the alternating direction implicit (ADI) finite-difference scheme on a variable mesh in v and θ, and the coefficients containing moments over both ion and electron distribution functions are recomputed at each ion time-step. The number density and mean energy characterizing the Maxwellian distribution of electrons are computed on a faster time-scale by solving a pair of ordinary differential equations that include the effects of ion–electron interactions and electron end losses and injection conditions. The ambipolar potential is determined by iteration on the electron equations in such a way as to equalize the positive and negative charge densities.

Marx (1970; Killeen and Marx, 1970) generalized the code to treat realistic spatial dependence of the magnitude of the confining magnetic field (z-dependence), as well as spatially distributed sources. In the z-dependent version of the code, the spatial variation of the ambipolar potential is a computed quantity.

A multispecies code (Killeen and Mirin, 1971; Futch et al., 1972) was developed in order to study D–T and D–^{3}He mirror reactors, including the effects of reaction products. The principal assumptions of this code are that the "Rosenbluth potentials" (see Section II) are isotropic and that the distribution functions can be represented by their lowest angular eigenfunction (see Section III). The resulting set of coupled equations for $f_a(v,t)$ are then solved numerically, using an implicit finite-difference scheme. An extensive parameter study (Futch et al., 1972) was conducted yielding values of the confinement parameter $n\tau$ and the figure of merit Q (the ratio of thermonuclear power to injected power) as a function of mirror ratio and injection energy.

In a Ph.D. thesis, Werkoff (1973) studied the effect of the assumptions that the Rosenbluth potentials are isotropic and that the distribution functions could be represented by the lowest normal mode. His initial conclusions were that including the anisotropic part of the ion–ion Rosenbluth potentials substantially increased the value of $n\tau$ for mirror containment. Recently, we have developed two independent multispecies codes which test this hypothesis and significantly improve the treatment of non-Maxwellian plasmas. It is these two codes which we describe in this article in Sections III and IV.

In Section II we describe the kinetic equations that are solved. Of particular importance is the inclusion of time-varying forces in the model for the simulation of specific experiments.

In Section III we give the method of solution of the Fokker–Planck equations using an arbitrary number of angular eigenfunctions. In particular, we study the mirror containment problem and find that including the anisotropic parts of the Rosenbluth potentials and using several normal modes to represent the ion distribution functions results in an increase of $n\tau$ by about 10%. This result is verified by use of the code described in Section IV.

In Section IV we describe a highly versatile, two-dimensional, multispecies code in which the ion kinetic equations are solved by finite-difference methods on a two-dimensional (v, θ) domain. We call this a "hybrid" code, because the electron distribution function is still represented using the lowest angular eigenfunction (isotropic in a full velocity space); hence, a one-dimensional electron kinetic equation is solved. We find that this method of solution for the ion equations is much faster than the method of Section III, so this has become our standard code for both mirror and toroidal plasma problems. In Section IV we give a variety of applications of this code and describe diagnostics that have been added for the special problems. For mirror systems with beam injection, we find values of the containment parameter $n\tau$ as much as 40% higher than those given by the one-dimensional Fokker–Planck code. We also consider two-component toroidal systems and calculate the energy multiplication factor for a number of different scenarios.

II. Mathematical Model

A. Fokker–Planck Equations

The appropriate kinetic equations are Boltzmann equations with Fokker–Planck collision terms, often referred to simply as Fokker–Planck equations:

$$\frac{\partial f_a}{\partial t} + \mathbf{v} \cdot \frac{\partial f_a}{\partial \mathbf{r}} + \frac{\mathbf{F}}{m_a} \cdot \frac{\partial f_a}{\partial \mathbf{v}} = \left(\frac{\partial f_a}{\partial t}\right)_c + S_a + L_a. \tag{1}$$

Here f_a is the distribution function in six-dimensional phase space for particles of species a, S_a is a source term, $(\partial f_a/\partial t)_c$ is the collision term, and L_a contains loss terms.

The Fokker–Planck collision term for an inverse-square force was derived by Rosenbluth et al. (1957) in the form

$$\frac{1}{\Gamma_a}\left(\frac{\partial f_a}{\partial t}\right)_c = -\frac{\partial}{\partial v_i}\left(f_a \frac{\partial h_a}{\partial v_i}\right) + \frac{1}{2}\frac{\partial^2}{\partial v_i\, \partial v_j}\left(f_a \frac{\partial^2 g_a}{\partial v_i\, \partial v_j}\right), \tag{2}$$

where $\Gamma_a = 4\pi Z_a^4 e^4/m_a^2$. In the present work we write the "Rosenbluth potentials"

$$g_a = \sum_b (Z_b/Z_a)^2 \ln \Lambda_{ab} \int f_b(\mathbf{v'})|\mathbf{v} - \mathbf{v'}|\, d\mathbf{v'}, \tag{3}$$

$$h_a = \sum_b [(m_a + m_b)/m_b](Z_b/Z_a)^2 \ln \Lambda_{ab} \int f_b(\mathbf{v'})|\mathbf{v} - \mathbf{v'}|^{-1}\, d\mathbf{v'}, \tag{4}$$

which differ from those given by Rosenbluth et al. (1957) in the dependence of the Coulomb logarithm on both interacting species and its consequent inclusion under the summations. The expression we use is

$$\ln \Lambda_{ab} = \ln\{[m_a m_b/(m_a + m_b)](2\alpha c\lambda_D/e^2) \max[(2\bar{E}/m)_{a,b}^{1/2}]\} - \tfrac{1}{2}, \tag{5}$$

where α is the fine structure constant, λ_D the Debye length, $\bar{E}$ the mean energy of particles of species a or b, and the other symbols have their usual meanings. The normalization of f_a in Eqs. (1) through (4) is such that the particle density is given by

$$n_a = \int f_a(\mathbf{v})\, d\mathbf{v}. \tag{6}$$

Since the collision term will be seen to contain velocity derivatives of f_a multiplied by velocity moments over f_a, Eq. (1) is a nonlinear, partial, integro-differential equation in seven independent variables. We choose a spherical coordinate system for velocity space (with $\theta = 0$ corresponding to the direction along a magnetic field line) and a cylindrical coordinate system for physical space (z along the magnetic axis). With these coordinate systems, the following assumptions are made:

1. The system is radially and azimuthally uniform in physical space, and hence, it is also azimuthally symmetric in velocity space. Equivalently, we neglect all gradients transverse to the magnetic field.

2. The system is axially uniform in physical space. For mirror systems this is equivalent to a magnetic square-well model; for toriodal systems this implies axisymmetry about the major axis of the device.

The transformation of Eq. (2) to spherical polar coordinates (v, θ, ϕ) in velocity space has been given by Rosenbluth *et al.* (1957). With our assumption of azimuthal symmetry, the resulting distribution functions are of the form $f_a(v, \mu, t)$, where $\mu = \cos\theta$ and $v = |\mathbf{v}|$. The equation for each species is

$$\frac{1}{\Gamma_a}\left(\frac{\partial f_a}{\partial t}\right) = -\frac{1}{v^2}\frac{\partial}{\partial v}\left(f_a v^2 \frac{\partial h_a}{\partial v}\right) - \frac{1}{v^2}\frac{\partial}{\partial \mu}\left(f_a(1-\mu^2)\frac{\partial h_a}{\partial \mu}\right)$$

$$+ \frac{1}{2v^2}\frac{\partial^2}{\partial v^2}\left(f_a v^2 \frac{\partial^2 g_a}{\partial v^2}\right) + \frac{1}{2v^2}\frac{\partial^2}{\partial \mu^2}\left\{f_a\left[\frac{1}{v^2}(1-\mu^2)^2\frac{\partial^2 g_a}{\partial \mu^2}\right.\right.$$

$$\left.\left. + \frac{1}{v}(1-\mu^2)\frac{\partial g_a}{\partial v} - \frac{1}{v^2}\mu(1-\mu^2)\frac{\partial g_a}{\partial \mu}\right]\right\} + \frac{1}{v^2}\frac{\partial^2}{\partial \mu\, \partial v}\left\{f_a(1-\mu^2)\right.$$

$$\left.\left[\frac{\partial^2 g_a}{\partial \mu\, \partial v} - \frac{1}{v}\frac{\partial g_a}{\partial \mu}\right]\right\} + \frac{1}{2v^2}\frac{\partial}{\partial v}\left\{f_a\left[-\frac{1}{v}(1-\mu^2)\frac{\partial^2 g_a}{\partial \mu^2} - 2\frac{\partial g_a}{\partial v}\right.\right.$$

$$\left.\left. + \frac{2\mu}{v}\frac{\partial g_a}{\partial \mu}\right]\right\} + \frac{1}{2v^2}\frac{\partial}{\partial \mu}\left\{f_a\left[\frac{1}{v^2}\mu(1-\mu^2)\frac{\partial^2 g_a}{\partial \mu^2} + \frac{2\mu}{v}\frac{\partial g_a}{\partial v}\right.\right.$$

$$\left.\left. + \frac{2}{v}(1-\mu^2)\frac{\partial^2 g_a}{\partial \mu\, \partial v} - \frac{2}{v^2}\frac{\partial g_a}{\partial \mu}\right]\right\}. \tag{7}$$

The functions g_a and h_a, defined by Eqs. (3) and (4), can be represented by expansions in Legendre polynomials (Rosenbluth *et al.*, 1957). For this purpose we let

$$f_a(v,\mu,t) = \sum_{j=0}^{\infty} V_j^a(v,t) P_j(\mu), \tag{8}$$

where

$$V_j^a(v,t) = \tfrac{1}{2}(2j+1) \int_{-1}^{+1} f_a(v,\mu,t) P_j(\mu)\, d\mu. \tag{9}$$

The expansions for g_a and h_a are

$$g_a(v,\mu,t) = \sum_{j=0}^{\infty} \sum_b (Z_b/Z_a)^2 \ln \Lambda_{ab}\, B_j^b(v,t) P_j(\mu), \tag{10}$$

$$h_a(v,\mu,t) = \sum_{j=0}^{\infty} \sum_b [(m_a+m_b)/m_b](Z_b/Z_a)^2 \ln \Lambda_{ab}\, A_j^b(v,t) P_j(\mu), \tag{11}$$

where

$$A_j^a = \frac{4\pi}{2j+1}\left[\int_0^v \frac{(v')^{j+2}}{v^{j+1}} V_j^a(v',t)\,dv' + \int_v^\infty \frac{v^j}{(v')^{j-1}} V_j^a(v',t)\,dv'\right] \tag{12}$$

$$B_j^a = -\frac{4\pi}{4j^2-1}\left[\int_0^v \frac{(v')^{j+2}}{v^{j-1}}\left(1 - \frac{j-1/2}{j+3/2}\frac{(v')^2}{v^2}\right) V_j^a(v')\,dv'\right.$$

$$\left. + \int_v^\infty \frac{v^j}{(v')^{j-3}}\left(1 - \frac{j-1/2}{j+3/2}\frac{v^2}{(v')^2}\right) V_j^a(v')\,dv'\right]. \tag{13}$$

In the computations we take a finite number (which can be varied) of terms in the Legendre expansions of g_a and h_a.

In Section III we shall describe the solution of Eq. (7) using an expansion in angular eigenfunctions $M_l^a(\mu)$.

In Section IV we describe the finite-difference solution in a two-dimensional velocity space. For this purpose we find it more convenient in differencing and applying boundary conditions to use (v,θ) coordinates rather than (v,μ) coordinates. Equation (2) in (v,θ) coordinates written in conservation form is

$$(1/\Gamma_a)(\partial f_a/\partial t)_c = (1/v^2)(\partial G_a/\partial v) + (1/v^2 \sin\theta)(\partial H_a/\partial\theta), \tag{14}$$

where

$$G_a = A_a f_a + B_a\, \partial f_a/\partial v + C_a\, \partial f_a/\partial\theta, \tag{15}$$

$$H_a = D_a f_a + E_a\, \partial f_a/\partial v + F_a\, \partial f_a/\partial\theta, \tag{16}$$

and

$$A_a = \frac{v^2}{2}\frac{\partial^3 g_a}{\partial v^3} + v\frac{\partial^2 g_a}{\partial v^2} - \frac{\partial g_a}{\partial v} - v^2\frac{\partial h_a}{\partial v} - \frac{1}{v}\frac{\partial^2 g_a}{\partial \theta^2} + \frac{1}{2}\frac{\partial^3 g_a}{\partial v\, \partial\theta^2}$$

$$- \frac{\cot\theta}{v}\frac{\partial g_a}{\partial \theta} + \frac{\cot\theta}{2}\frac{\partial^2 g_a}{\partial v\, \partial\theta}, \tag{17}$$

$$B_a = (v^2/2)(\partial^2 g_a/\partial v^2), \tag{18}$$

$$C_a = -(1/2v)(\partial g_a/\partial \theta) + \tfrac{1}{2}\partial^2 g_a/\partial v\, \partial\theta, \tag{19}$$

$$D_a = \frac{\sin\theta}{2v^2}\frac{\partial^3 g_a}{\partial \theta^3} + \frac{\sin\theta}{2}\frac{\partial^3 g_a}{\partial v^2\, \partial\theta} + \frac{\sin\theta}{v}\frac{\partial^2 g_a}{\partial v\, \partial\theta} - \frac{1}{2v^2 \sin\theta}\frac{\partial g_a}{\partial \theta}$$

$$+ \frac{\cos\theta}{2v^2}\frac{\partial^2 g_a}{\partial \theta^2} - \sin\theta\frac{\partial h_a}{\partial \theta}, \tag{20}$$

$$E_a = \sin\theta[-(1/2v)\,\partial g_a/\partial\theta + \tfrac{1}{2}\partial^2 g_a/\partial v\, \partial\theta], \tag{21}$$

$$F_a = \frac{\sin\theta}{2v^2}\frac{\partial^2 g_a}{\partial \theta^2} + \frac{\sin\theta}{2v}\frac{\partial g_a}{\partial v}. \tag{22}$$

Equation (14) is used for each ion species present. For electrons we assume that the distribution function is isotropic, so Eq. (2) becomes

$$(1/\Gamma_e)(\partial f_e/\partial t)_c = (1/v^2)(\partial G_e/\partial v), \tag{23}$$

where

$$G_e = A_e f_e + B_e\, \partial f_e/\partial v \tag{24}$$

and

$$A_e = 4\pi \sum_b \left[(Z_b/Z_e)^2 m_e/m_b \ln \Lambda_{eb} \int_0^v V_0^b(v',t) v'^2\, dv' \right] \tag{25}$$

$$B_e = 4\pi \sum_b \left\{ (Z_b/Z_e)^2 \ln \Lambda_{eb} \left[(1/3v) \int_0^v V_0^b(v',t) v'^4\, dv' \right.\right.$$

$$\left.\left. + (v^2/3) \int_v^\infty V_0^b(v',t) v'\, dv' \right] \right\}. \tag{26}$$

B. Time-Varying Forces

Within the context of our idealized homogeneous plasma model the form of the kinetic equation with time-varying magnetic field is

$$df/dt = \partial f/\partial t + \dot{v}_\parallel\, \partial f/\partial v_\parallel + \dot{v}_\perp\, \partial f/\partial v_\perp = (\partial f/\partial t)_c + S + L, \tag{27}$$

where $(\dot{v}_\parallel, \dot{v}_\perp)$ is the cyclotron-averaged inductive acceleration. For slowly varying magnetic fields $(\dot{v}_\parallel, \dot{v}_\perp)$ can be derived from the adiabatic constants of the particle motion.

1. *Magnetic Field Compression in a Mirror Machine*

In linear mirror systems the constants of the particle motion are

$$mv_\perp^2/2B = \text{magnetic moment}, \tag{28}$$

$$2mv_\parallel L = \text{longitudinal invariant}, \tag{29}$$

where L is the distance between mirrors. To derive $(\dot{v}_\parallel, \dot{v}_\perp)$ we take the time derivative of these equations, obtaining

$$2v_\perp \dot{v}_\perp/B - v_\perp^2 \dot{B}/B^2 = 0, \tag{30}$$

$$\dot{v}_\parallel L + v_\parallel \dot{L} = 0, \tag{31}$$

and subsequently

$$\dot{v}_\perp = \tfrac{1}{2} v_\perp \dot{B}/B, \tag{32}$$

$$\dot{v}_\parallel = -v_\parallel \dot{L}/L. \tag{33}$$

When these results for $\dot{v}_\parallel$ and $\dot{v}_\perp$ are inserted in Eq. (27) the Fokker–Planck equation becomes

$$\partial f/\partial t + (\tfrac{1}{2} v \sin^2\theta \, \partial f/\partial v + \tfrac{1}{2} \sin\theta \cos\theta \, \partial f/\partial \theta) \dot{B}/B = (\partial f/\partial t)_c + S + L, \tag{34}$$

where we have used

$$v_\perp = v \sin\theta, \tag{35}$$

$$v_\parallel = v \cos\theta, \tag{36}$$

and longitudinal compression has been neglected. For isotropic electrons $\partial f/\partial \theta$ is zero and $\sin^2\theta$ is replaced by its average value $\langle \sin^2\theta \rangle = 2/3$ so the electron Fokker–Planck equation becomes

$$\partial f/\partial t + (\tfrac{1}{3} v \, \partial f/\partial v) \dot{B}/B = (\partial f/\partial t)_c + S + L. \tag{37}$$

2. Major Radius Compression in a Tokamak

In axisymmetric toroidal systems the constants of the particle motion are

$$mv_\perp^2/2B = \text{magnetic moment,} \tag{38}$$

$$mv_\parallel R = \text{toroidal angular momentum,} \tag{39}$$

where R is the major radius of the torus. For a time-varying major radius (Furth and Yoshikawa, 1970) we derive $(\dot{v}_\parallel, \dot{v}_\perp)$ by taking the time derivative of these equations, obtaining

$$2v_\perp \dot{v}_\perp/B - v_\perp^2 \dot{B}/B^2 = 0, \tag{40}$$

$$\dot{v}_\parallel R + v_\parallel \dot{R} = 0, \tag{41}$$

and subsequently

$$\dot{v}_\perp = +\tfrac{1}{2} v_\perp \dot{B}/B, \tag{42}$$

$$\dot{v}_\parallel = -v_\parallel \dot{R}/R. \tag{43}$$

Since B is essentially just the toroidal field strength, it varies inversely with R, yielding

$$\dot{B}/B = -\dot{R}/R. \tag{44}$$

When these results are inserted in Eq. (27) the Fokker–Planck equation becomes

$$\frac{\partial f}{\partial t} + \frac{\dot{R}}{R}\left[-\left(1 - \tfrac{1}{2}\sin^2\theta\right)v\frac{\partial f}{\partial v} + \tfrac{1}{2}\sin\theta\cos\theta\frac{\partial f}{\partial \theta}\right] = \left(\frac{\partial f}{\partial t}\right)_c + S + L, \tag{45}$$

where $(\partial f/\partial t)_c$ is given by Eq. (14). For isotropic electrons we have

$$\partial f_e/\partial t + (-\tfrac{2}{3} v\, \partial f_e/\partial v)\dot{R}/R = (\partial f_e/\partial t)_c + S_e + L_e. \tag{46}$$

C. Boundary Conditions

1. Loss Cone Domain, Ambipolar Potential

Since several examples discussed here are devoted to the problem of plasma confinement within systems of magnetic mirrors, in this section we consider the mathematical description of such systems.

The loss cone angle is (Spitzer, 1962)

$$\sin^2 \theta_{LC} = 1/R_m, \tag{47}$$

where $R_m = B_m/B(z)$; B_m is the magnetic field at the mirror, and $B(z)$ is the magnetic field at the interior point being considered. The orientation of the loss cone in velocity space is displayed in Fig. 1. A particle whose angle in velocity space is less than θ_{LC} will be immediately lost from the mirror system. θ_{LC} is independent of velocity as well as particle mass and charge. Equation (47) is derived under the assumption that no electrostatic potential exists, and θ_{LC} is the actual loss angle only under that condition.

However, because of their greater mobility, the scattering rate of electrons will be greater than that of ions, and more electrons than ions will tend to leak out of the ends of the device. Hence, an ambipolar potential will build up, being greatest at the center and decreasing toward the ends. The fact that this potential is established leads to a fundamental change in the loss characteristics for the two types of particles. The loss regions are then defined by a loss angle which is a function of speed and charge (Kaufman, 1956; Post, 1961; BenDaniel, 1961; Yushmanov, 1966). If $Z_a e$ is charge and $-\phi$ is electrostatic potential relative to the midplane, the loss angle is given by

$$\sin^2 \theta_{LC} = 1/R_a, \tag{48}$$

where

$$R_a = [(1 + Z_a e\phi/\tfrac{1}{2} m_a v^2)/R_m]^{-1} \tag{49}$$

is the "effective" mirror ratio.

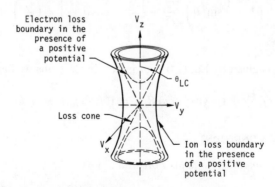

FIG. 1. Velocity-space loss-cone boundaries.

Equation (48) approaches Eq. (47) asymptotically as $v \to \infty$. For ions, the right-hand side of Eq. (48) can exceed unity; no ion in such a velocity regime can be contained. Conversely, for electrons, this term can be less than zero; all electrons at such velocities will be electrostatically trapped. These regions are shown in Fig. 1. The loss region for ions is transformed from a cone into a hyperboloid of one sheet. Its minimum radius occurs at $\theta = \pi/2$, and is equal to the minimum ion velocity possible for confinement. The electron loss region is transformed into a hyperboloid of two sheets.

In some situations, the potential acting on the electrons may be different from that acting on the ions, so we include a provision for specifying the ratio of these two potentials. There is also the option of choosing purely electrostatic confinement for electrons, in which case one uses the boundary condition $f_e(v \geqslant v_\phi) = 0$ where $\tfrac{1}{2} m_e v_\phi^2 = e\phi$.

2. *Full Velocity Space*

In problems where one wants to observe the relaxation of a distribution in the absence of a loss cone, or where one assumes that ions in a loss cone domain are not lost instantly, full velocity-space boundary conditions are applied, namely:

(a) $\quad f(v = 0, \theta)$ is independent of θ, $\hfill$ (50)

(b) $\quad \partial f/\partial v (v = 0, \theta = \pi/2) = 0$, $\hfill$ (51)

(c) $\quad \partial f/\partial \theta (v, \theta = 0) = \partial f/\partial \theta (v, \theta = \pi) = 0$. $\hfill$ (52)

In devices with symmetry about the midplane, $\theta = \pi/2$, it is assumed that

(d) $\quad \partial f/\partial \theta (v, \theta = \pi/2) = 0$. $\hfill$ (53)

D. Source and Loss Terms

The source term S_a in Eq. (1) is of the form

$$S_a(v, \theta, t) = \sum_l J_a^{\,l}(t)\, S_a^{\,l}(v, \theta)\, \delta_s^{a,l}(t), \qquad (54)$$

where the shape function $S_a^{\,l}(v, \theta)$ is a Gaussian in v and $\cos\theta$ of density 1, $\delta_s^{a,l}(t)$ is either 0 or 1, and $J_a^{\,l}(t)$ is a current of the form

$$J_a^l = \sum_b [A_{ab}^l + B_{ab}^l (n_b)^{1/2} + C_{ab}^l n_b]. \tag{55}$$

The quantities A_{ab}^l, B_{ab}^l, and C_{ab}^l are parameters independent of time, and n_b is the density of species "b."

The loss term L_a in Eq. (1) may combine several loss processes. Losses due to charge exchange with the beam are expressed as

$$L_a^c = -\left[\sum_{b,l} D_{ab}^l \delta_s^{b,l}(t)\right] f_a(v,\theta,t), \tag{56}$$

where the quantities D_{ab}^l are constant parameters. The presence of the factors $\delta_s^{b,l}$ allows the whole charge exchange process to be implemented as a unit; that is, a given charge exchange term along with its corresponding source term depends on the same temporal function $\delta_s^{b,l}(t)$.

In mirror applications it is often assumed that ions in the loss cone domain are lost instantly. However, it takes a finite length of time for such ions to make one pass between the mirrors before they are actually lost. Thus, the escape time is finite, and for cold ions (as in a target plasma) this time can be quite long. To account for this, we consider a contribution to the loss term L_a of the form

$$L_a^m = -f_a(v,\theta,t)(v_\parallel/L)\delta^m(v,\theta), \tag{57}$$

where L is the axial length of the mirror system, and $\delta^m(v,\theta)$ is either 1 or 0, depending on whether or not (v,θ) is in the loss cone domain.

The effects of finite particle and energy confinement times in a Tokamak are incorporated by adding a contribution to L_a of the form

$$L_a^\tau = -\frac{f_a(v,\theta,t)}{\tau_p} + \frac{1}{v^2}\frac{\partial}{\partial v}\left[\left(\frac{1}{\tau_e} - \frac{1}{\tau_p}\right)\frac{v^3 f_a(v,\theta,t)}{2}\right], \tag{58}$$

where τ_p and τ_e are particle and energy confinement times, respectively. If we ignore all other terms in Eq. (1) and compute moments, we find

$$\partial n_a/\partial t = -n_a/\tau_p, \tag{59}$$

$$\partial(n_a \bar{E}_a)/\partial t = -n_a \bar{E}_a/\tau_e. \tag{60}$$

The quantity $\bar{E}_a$ is the mean energy of species "a."

III. Solutions Using Angular Eigenfunctions

A. Numerical Methods

It will be assumed throughout this section that our model is an open-ended device with mirror ratio R_m. The technique of eigenfunction expansion, however, is not restricted to this case.

We represent the solutions of Eq. (7) by the orthogonal series

$$f_a(v, \mu, t) = \sum_{l=1}^{\infty} U_l^a(v, t) M_l^a(\mu), \tag{61}$$

where $M_l^a(\mu)$ are eigenfunctions of the equation

$$(1 - \mu^2) d^2 M_l^a/d\mu^2 - 2\mu \, dM_l^a/d\mu + \lambda_l^a M_l^a = 0 \tag{62}$$

which is Legendre's equation on the domain $-\cos\theta_{LC} \leq \mu \leq \cos\theta_{LC}$, where θ_{LC} is the loss cone angle defined by Eq. (47).

In this treatment we do not assume that g_a and h_a are isotropic, so Eq. (7) is not separable. However, we do use the fact that the functions $M_l^a(\mu)$ form a complete orthonormal set on the domain $-\mu_{LC} \leq \mu \leq \mu_{LC} = \cos\theta_{LC}$, i.e., for $k \neq l$

$$\int_{-\mu_{LC}}^{\mu_{LC}} M_l^a(\mu) M_k^a(\mu) \, d\mu = 0, \quad \int_{-\mu_{LC}}^{\mu_{LC}} [M_l^a(\mu)]^2 \, d\mu = 1. \tag{63}$$

For the purposes of computation, we consider a finite number, N, of normal modes, $M_k^a(\mu)$, for each species. We obtain these functions by the numerical solution of Eq. (62), as a two-point eigenvalue problem, for $k = 1, 2, ..., N$.

If we substitute the right-hand side of Eq. (61) into Eq. (7), multiply by $M_k^a(\mu)$ and integrate with respect to μ, making use of Eq. (63), we obtain the following equation for $U_k^a(v, t)$

$$\frac{\partial U_k^a}{\partial t} = \sum_{l=1}^{N} \{(1/v^2)(\partial/\partial v)[\alpha_{kl}^a U_l^a + \beta_{kl}^a (\partial U_l^a/\partial v)] - (\gamma_{kl}^a/v^2) U_l^a\} \tag{64}$$

for $k = 1, ..., N$, where

$$\alpha_{kl}^a = \Gamma_a \int_{-\mu_{LC}}^{\mu_{LC}} d\mu \, M_k^a(\mu) \left\{ M_l^a(\mu) \left[-v^2 \frac{\partial h_a}{\partial v} + \frac{v^2}{2} \frac{\partial^3 g_a}{\partial v^3} + v \frac{\partial^2 g_a}{\partial v^2} \right. \right.$$
$$\left. - \frac{1}{2v}(1 - \mu^2) \frac{\partial^2 g_a}{\partial \mu^2} - \frac{\partial g_a}{\partial v} + \frac{\mu}{v} \frac{\partial g_a}{\partial \mu} \right] +$$

$$+ \frac{\partial}{\partial \mu}\left[(1-\mu^2) M_l^a \left(\frac{\partial^2 g_a}{\partial \mu \, \partial v} - \frac{1}{v}\frac{\partial g_a}{\partial \mu}\right)\right]\Bigg\}, \tag{65}$$

$$\beta_{kl}^a = \Gamma_a \int_{-\mu_{LC}}^{\mu_{LC}} d\mu \, [M_k^a(\mu) \, M_l^a(\mu) \tfrac{1}{2} v^2 \, (\partial^2 g_a / \partial v^2)], \tag{66}$$

$$\gamma_{kl}^a = -\Gamma_a \int_{-\mu_{LC}}^{\mu_{LC}} d\mu \, M_k^a(\mu) \Bigg\{ \frac{\partial}{\partial \mu}\left[M_l^a(\mu)\left(-(1-\mu^2)\frac{\partial h_a}{\partial \mu}\right.\right.$$
$$+ \frac{\mu}{2v^2}(1-\mu^2)\frac{\partial^2 g_a}{\partial \mu^2} + \frac{\mu}{v}\frac{\partial g_a}{\partial v} + \frac{(1-\mu^2)}{v}\frac{\partial^2 g_a}{\partial \mu \, \partial v} - \frac{1}{v^2}\frac{\partial g_a}{\partial \mu}\bigg)\bigg]$$
$$+ \frac{\partial^2}{\partial \mu^2}\left[M_l^a(\mu)\left(\frac{(1-\mu^2)^2}{2v^2}\frac{\partial^2 g_a}{\partial \mu^2} + \frac{(1-\mu^2)}{2v}\frac{\partial g_a}{\partial v} - \frac{\mu(1-\mu^2)}{2v^2}\frac{\partial g_a}{\partial \mu}\right)\right]\Bigg\}. \tag{67}$$

We simplify Eq. (67) by making use of Eq. (62); hence

$$\gamma_{kl}^a = -\Gamma_a \int_{-\mu_{LC}}^{\mu_{LC}} d\mu \, M_k^a(\mu) \Bigg\{ \frac{\partial M_l^a}{\partial \mu}\left(\frac{2(1-\mu^2)}{v}\frac{\partial^2 g_a}{\partial \mu \, \partial v} - \frac{7\mu(1-\mu^2)}{2v^2}\frac{\partial^2 g_a}{\partial \mu^2}\right.$$
$$+ \frac{(1-\mu^2)^2}{v^2}\frac{\partial^3 g_a}{\partial \mu^3} + \frac{2(\mu^2-1)}{v^2}\frac{\partial g_a}{\partial \mu} - (1-\mu^2)\frac{\partial h_a}{\partial \mu}\bigg)$$
$$+ M_l^a\Bigg[-\frac{3\mu}{v}\frac{\partial^2 g_a}{\partial \mu \, \partial v} + \left(\frac{15\mu^2 - 7}{2v^2} - \frac{\lambda_l^a(1-\mu^2)}{2v^2}\right)\frac{\partial^2 g_a}{\partial \mu^2}$$
$$- \frac{4\mu(1-\mu^2)}{v^2}\frac{\partial^3 g_a}{\partial \mu^3} + \frac{3(1-\mu^2)}{2v}\frac{\partial^3 g_a}{\partial \mu^2 \, \partial v} + \frac{(1-\mu^2)^2}{2v^2}\frac{\partial^4 g_a}{\partial \mu^4}$$
$$+ \left(\frac{3\mu}{v^2} + \lambda_l^a \frac{\mu}{2v^2}\right)\frac{\partial g_a}{\partial \mu} - \frac{\lambda_l^a}{2v}\frac{\partial g_a}{\partial v} + 2\mu \frac{\partial h_a}{\partial \mu} - (1-\mu^2)\frac{\partial^2 h_a}{\partial \mu^2}\Bigg]\Bigg\}. \tag{68}$$

The functions g_a and h_a in Eqs. (65)–(68) are represented by the expansions (10) and (11), where Eqs. (12) and (13) make use of

$$V_j^a(v,t) = \sum_{l=1}^{N} U_l^a(v,t) \int_{-1}^{+1} M_l^a(\mu) P_j(\mu) \, d\mu \Big/ \int_{-1}^{+1} [P_j(\mu)]^2 \, d\mu. \tag{69}$$

All of the derivatives of g_a and h_a with respect to v and μ can be carried out analytically.

The coefficients (65)–(68) of Eq. (64) can be expressed analytically in terms of four time-dependent functionals

$$M_j(V_j^a) = \int_v^\infty V_j^a(v',t)(v')^{1-j} \, dv', \tag{70}$$

$$N_j(V_j^a) = \int_0^v V_j^a(v',t)(v')^{2+j}\,dv', \tag{71}$$

$$E_j(V_j^a) = \int_0^v V_j^a(v',t)(v')^{4+j}\,dv', \tag{72}$$

$$R_j(V_j^a) = \int_v^\infty V_j^a(v',t)(v')^{3-j}\,dv', \tag{73}$$

and five definite integrals

$$(S_1)_{jkl}^{ab} = \int_{-\mu_{LC}}^{+\mu_{LC}} \frac{dP_j}{d\mu} M_k^a(\mu) \frac{dM_l^b}{d\mu}(1-\mu^2)\,d\mu, \tag{74}$$

$$(S_2)_{jkl}^{ab} = \int_{-\mu_{LC}}^{+\mu_{LC}} P_j(\mu) M_k^a(\mu) M_l^b(\mu)\,d\mu, \tag{75}$$

$$(S_3)_{jkl}^{ab} = \int_{-\mu_{LC}}^{+\mu_{LC}} \frac{dP_j}{d\mu} M_k^a(\mu) \frac{dM_l^b}{d\mu}\,d\mu, \tag{76}$$

$$(S_4)_{jkl}^{ab} = \int_{-\mu_{LC}}^{+\mu_{LC}} P_j(\mu) M_k^a(\mu) \frac{dM_l^b}{d\mu} \mu\,d\mu, \tag{77}$$

$$(S_5)_{jkl}^{ab} = \int_{-\mu_{LC}}^{+\mu_{LC}} \frac{dP_j}{d\mu} M_k^a(\mu) M_l^b(\mu) \mu\,d\mu. \tag{78}$$

The expressions for the coefficients in Eqs. (64)–(68) are available from the authors. The integrals (74)–(78) are computed using Eq. (47) for the loss cone angle; hence they are time-independent. It turns out, though, that we must ignore all anisotropic components of the Rosenbluth potentials for all non-ion–ion interactions; i.e., we set $(S_{1-5})_{jkl}^{ab} = 0$ if $j > 0$ and "a" or "b" equals "e."

The eigenvalues λ_l^a of Eq. (62) which appear in Eq. (68) do depend on the "effective" mirror ratio, R_a, given in Eq. (49). Note that this formula applies only if "a" is an ion and $v^2 \geq v_{ca}^2 = Z_a e\phi/\tfrac{1}{2}m_a(R_m - 1)$ or if "a" is an electron and $v^2 \geq v_{ce}^2 = e\phi/\tfrac{1}{2}m_e$. We extend the definition of R_a to all velocities as $R_a = 1$ if "a" is an ion and $v^2 \leq v_{ca}^2$, and $R_e = \infty$ if $v^2 \geq v_{ce}^2$. The smallest eigenvalue λ_1^a is given by

$$\lambda_1^a \approx 1/\log_{10} R_a, \tag{79}$$

and the larger eigenvalues $\lambda_l^a (l \neq 1)$ are obtained through piecewise linear approximations of λ_l^a/λ_1^a as functions of $\log_{10} R_a$. The distribution function U_l^a is set to 0 at those points where λ_l^a is infinite.

Equations (64)–(68) are solved on a finite difference mesh $\{v_j\}_{j=1}^J$, where $v_1 = 0$ and $v_2 \leq v_J/(J-1)$. The value of v_2 determines a unique mesh ratio

$r \geqslant 1$ satisfying

$$v_{j+1} - v_j = r(v_j - v_{j-1}), \qquad j = 2, ..., J - 1. \tag{80}$$

The purpose of incorporating a nonuniform mesh of this type is to assure adequate representation of the low velocity ions and the high velocity electrons.

An implicit algorithm is used to integrate the vector system (64). The spatial and temporal difference approximations are analogous to those of Futch et al. (1972). For each species "a" we obtain a tridiagonal system of the form

$$-A_j U_{j+1}^{n+1} + B_j U_j^{n+1} - C_j U_{j-1}^{n+1} = D_j, \tag{81}$$

where A_j, B_j, and C_j are N by N matrices, D_j is an N-vector, and U_j^n represents $U_k^a(v_j, n\Delta t)$. The method of solution is described in Richtmyer and Morton (1967).

For various models we compute a self-consistent ambipolar potential as a function of time. We let

$$n_+ = \sum_{\text{ions}} n_a Z_a, \tag{82}$$

$$n_- = n_e, \tag{83}$$

$$d_\pm(t) = n_\pm(t + \Delta t) - n_\pm(t). \tag{84}$$

If $n_+(t)\{\lessgtr\}n_-(t)$ and $d_+(t)\{\lessgtr\}d_-(t)$, we {lower/raise} the ambipolar potential by an amount $\Delta\phi$ and repeat the last time-step. This iterative procedure continues until $d_+(t)\{\gtrless\}d_-(t)$.

It will be shown in Section III, B that for problems in which the source is proportional to the lowest normal mode, reasonably accurate results may be obtained by considering only one normal mode (i.e., $N = 1$) and by considering only the terms in Eqs. (10) and (11) proportional to $P_0(\mu)$ and $P_2(\mu)$. [The term corresponding to $P_1(\mu)$ vanishes due to symmetry about $\mu = 0$.] The lowest normal mode model in Futch et al. (1972) has been subsequently improved to consider the projections of the Rosenbluth potentials on $P_2(\mu)$ in addition to the isotropic components.

B. APPLICATIONS

1. *Several Normal Modes*

Here we consider a mirror-confined deuterium plasma in which the injection energy of the deuterons is 100 keV. The ambipolar potential is computed self-consistently as described earlier. We calculate the confinement figure of merit

$$n\tau = n_i^2/J_i \tag{85}$$

as a function of the angular source shape, the mirror ratio (R), the number of even Legendre polynomials in the Rosenbluth potential expansions (M), and the number of normal modes used to represent the distribution function (N). Results are displayed in Table I. Here we also include results from the two-dimensional hybrid model described in Section IV.

Two conclusions may be drawn:

1. The Rosenbluth potentials may be expressed in terms of their projections on just $P_0(\mu)$ and $P_2(\mu)$; i.e., Eqs. (10) and (11) may be truncated after $j = 2$.

2. For narrow source problems, there is excellent agreement between the expansion code and the hybrid code.

TABLE I

Comparisons of Confinement Parameter Values

Angular source	R	M	N	Code	$n\tau(\times 10^{13})$
Normal mode	3	1	1	Expansion	1.03
Normal mode	3	2	1	Expansion	1.13
Normal mode	3	3	1	Expansion	1.14
Normal mode	3	4	1	Expansion	1.14
Normal mode	3	5	1	Expansion	1.14
Normal mode	3	2	4	Expansion	1.09
Normal mode	3	5	—	Hybrid	1.02
Narrow ($e^{-10\mu^2}$)	3	2	4	Expansion	1.24
Narrow	3	5	—	Hybrid	1.22
Normal mode	10	1	1	Expansion	2.07
Normal mode	10	2	1	Expansion	2.34
Normal mode	10	3	1	Expansion	2.34
Normal mode	10	4	1	Expansion	2.34
Normal mode	10	5	1	Expansion	2.34
Normal mode	10	2	4	Expansion	2.29
Normal mode	10	5	—	Hybrid	2.24
Narrow ($e^{-10\mu^2}$)	10	1	2	Expansion	2.48
Narrow	10	1	3	Expansion	2.34
Narrow	10	1	4	Expansion	2.36
Narrow	10	2	2	Expansion	2.86
Narrow	10	2	3	Expansion	2.58
Narrow	10	2	4	Expansion	2.61
Narrow	10	3	2	Expansion	2.86
Narrow	10	5	—	Hybrid	2.65

2. Lowest Mode

a. D–T and D–^{3}He Mirror Reactors. The lowest-normal-mode representation has been used in a parametric study of the figure of merit Q for mirror reactors employing D–T and D–^{3}He fuel cycles (Futch *et al.*, 1972). In these investigations one considers a steady-state system in which the mirror-confined plasma is sustained by energetic neutral beam injection. The Q of the system is defined by

$$Q = \frac{\text{(thermonuclear power)}}{\text{(injection power required to maintain the plasma)}}$$

$$= \tfrac{1}{2} \sum_a \sum_b n_a n_b (\overline{\sigma v})_{ab} E_{ab} / \sum_a J_a E_a, \tag{86}$$

where n_a is the number density (particles/cm^3), J_a is the injected source current density (particles/cm^3/sec), and E_a is the source energy for particles of species "a." The quantity $(\overline{\sigma v})_{ab}$ is the product of the fusion cross section and relative speed, averaged over the distribution functions of the interacting particles, and E_{ab} is the energy released per fusion reaction. For given source parameters (J_a, E_a) and mirror ratio R_m, the velocity distribution functions for all plasma species are obtained from Eq. (64) with source terms added and are used to calculate the thermonuclear reaction rate

$$n_a n_b (\overline{\sigma v})_{ab} = \int d\mathbf{v}_a \int d\mathbf{v}_b \, f_a(v_a) f_b(v_b) \, \sigma_{ab}(v_{ab}) v_{ab}, \tag{87}$$

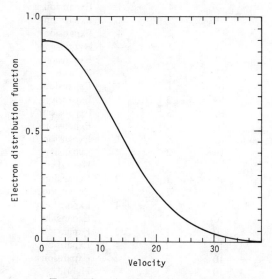

FIG. 2. Electron velocity distribution.

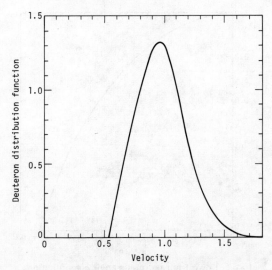

FIG. 3. Deuteron velocity distribution.

where $v_{ab} = |\mathbf{v}_a - \mathbf{v}_b|$ is the relative speed. We neglect the anisotropic nature of the distribution functions in performing these Q calculations so the explicit form of the normal mode eigenfunction $M_a(\mu)$ is not needed.

To illustrate the nature of the solutions, we show some typical velocity distribution functions for electrons, deuterons, and tritons in Figs. 2–4. A

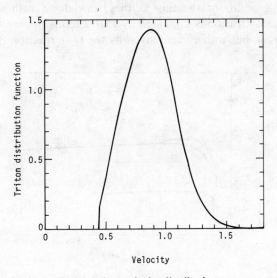

FIG. 4. Triton velocity distribution.

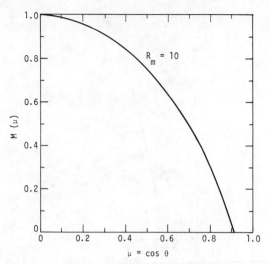

FIG. 5. Lowest-normal-mode angle eigenfunction.

typical normal mode angular distribution is shown in Fig. 5. The results of a parameter survey for the D–T mirror reactor fuel cycle are summarized in Fig. 6 where Q is given as a function of injection energy ($E_D = E_T = E_0$) and mirror ratio, with source currents adjusted so as to yield equal densities for the deuterons and tritons. Charged fusion reaction products (alpha particles) in the four-species runs (D–T–E–α) are assumed to be adiabatically confined, thus producing additional heating as they slow down within the plasma. Their continuous creation is simulated by source terms similar in form to the ion source terms, but with current given by the D–T reaction rate, and with

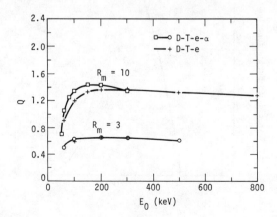

FIG. 6. Mirror system Q versus injection energy ($E_{DT} = 22.4$ MeV).

E_α equal to 3.5 MeV. The loss due to burnup is included by subtracting the reaction rate from the D and T source terms. The results in Fig. 6 show that there is a very broad maximum in Q as a function of the D–T injection energy E_0, and alpha-particle confinement ceases to be beneficial beyond about $E_0 = 300$ keV. More details on these and other results can be found in Futch et al. (1972).

The Q values obtained here are sensitive to the number of Legendre polynomials retained in the expansion of the Rosenbluth potentials [see Eqs. (10) and (11)]. The results in Fig. 6 and Futch et al. (1972) were derived using only the $P_0(\mu)$ term. The effect of including the next nonzero term, proportional to $P_2(\mu)$, is shown in Table II for a few typical cases. One sees that Q increases by about 20%. Additional terms in the expansion yield smaller corrections. Results from the two-dimensional code described in Section IV are also included in this table and good agreement is obtained when the $P_2(\mu)$ corrections are included in the one-dimensional model.

TABLE II

COMPARISON OF Q VALUES ($R_m = 10$)

Code description	Plasma species			
	D-e-T	D-e-T-α	D-e-^{3}He	D-e-^{3}He-α-p
One-dimensional code; P_0 only	1.22	1.39	0.244	0.264
One-dimensional code; P_0 and P_2	1.44	1.68	0.289	0.312
Two-dimensional code; normal mode source	1.38	1.61	0.278	0.301
Two-dimensional code; narrow source	1.71	1.99	0.337	0.365

The lowest-normal-mode Fokker–Planck model has also been used in a study of two-component mirror reactors, as described by Post et al. (1973). A modified loss term accounted for deuteron losses due to axial diffusion at low energies and mirror end losses at high energy. The bulk tritium and electron plasma was assumed to form a fixed Maxwellian background for the injected deuterons. Q values comparable to those for toroidal two-component systems (Dawson et al., 1971) are obtained.

b. Two-Component Toroidal Reactors. In this section we illustrate the time-dependent or transient aspect of the solution to the kinetic equations for a multispecies plasma. We apply the code to the study of the energy multiplication factor in the two-component toroidal reactor (Dawson et al., 1971;

Berk et al., 1974). In this application there are three plasma species (e–D–T), each of which is assumed to have an isotropic distribution in velocity space. We can also include the alpha-particles produced by D–T fusion reactions. Energetic deuterons are injected into an ohmic-heated tritium plasma in a Tokamak. We follow the evolution of the initially peaked deuteron distribution and calculate the energy produced via D–T fusion reactions as the deuterons slow down in the background plasma. In this instance there are no loss terms in the kinetic equations (i.e., no loss cone in velocity space) so we effectively assume that the particle and energy confinement times are long compared to the slowing-down time of the deuterons. The figure of merit for the system is the energy multiplication factor, F, defined by

$$F = \text{(Fusion Energy Produced)}/\text{(Initial Energy in the Deuterons)}$$
$$= \int_0^t dt\, n_D n_T \overline{\sigma v}(t)\, E_F / n_D \bar{E}_D(0), \tag{88}$$

where $\bar{E}_D(0)$ is the mean energy per deuteron at $t = 0$ and $\overline{\sigma v}$ is the D–T fusion rate parameter which depends on the detailed shape of the distributions.

Results for the energy multiplication factor F in several illustrative cases are given in Fig. 7. For times long compared to the slowing-down time of the deuterons, the system approaches thermal equilibrium so that $\overline{\sigma v}$ becomes constant in time and F increases linearly. This is clearly seen in Fig. 7. To obtain an F value which does *not* depend on the time interval chosen in Eq. (88), one can define a "transient" contribution to F by subtracting the asymp-

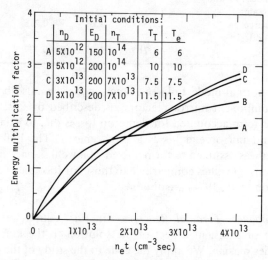

FIG. 7. TCT energy multiplication (curve D includes alphas).

totic steady-state component of the fusion power from the integrand in Eq. (88). This procedure yields well-defined F values which are more nearly comparable to those obtained by Dawson et al. (1971).

IV. Finite-Difference Solution in a Two-Dimensional Velocity Space

A. NUMERICAL METHODS

In this section we consider a "hybrid" treatment, in which the ions are represented and differenced in v–θ coordinates and the electrons are represented by their lowest normal mode. When applying this model to mirror devices, an electron loss term

$$-\tfrac{1}{2}\lambda_1^e \cdot (1/v^3)(\partial g_e/\partial v) f_e(v, t) \tag{89}$$

is added to the right-hand side of Eq. (23), where λ_1^e is defined by Eq. (79).

Equation (14) is integrated using the method of splitting, or fractional time-steps. We first advance

$$(1/\Gamma_a)(\partial f_a/\partial t) = (1/v^2)(\partial G_a/\partial v) \tag{90}$$

using a semi-implicit difference algorithm, and then advance

$$(1/\Gamma_a)(\partial f_a/\partial t) = (1/v^2 \sin\theta)(\partial H_a/\partial \theta) \tag{91}$$

in an analogous manner.

Equation (90) is discretized as follows:

$$\frac{f_{i,j}^{n+1} - f_{i,j}^n}{\Gamma_a \Delta t} = \frac{A_{i,j+1}^n f_{i,j+1}^{n+1} - A_{i,j-1}^n f_{i,j-1}^{n+1}}{2v_j^2 \Delta v_j}$$

$$+ \frac{1}{v_j^2 \Delta v_j}\left(\frac{B_{i,j+1/2}^n (f_{i,j+1}^{n+1} - f_{i,j}^{n+1})}{\Delta v_{j+1/2}} - \frac{B_{i,j-1/2}^n (f_{i,j}^{n+1} - f_{i,j-1}^{n+1})}{\Delta v_{j-1/2}}\right)$$

$$+ \frac{1}{2v_j^2 \Delta v_j}\left(\frac{C_{i,j+1}^n (f_{i+1,j+1}^n - f_{i-1,j+1}^n)}{2\Delta\theta_i}\right.$$

$$\left. - \frac{C_{i,j-1}^n (f_{i+1,j-1}^n - f_{i-1,j-1}^n)}{2\Delta\theta_i}\right). \tag{92}$$

Here, $f_{i,j}^n = f(v_j, \theta_i, n\Delta t)$, and the coefficient $B_{i,j+1/2}^n$ is a simple average of $B_{i,j}^n$ and $B_{i,j\pm 1}^n$. The subscript "a" has also been dropped.

By rearranging terms, Eq. (92) may be put into the tridiagonal form:

$$-\alpha_{i,j}^n f_{i,j+1}^{n+1} + \beta_{i,j}^n f_{i,j}^{n+1} - \gamma_{i,j}^n f_{i,j-1}^{n+1} = \delta_{i,j}^n. \qquad (93)$$

We see that the terms of mixed second derivative type may not be written fully implicitly if we wish to maintain a tridiagonal form. Equation (91) is integrated in a similar manner, with the roles of v and θ reversed.

The electron equation is integrated using the method described in Section III. Here we allow the time-step Δt to be divided into several smaller time-steps, $\Delta t/K = \Delta t_e$, where for each small time-step the electron coefficients are recomputed. This procedure takes very little extra computer time, and is quite useful in time-dependent problems, where the electrons are usually on a faster time-scale.

B. Applications

1. *Mirror-Confined Plasmas*

The direct two-dimensional approach for obtaining the ion velocity-space distributions in mirror-confined plasmas can be used to analyze many different physical situations which cannot be studied with the one-dimensional lowest-normal-mode approach described in Section III. In particular, transient plasmas typically encountered in mirror experiments such as 2XII (Coensgen *et al.*, 1974) may undergo rapid changes which would not allow the establishment of a lowest-normal-mode ion distribution. Furthermore, the injection of well-collimated neutral beams, as in BBII (Anderson *et al.*, 1974; Berkner *et al.*, 1971) and 2XIIB (Coensgen *et al.*, 1974) gives rise to ion sources that are sharply peaked in velocity space so that many normal modes would be required for adequate representation.

In the following paragraphs we give examples of both transient and steady-state mirror problems with neutral beam injection. We also show how the solutions of the Fokker–Planck equations can be used to obtain information about MHD equilibrium and stability for mirror-confined plasmas.

a. The 2XII Experiment. One of the first applications of the two-dimensional Fokker–Planck code was the study of the 2XII mirror experiment (Coensgen *et al.*, 1974). A plasma pulse is injected into a mirror trap, magnetically compressed to increase the density and energy, and decays slowly as ions and electrons escape through the mirrors due to collisional processes. Experimental data on the time history of various plasma properties can be compared with the time-dependent solutions of the Fokker–Planck equations for electrons and ions (Coensgen *et al.*, 1974). Typically, the form of the

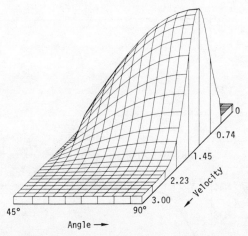

FIG. 8. Initial ion distribution in 2XII.

experimentally trapped ion distribution function is fed into the code as an initial condition, and the time dependence of the magnetic field compression term is set up to model the time-dependent magnetic field used in the experiment. An example of an initial ion distribution is shown in Fig. 8. In this three-dimensional display, the vertical distance represents the magnitude of the distribution function at each (v, θ) point in the horizontal plane. The form of the time-dependent magnetic field is illustrated in Fig. 9. Some typical code results are given in Figs. 10–12 where the density, mean ion energy, and mean electron energy are plotted as a function of time. The magnetic compression stage occurs for $0 \leqslant t \leqslant 535$ μsec. After compression one sees that the density decays while the mean ion energy increases due to the preferential loss of low

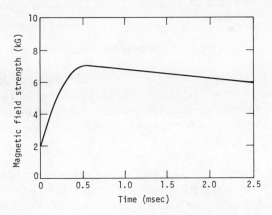

FIG. 9. 2XIIB magnetic field strength.

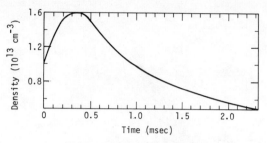

FIG. 10. 2XIIB plasma density.

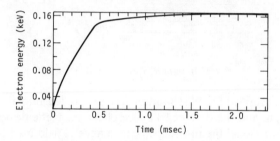

FIG. 11. 2XIIB mean electron energy.

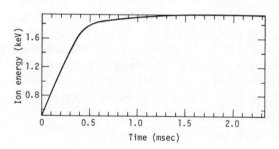

FIG. 12. 2XIIB mean ion energy.

energy ions by Coulomb scattering. Experimentally observed density decays are typically two or three times faster than that given by the Fokker–Planck code, indicating that nonclassical processes are occurring in the experiment. Methods for modeling such processes in the code are currently being investigated.

b. The BBII Experiment. The BBII mirror experiment (Anderson *et al.*, 1974) features plasma buildup from vacuum conditions via continuous neutral beam injection. A steady-state plasma is established as mirror end losses are balanced by the continuous ionization of the neutral beam within the mirror trap. The beam is highly collimated and monoenergetic so it tends to produce an ion distribution which is sharply peaked in velocity space. The most appropriate method for numerically modeling this particular feature of the experiment is the direct two-dimensional solution of the Fokker–Planck equations.

Source and loss terms which approximate the experimental conditions are inserted into the code and the steady-state ion and electron distribution functions are computed. Examples are shown in Figs. 13 and 14. The low density nature of this experiment makes it imperative that many of the important atomic processes occurring within the plasma be adequately represented in the code. The forms of the three terms in the ion and electron source currents, Eq. (55), have been chosen so as to represent Lorentz ionization, inverted cascade ionization (Hiskes, 1963), and electron and ion impact ionization (Riviere, 1971) of the neutral beam. The presence of significant amounts of background gas, usually neon and nitrogen, gives rise to an additional electron source because when these gases are ionized by the plasma the electrons are electrostatically confined by the ambipolar potential. The cold impurity ion that is produced is immediately pushed out of the system by this same potential. The net effect is a "cooling" of the electrons so that the mean electron energy is lower than in the absence of the gas. The effect on the form of the ion distribution is also quite dramatic, as shown in Figs. 13 and 14. The computed ion distribution function shown in Fig. 13 is representative of the BBII experimental conditions at steady state. Figure 14 shows how this distribution would change if the background gas were completely eliminated. In both cases, the sharp peak in the ion distribution coincides with the ion source energy and injection angle. Charge exchange

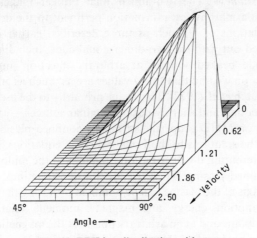

FIG. 13. BBII ion distribution with gas.

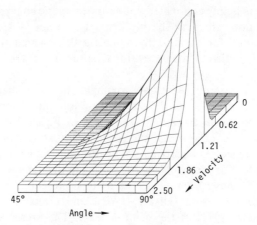

FIG. 14. BBII ion distribution without gas.

processes are important in BBII under certain operating conditions, and these have also been included in our Fokker–Planck runs.

Several quantitative features of the experiment, such as the density and mean energies and ambipolar potential, are reproduced quite accurately in the code results when the proper atomic physics is included in the model. A major unsolved problem is the identification of the density-limiting instability seen in the experiment. This may be caused by some peculiarity in the shape of the ion distribution, but an examination of the Fokker–Planck results has not yet produced a definitive answer to this question.

c. Reactor Studies. The two-dimensional Fokker–Planck code is also used to obtain plasma physics information pertinent to the design of mirror reactors. Calculations of Q, such as those described earlier in Section III, have been carried out with the two-dimensional code, including the effect of non-normal-mode peaked sources at arbitrary injection angles. For perpendicular injection we find that the Q values are as much as 50% higher than those given in Futch *et al.* (1972). This is due primarily to the use of anisotropic Rosenbluth potentials and sharply peaked angular sources.

Information on MHD properties of mirror-confined plasmas can also be extracted from the solutions to the Fokker–Planck equations. We regard the solutions as representing the distribution functions at the midplane of a long, thin mirror-confined plasma. Then the distribution functions at every other point along the axis of the plasma can be obtained by noting that the constants of the particle motion, energy, and magnetic moment, relate the velocity vector at the midplane and at an arbitrary point via the magnetic field strength B and ambipolar potential ϕ. Denoting midplane values by a zero subscript, we have:

$$\text{magnetic moment} = mv^2 \sin^2\theta/2B = mv_0^2 \sin^2\theta_0/2B_0, \quad (94)$$

$$\text{total energy} = \tfrac{1}{2}mv^2 + eZ\phi = \tfrac{1}{2}mv_0^2. \quad (95)$$

If we assume that the particle mean free path is long compared to the distance between mirrors, then at some distance l from the midplane, where the value of the magnetic field strength is $B(l) = B$ and the ambipolar potential is $\phi(l) = \phi$, the distribution function is given by

$$f_a[v, \theta, l] = f_a[v_0(v, \phi), \theta_0(v, \theta, \phi, B), 0], \quad (96)$$

where $v_0(v, \phi)$ and $\theta_0(v, \theta, \phi, B)$ are obtained from Eqs. (94) and (95), and $f[v_0, \theta_0, 0]$ is the solution of the two-dimensional Fokker–Planck equation. Computing the density for each plasma species,

$$n_a(B, \phi) = 4\pi \int_0^{\pi/2} d\theta \sin\theta \int_0^\infty dv\, v^2 f_a[v, \theta, l], \quad (97)$$

we obtain the ambipolar potential as a function of magnetic field strength, $\phi = \phi(B)$, by imposing charge neutrality at each point in the plasma,

$$n_e(B, \phi) = \sum_a Z_a n_a(B, \phi), \quad (98)$$

and solving for ϕ in terms of B. A typical ion distribution function, $f(v_0, \theta_0, 0)$ is shown in Fig. 15, and the potential profile derived from it is given in Fig. 16.

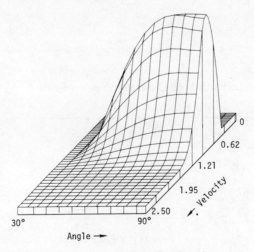

FIG. 15. Deuteron velocity distribution for mirror reactor.

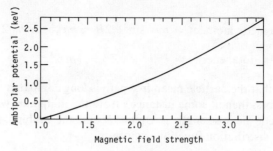

FIG. 16. Ambipolar potential profile along a magnetic field line.

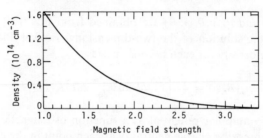

FIG. 17. Density profile along a magnetic field line.

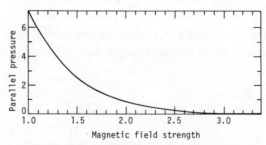

FIG. 18. Parallel pressure profile along a magnetic field line.

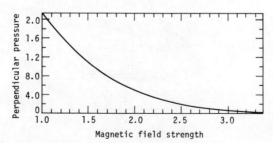

FIG. 19. Perpendicular pressure profile along a magnetic field line.

When $\phi(B)$ is substituted into Eq. (96), one obtains the distribution function at an arbitrary point parameterized only by the normalized magnetic field strength, $\psi \equiv B/B_0$ where $\psi = 1$ at the midplane and $\psi = R_m$ at the mirror. The density $n(\psi)$ and pressure tensor components $[p_\perp(\psi), p_\parallel(\psi)]$ can then be computed and used in an MHD description of the plasma. Examples of density and pressure profiles obtained from the numerically computed distribution functions are given in Figs. 17–19.

For MHD stability the plasma pressure must satisfy the two conditions (Hall, 1972)

$$B + 4\pi \, dp_\perp/dB > 0, \tag{99}$$

$$B - 4\pi \, dp_\parallel/dB > 0, \tag{100}$$

which are imposed by the mirror mode and firehose mode, respectively. When these conditions are applied to the pressure profiles computed from the Fokker–Planck solutions, one can determine the midplane magnetic field strength B_0 for which the plasma would be marginally stable. Then one can also obtain the limiting plasma beta, defined at the midplane by

$$\beta_0 = p_\perp(B_0)/[p_\perp(B_0) + B_0^2/8\pi]. \tag{101}$$

For the example given in Figs. 15–19, the maximum beta was found to be $\beta_0 = 0.65$; input data for this example is given in Table III.

TABLE III

INPUT DATA FOR MIRROR REACTOR EXAMPLE

Mirror ratio	3.381
Deuteron source energy	65 keV
Deuteron source current	2×10^{15} cm^{-3} sec^{-1}
Triton source energy	97.5 keV
Triton source current	1×10^{15} cm^{-3} sec^{-1}
Injection angles (4 sources)	55°, 63°, 70°, 78°

Certain aspects of the equilibrium configuration for a mirror-confined plasma can also be deduced using the pressure tensor derived from the Fokker–Planck distributions. In particular, the modification of the vacuum magnetic field due to the presence of a high-beta plasma can be approximately evaluated for a long, thin mirror system. The basic equation is that giving the radial pressure balance at each point along the axis of the plasma:

$$B_{vac}^2/8\pi = B^2/8\pi + p_\perp(B). \tag{102}$$

This equation can be solved for B in terms of B_{vac} and, assuming that B_{vac} is approximately uniform over the cross-sectional area of the plasma, this gives the modification of the vacuum magnetic field due to a plasma with pressure $p_\perp(B)$. In particular, the vacuum mirror ratio, R_{vac}, is enhanced due to the high-beta plasma:

$$R_m = R_{vac}/(1 - \beta_0)^{1/2}. \tag{103}$$

For the example described in Figs. 15–19 and Table III, the relation between B_{vac} and B derived from Eq. (102) is shown in Fig. 20. Furthermore, if the spatial dependence of the vacuum magnetic field, $B_{vac}(l)$ is known, then the spatial form of the magnetic field with plasma can also be deduced. For example, if one assumes the form

$$B_{vac}(l) = B_0[\tfrac{1}{2}(R_{vac} + 1) - \tfrac{1}{2}(R_{vac} - 1)\cos(\pi l/L_T)], \tag{104}$$

the spatial profile is modified as shown in Fig. 21 for $\beta = \beta_0$.

An extensive Fokker–Planck parameter study of mirror-confined plasmas with neutral injection is in progress. The motivation is to provide a data base which can serve to identify interesting parameter regimes and form a basis for starting more detailed design calculations. The input parameters which can be varied in this study are the mirror ratio, the number and type of ion species, the number of sources for each ion species, and the injection energy, injection angle, and relative strength of each source. Some results for single-species,

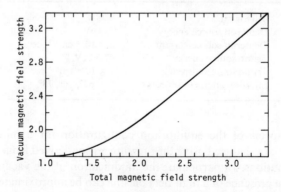

FIG. 20. Magnetic field strength in vacuum and in high-beta plasma.

single-source systems are shown in Figs. 22 and 23 where the particle confinement parameter $\langle n\tau \rangle$ and maximum beta are given as a function of the vacuum mirror ratio and injection angle (Rensink et al., 1974). These results indicate that the limiting beta can be significantly increased without degrading $\langle n\tau \rangle$ by injecting with $\theta_0 \approx 50°$ rather than the conventional $\theta_0 = 90°$.

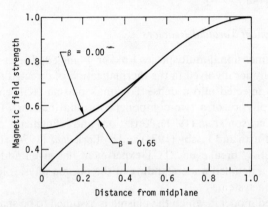

FIG. 21. Magnetic field profile in vacuum and at MHD stability limit.

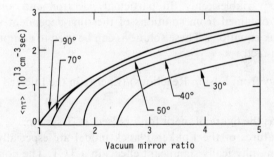

FIG. 22. Plasma confinement parameter versus vacuum mirror ratio and injection angle.

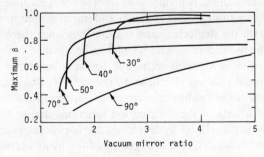

FIG. 23. Maximum stable beta versus vacuum mirror ratio and injection angle.

The more general problem of optimizing injection with many sources present has been investigated by Hall *et al.* (1975) using simplified Fokker–Planck and reactor models. Their results are consistent with our observations on the use of various injection angles to maximize beta. Two-component mirror reactors are also being investigated with the two-dimensional Fokker–Planck model (Byers *et al.*, 1974).

2. *Two-Component Toroidal Reactors*

The two-dimensional multispecies Fokker–Planck code is a useful tool for studying the physics involved in the thermalization of directed monoenergetic neutral beams injected into a dense Tokamak plasma (Killeen *et al.*, 1974). The essential physics of a two-component toroidal reactor (TCT) was first described by Dawson *et al.* (1971). Variations and refinements of this concept were given by Furth and Jassby (1974) and by Berk *et al.* (1974). The authorized construction of a "breakeven" TCT experiment has given added impetus to the search for a detailed understanding of the plasma physics involved in the design of such a system.

An idealized model in which the plasma is assumed to be spatially uniform over some finite toroidal volume allows one to analyze the system in terms of velocity-space variables only. In particular, electron and ion distribution functions are obtained from solutions of the time-dependent velocity-space Fokker–Planck equations. These solutions can be used to compute the energy multiplication factor

$$Q = \text{(Energy from Fusion Reactions)}/\text{(Energy in the Injected Deuterons)}$$
(105)

which serves as the figure of merit for a pulsed TCT system.

Several features of the Fokker–Planck model are especially appropriate for representing physically significant effects in a TCT. The nonlinear nature of the kinetic equations ensures that the collisional interaction of all species, including self-interactions, is properly accounted for regardless of the form of the distribution functions. Alpha particles and impurity ions are treated on an equal footing with the deuterium and tritium ions since there is no inherent restriction on the number of species which can be handled in the code. Major radius compression is a useful technique for supplementing neutral beam injection in toroidal plasmas (Furth and Jassby, 1974; Bol *et al.*, 1972, 1974) and the two-dimensional nature of our velocity space allows us to accurately account for distortions of the distribution functions due to this anisotropic driving force (see Section II, B). System losses due to processes such as diffusion, charge exchange, radiation, etc., are accounted for by loss terms described in Section II, D which parameterize these losses via particle and energy confinement times for each plasma species. These confinement times are read into the code as part of the input data for each run. We neglect any loss boundaries in velocity space due, for example, to banana orbits which might extend outside the plasma.

Some special diagnostics are available for analyzing TCT problems and examples of these are given in Figs. 24–27. In addition to our standard three-dimensional display of the ion distribution functions in (v, θ) space shown in

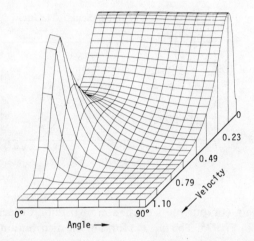

FIG. 24. Deuteron velocity distribution for steady-state tangential injection in TCT.

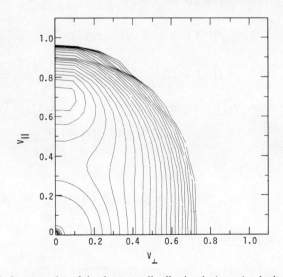

FIG. 25. Contour plot of the deuteron distribution in $(v_\perp, v_\parallel)$ velocity space.

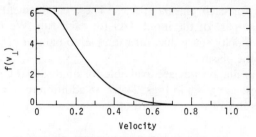

FIG. 26. Distribution of perpendicular velocities.

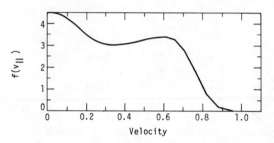

FIG. 27. Distribution of parallel velocities.

Fig. 24, we obtain contour plots of the ion distribution functions in $(v_\perp, v_\parallel)$ space as shown in Fig. 25. The $(v_\perp, v_\parallel)$ form of the distribution function, when integrated over one of the velocity space components, is shown in Figs. 26 and 27. These distributions are defined by

$$f(v_\perp) = \int_{-\infty}^{+\infty} dv_\parallel \, f(v_\perp, v_\parallel), \tag{106}$$

$$f(v_\parallel) = 2\pi \int_0^\infty v_\perp \, dv_\perp \, f(v_\perp, v_\parallel). \tag{107}$$

Other useful diagnostic quantities are the total system energy

$$W_{\text{SYST}}(t) = \sum_a n_a(t) \bar{E}_a(t) V(t), \tag{108}$$

the cumulative energy loss

$$W_{\text{LOSS}}(t) = \sum_a \int_0^t dt \, n_a(t) \bar{E}_a(t) V(t)/\tau_a, \tag{109}$$

and the cumulative fusion energy produced

$$W_{\text{F}}(t) = \int_0^t dt \, n_{\text{D}}(t) n_{\text{T}}(t) \overline{\sigma_{\text{DT}} v}(t) V(t) E_{\text{DT}}, \tag{110}$$

where $E_{DT} = 17.6$ MeV and $\overline{\sigma_{DT} v}(t)$ is the instantaneous value of the fusion reaction rate parameter, averaged over the velocity distribution functions of the deuteron and triton components. The plasma "volume" $V(t)$ is normalized to unity at $t = 0$ and decreases with major radius compression so as to conserve the total number of particles, $N_a = n_a(t)V(t)$, in the absence of losses.

a. Energy Multiplication Factor. The essential feature of a TCT reactor is the production of fusion energy by an initially energetic deuteron as it slows down in a background tritium plasma. In the simplest application of the Fokker–Planck code we initially set up a deuteron velocity distribution which approximates a delta-function in energy and angle (θ is defined relative to the direction of the toroidal field) together with a Maxwellian distribution for tritons and electrons. Then the evolution of each of the plasma species is followed until the system approaches thermal equilibrium, and the total energy produced by D–T fusion reactions is noted [see Eq. (88)]. In thermal equilibrium, fusion energy is produced at a constant rate (corresponding to the D–T plasma temperature) and the total fusion energy $W_F(t)$ increases linearly with time as $t \to \infty$. For two-component systems we are primarily interested in the *transient* contribution to the fusion energy so we subtract the steady-state contribution, defining

$$W_F'(t) = W_F(t) - n_D n_T (\sigma_{DT} v)_{\text{eq.}} E_{DT} t, \qquad (111)$$

where $(\overline{\sigma_{DT} v})_{\text{eq.}}$ is the equilibrium rate parameter for a Maxwellian plasma. Then the energy multiplication factor is defined by

$$Q' = \lim_{t \to \infty} W_F'/n_D(0)\bar{E}_D(0). \qquad (112)$$

Some typical results are given in Table IV, where we have used a background tritium plasma with $n = 10^{14}$ cm^{-3} and $T = 5$ keV. In this set of runs the deuteron particle density is chosen very small compared to that of the background plasma so that self-interaction of the "test particle" deuterons is

TABLE IV

COMPARISON OF ENERGY MULTIPLICATION FACTORS

Deuteron energy (keV)	Q'	Q''	F
70	0.612	0.487	0.301
110	0.996	0.946	0.644
150	1.187	1.191	1.096
190	1.244	1.264	1.151

negligible and the background plasma is not heated significantly. The energy multiplication factors Q' can then be compared with the calculations of Dawson et al. (1971) denoted by F in Table IV. The Q' values are larger than the corresponding F values primarily because Q' includes the effect of deuteron energy dispersion and finite tritium temperature in the evaluation of $\overline{\sigma v}$:

$$\overline{\sigma v} \equiv \int d\mathbf{v}_1 \int d\mathbf{v}_2 \, f_D(\mathbf{v}_1) f_T(\mathbf{v}_2) \, \sigma(v_{12}) v_{12} / n_D n_T. \tag{113}$$

For comparison we show results, denoted by Q'', obtained by neglecting the finite tritium temperature in evaluating $\overline{\sigma v}$. The effects of deuteron energy dispersion and finite tritium temperature are seen to be most pronounced at low energies.

b. Pulsed Reactors. Instead of simply postulating the existence of specially prepared ion and electron components in the initial conditions for our Fokker–Planck runs as described in the last paragraph, it is possible to more realistically model the sequence of events which might occur in a pulsed TCT reactor by using the various source and loss and magnetic compression terms described in Section II. Several scenarios of varying degrees of complexity have been suggested. A few examples of these will be described here.

First, it is necessary to precisely define the figure of merit Q_b for a pulsed system. The "breakeven experiment" in a pulsed TCT is defined to occur during a time interval which begins when the energetic deuteron component and thermal tritium plasma have reached some optimum condition for energy multiplication, and ends at such time as the thermal energy in the tritium plasma falls below its "initial" value, i.e., its value at the beginning of the breakeven experiment. The breakeven factor Q_b is then defined as the ratio of the fusion energy produced during the breakeven experiment and the "initial" energy in the hot deuteron component of the plasma.

In the simplest scenario for a TCT we assume the existence of an ohmic-heated 10 keV tritium plasma into which we inject 200 keV deuterons for 30 msec. The deuteron beams are assumed to be injected tangentially into the torus. We assume symmetry about $\theta = \pi/2$ so the use of both co- and counter-injected beams is implicit in our model. The source strength or beam current is chosen such that the energy transfer from the deuterons to the tritium plasma during the breakeven experiment is sufficient to balance any energy losses from the tritium plasma. The breakeven factor in this case is $Q_b = 1.08$. The effect on Q_b of explicitly including additional ion species, such as alpha particles from fusion reactions and impurity ions (carbon and/or oxygen), is shown in Table V. The alphas provide a small amount of additional plasma heating, whereas the impurities allow the system to lose energy faster. The

TABLE V

EFECT OF IMPURITIES AND ALPHAS ON Q

Plasma species	Z_{eff}	Q
e-D-T	1	1.08
e-D-T-α	1	1.09
e-D-T-α-C	2	0.92
e-D-T-α-C-O	2.7	0.84

effective Z of the plasma is defined by

$$Z_{\text{eff}} = \sum_a n_a Z_a^2 \bigg/ \sum_a n_a Z_a, \tag{114}$$

where the index "a" runs over all ion species in the plasma.

A somewhat more complex scenario for TCT operation is indicated schematically in Fig. 28. Here one assumes that the beam energies and plasma temperatures necessary for breakeven ($Q_b \geq 1$) are not achievable by "conventional" means. Major radius compression is used to boost the energy of both the injected deuterons and the tritium plasma. When injection has been completed, the plasma system is compressed for 10 msec at a rate given by $\dot{R}/R = -40.5 \text{ sec}^{-1}$ so that the final compression ratio is $C = 1.5$. The breakeven experiment begins when the compression phase is completed, $t = 40$ msec, and ends at $t = 200$ msec. The time histories of the deuteron density and mean energy are shown in Figs. 29 and 30 where the effects of injection and compression can be clearly seen. The breakeven factor was $Q_b = 1.28$ in this particular example.

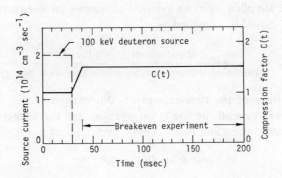

FIG 28. Schematic of simple scenario for a pulsed TCT.

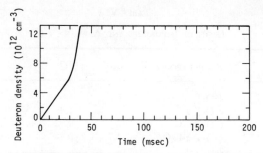

FIG. 29. Deuteron density behavior in simple TCT scenario.

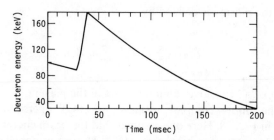

FIG. 30. Deuteron mean energy in simple TCT scenario.

Another scenario for TCT operation uses the major radius compression capability to achieve "energy clamping" of the fast deuterons. The basic idea is to inject deuterons at an energy which maximizes the instantaneous D–T fusion reaction rate ($E_D \approx 125$ keV), and then supply enough energy to the deuterons via magnetic compression to make up for the energy transfer from the deuterons to the bulk tritium plasma. In this way the deuteron energy is "clamped" at the optimum value for fusion energy production. The price one has to pay for this added feature is that the compression energy must ultimately be supplied from an external source so the breakeven factor for this scenario should be redefined as

$$Q_b = \frac{\text{(fusion energy produced)}}{\text{(initial deuteron energy)} + \text{(compression energy)}}. \tag{115}$$

The precise form of the time-dependent compression needed to clamp the deuteron energy depends on the beam intensity and the background plasma energy loss rate (Jassby and Furth, 1974). Typically, we use the form

$$\dot{R}/R = A/[1 + Be^{-t/\tau}], \tag{116}$$

with parameters (A, B, τ) specified separately for each phase of the TCT scenario. A possible sequence of events for a TCT with energy clamping is indicated schematically in Fig. 31, and results are shown in Figs. 32 and 33. A breakeven factor $Q_b = 0.86$ was achieved in this example.

c. Steady-State Reactors. It may be possible for a two-component toroidal reactor to operate as a steady-state device. However, this would require significant technological progress in the areas of superconducting magnets and neutral beam sources, as well as an understanding and control of impurities in Tokamaks. We have modeled steady-state TCT systems with our Fokker–Planck code and we give some sample results in this section. This investigation of steady-state operation is also relevant to pulsed systems in which the duration of the injected deuteron beam pulse is long compared to the slowing-down time for the deuterons.

In the example given here we assume that the bulk tritium plasma density remains constant, $n_T = 10^{14}$ cm^{-3}; we do not attempt to model the processes by which this bulk plasma is maintained against particle losses. However, we do allow for a finite plasma *energy* confinement time, typically 200 msec.

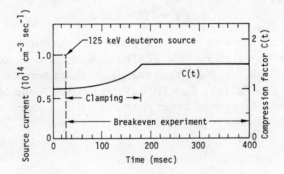

FIG. 31. Schematic of TCT scenario with energy clamping.

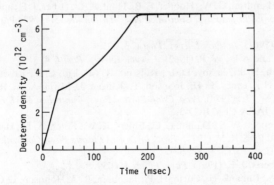

FIG. 32. Deuteron density behavior in clamping scenario.

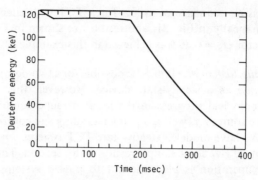

FIG. 33. Deuteron mean energy in clamping scenario.

The steady-state injection of energetic 200 keV deuterons is modeled with a constant current source term, $J_D = 1.5 \times 10^{14}$ cm^{-3} sec^{-1}. As the deuterons slow down, they are likely to be transported out of the system at low energy so we have used an energy-dependent deuteron particle confinement time,

$$\tau_D(E) = 3(E)^{3/2} \text{ msec}, \tag{117}$$

where E is in keV. The form of the steady-state deuteron distribution obtained in this case is shown in Figs. 24–27. The equilibrium deuteron density is $n_D = 3.8 \times 10^{13}$ cm^{-3} and the mean energies of the system components are $\bar{E}_e = 15$ keV, $\bar{E}_T = 26$ keV, $\bar{E}_D = 100$ keV. The system figure of merit Q is simply the ratio of the fusion power produced and the injected beam power, and in this case its value is $Q = 1.85$.

References

Anderson, O. A., Birdsall, D. H., Damm, C. C., Foote, J. H., Futch, A. H., Goodman, R. K., Gordon, F. J., Hamilton, G. W., Hooper, E. B., Hunt, A. L., Osher, J. E., and Porter, G. D. (1974). *Proc. Conf. Plasma Phys. Controlled Nucl. Fusion Res. 5th, 1974* Paper No. IAEA-CN-33/D5-2.

BenDaniel, D. J. (1961). *J. Nucl. Energy, Part C* **3**, 235.

BenDaniel, D. J., and Allis, W. P. (1962). *J. Nucl. Energy, Part C* **4**, 31.

Berk, H. L., Furth, H. P., Jassby, D. L., Kulsrud, R. M., Liu, C. S., Rosenbluth, M. N., Rutherford, P. H., Tenney, F. H., Johnson, T., Killeen, J., Mirin, A. A., Rensink, M. E., and Horton, C. W., Jr. (1974). *Proc. Conf. Plasma Phys. Controlled Nucl. Fusion Res. 5th, 1974* Paper No. IAEA-CN-33/G2-3.

Berkner, K. H., Cooper, W. S., Damm, C. C., Ehlers, K. W., Futch, A. H., Hamilton, G. W., Osher, J. E., and Pyle, R. V. (1971). *Proc. Int. Conf. Plasma Phys. Controlled Nucl. Fusion Res. 1971* Vol. II, p. 707.

Bing, G., and Roberts, J. E. (1961). *Phys. Fluids* **4**, 1039.

Bol, K., Ellis, R. A., Eubank, H., Furth, H. P., Jacobsen, R. A., Johnson, L. C.. Mazzucato, E., Stodiek, W., and Tolnas, E. L. (1972). *Phys. Rev. Lett.* **29**, 1495.

Bol, K., Cecchi, J. L., Daughney, C. C., Ellis, R. A., Jr., Eubank, H. P., Furth, H. P., Goldston, R. J., Hsuan, H., Jacobsen, R. A., Mazzucato, E., Smith, R. R., and Stix, T. H. (1974). *Phys. Rev. Lett.* **32**, 661.

Byers, J. A., Freis, R. P., Killeen, J., Lee, E. P., McNamara, B., Mirin, A. A., and Rensink, M. E. (1974). *Proc. Conf. Plasma Phys. Controlled Nucl. Fusion Res.*, *5th, 1974* Paper No. IAEA-CN-33/G2-1.

Coensgen, F. H., Cummins, W. F., Molvik, A. W., Nexsen, W. E., Jr., Simonen, T. C., and Stallard, B. W. (1974). *Proc. Conf. Plasma Phys. Controlled Nucl. Fusion Res.*, *5th, 1974* Paper No. IAEA-CN-33/D2-1.

Dawson, J. M., Furth, H. P., and Tenney, F. H. (1971). *Phys. Rev. Lett.* **26**, 1156.

Fowler, T. K., and Rankin, M. (1966). *J. Nucl. Energy*, Part C **8**, 121.

Furth, H. P., and Jassby, D. L. (1974). *Phys. Rev. Lett.* **32**, 1176.

Furth, H. P., and Yoshikawa, S. (1970). *Phys Fluids* **13**, 2593.

Futch, A. H., Jr., Holdren, J. P., Killeen, J., and Mirin, A. A. (1972). *Plasma Phys.* **14**, 211.

Hall, L. S. (1972). *Phys. Fluids* **15**, 882.

Hall, L. S., Hooper, E. B., Jr., and Newcomb, W. A. (1975). "Optimal Injection of Beams in a Mirror Well," Report UCRL-75561. Lawrence Livermore Laboratory, Livermore, California.

Hiskes, J. R. (1963). *Phys. Rev. Lett.* **10**, 102.

Jassby, D. L., and Furth, H. P. (1974). "Two-Energy-Component Fusion Reactors with Sustained Suprathermal-Ion Energy," Report MATT-1048. Princeton Plasma Phys. Lab., Princeton, New Jersey.

Kaufman, A. N. (1956). *US At. Energy Comm.*, *Doc.* **TID-7520**, Part 2, 387.

Killeen, J., and Futch, A. H., Jr. (1968). *J. Comput. Phys.* **2**, 236.

Killeen, J., and Marx, K. D. (1970). *Methods Comput. Phys.* **9**, 422.

Killeen, J., and Mirin, A. A. (1971). In "Proceedings of the Fourth Conference on the Numerical Simulation of Plasmas" (J. P. Boris and R. A. Shanny, eds.). Stock 0851 00059, 685. US Govt. Printing Office, Washington, D.C.

Killeen, J., Johnson, T. H., Mirin, A. A., and Rensink, M. E. (1974). "Computational Studies of the Two-Component Toroidal Fusion Test Reactor," Report UCID-16530. Lawrence Livermore Laboratory, Livermore, California.

Marx, K. D. (1970). *Phys. Fluids* **13**, 1355.

Post, R. F. (1961). *Phys. Fluids* **4**, 902.

Post, R. F., Fowler, T. K., Killeen, J., and Mirin, A. A. (1973). *Phys. Rev. Lett.* **31**, 280.

Rensink, M. E., Fowler, T. K., Freis, R. P., Killeen, J., Mirin, A. A., Moir, R. W., Pearlstein, L. D., Post, R. F., Tull, C. G., Hall, L. S., McNamara, B., Boyd, J. K., Finan, C. A., III, Fuss, D., and Wilgus, C. A. (1974). *Proc. Conf. Plasma Phys. Controlled Nucl. Fusion Res.* *5th, 1974* Paper No. IAEA-CN-33/D1-2.

Richtmyer, R. D., and Morton, K. W. (1967). "Difference Methods for Initial Value Problems." Wiley (Interscience), New York.

Riviere, A. C. (1971). *Nucl. Fusion* **11**, 363.

Roberts, J. E., and Carr, M. L. (1960). "End-Losses from Mirror Machines," Report UCRL-5651-T. Lawrence Livermore Laboratory, Livermore, California.

Rosenbluth, M. N., MacDonald, W. M., and Judd, D. L. (1957). *Phys. Rev.* **107**, 1.

Spitzer, L., Jr. (1962). "Physics of Fully Ionized Gases," 2nd ed. Wiley (Interscience), New York.

Werkoff, F. (1973). "Equations de Fokker-Planck avec des Coéfficients Anisotropes et Bilan Energétique d'un Réacteur à Miroirs." Association Euratom-CEA, Grenoble, France.

Yushmanov, E. E. (1966). *Zh. Eksp. Teor. Fiz.* **49**, 588; *Sov. Phys.—JETP* **22**, 409 (1966).

Author Index

Numbers in italics refer to the pages on which the complete references are listed.

A

Adam, J., *164*
Ahlberg, J. H., 86, *128*
Aldcroft, D., 204, *207*
Allis, W. P., 390, *430*
Amsden, A. A., 7, 24, 25, 26, 27, 31, 32, *39*, *40*
Anashin, A. M., 166, *207*
Anderson, D. A., 118, 127, *128*
Anderson, D. V., 10, *39*, 241, 242, *249*
Anderson, O. A., 233, *249*, 412, 414, *430*
Appert, K., 255, 257, 264, *278*
Arai, T., 166, *208*
Armentrout, C. J., 282, 304, *325*
Artsimovich, L. A., 50, *82*, 131, 145, *162*, 166, 168, 196, *207*, 219, *250*, 254, *279*
Ashby, D. E. T. F., 169, *207*
Auer, P. L., 368, *387*

B

Baker, D. A., 256, *279*
Baldwin, D. E., 232, *250*
Bariaud, A., 166, 197, *209*
Barnett, C. F., 166, *208*
Barter, J. D., 282, 304, *325*
Bateman, G., 9, 36, *39*, *41*, 166, 167, *207*, *209*, 255, *279*
Beasley, C. O., 232, *250*
Becker, G. 219, *250*
Bengston, R., 166, *208*
BenDaniel, D. J., 390, *430*
Berger, D., 257, 264, *278*
Berk, H. L., 48, *82*, 232, *250*, 410, 422, *430*
Berkner, K. H., 412, *430*
Bernstein, I. B., 277, *279*
Berry, L. A., 152, 154, *162*, 166, 190, *207*, *208*
Bertrand, P., 48, *82*

Betancourt, O. L., 51, 55, 56, 57, 59, *82*, 214, *250*
Bickerton, R. J., 166, *208*
Bing, G., 390, *430*
Birdsall, C. K., 328, 329, 330, *364*, *365*, *366*, 368, *387*
Birdsall, D. H., 233, *249*, 412, 414, *430*
Bishop, A. S., 132, *162*
Biskamp, D., 351, *364*
Blank, A. A., 214, *250*
Bleach, R. D., 116, *129*
Bloch, E., 214, *250*
Bodin, N. A. B., 2, 3, *39*, 218, *250*
Bodner, S. E., 116, *129*
Bohm, D., 282, *325*
Bol, K., 166, *207*, 422, *431*
Book, D. L., 10, 16, *39*, 86, 87, 88, 91, 94, 95, 99, 100, 102, 103, 105, 106, 115, 123, 126, 127, *128*, *129*
Boris, J. P., 3, 8, 10, 16, *39*, *40*, 86, 87, 88, 91, 94, 95, 99, 100, 101, 102, 103, 104, 105, 106, 115, 122, 123, 124, 126, 127, *128*, *128*, *129*, 331, 335, 336, 337, 339, 359, 361, 362, *364*, *365*, 368, 371, 378, *387*
Boujot, J. P., 138, *164*
Boyd, J. K., 233, 244, *250*, 421, *431*
Boyd, T. J. M., 257, *279*
Brackbill, J. U., 10, 18, 27, *40*
Braginskii, S. I., 116, *129*, 136, *162*, 170, *207*
Brennan, M. H., 306, *325*
Breton, C., 166, 197, *208*
Brettschneider, M., 363, *364*
Bruin, R. A., 282, 304, *325*
Buneman, O., 329, 331, 333, 334, 336, 339, 340, *364*, *366*, 286, *325*, 371, *387*
Burcham, J., 204, *207*
Burger, P., 329, *365*
Burke, V. M., 179, *207*
Burkhalter, P. G., 116, *129*
Bush, C. E., 166, *208*
Bussac, J. P., 166, 197, *209*

433

Butler, J. D., 13, 16, 27, *40*
Butler, T. D., 26, *40*
Butt, E. P., 218, *250*
Buzbee, B. L., 379, 380, *387, 388*
Byers, J. A., 363, *365*, 384, *387*, 422, *431*

C

Callen, J. D., 50, 61, *82*, 135, 152, *162*, 166, 204, *207, 208, 209*, 221, 222, *250*, 323, 324, *325*
Canosa, J., 308, 316, *325*
Carlile, R. M., 287, *325*
Carr, M. L., 390, *431*
Cavallo, A. J., 282, 304, *325*
Cecchi, J. L., 422, *431*
Chance, M. S., 254, 258, *279*
Chapman, S., 169, *207*
Chen, L., 286, *325*
Chevalier, R. A., 127, *129*
Christiansen, J. P., 168, 169, 179, 180, *207*
Chu, C., 282, 294, 297, 298, 300, 308, *325*
Chu, C. K., 2, 3, 8, *40*, 44, *82*
Chu, M. S., 222, 223, *250*
Clark, J. F., 152, 154, 155, *162, 163*
Clark, M., 198, *209*
Clark, R. W., 127, *128, 129*
Clarke, J. F., 160, *163*, 166, 190, 204, *207, 208, 209*
Coensgen, F. H., 239, *251*, 412, *431*
Coffey, T. C., 128, *128*
Cohen, B. I., 361, *365*
Colchin, R. J., 166, 204, *207, 208*
Cole, H. C., 166, 204, *207*
Columbant, D. G., 115, 116, *129*
Concus, P., 380, *387*
Connor, J. W., 204, *207*, 277, *279*
Cook, J. L., 7, 25, 26, *40*
Cooley, J. W., 341, *365*
Cooper, W. S., 412, *430*
Cordey, J. G., 135, 152, *162*, 204, *207*
Core, W. G. F., 135, 152, *162*, 204, *207*
Cowlin, M., 204, *207*
Cowling, T. G., 169, *207*
Courant, R., 91, 96, *129*
Crenn, J. P., 166, 197, *209*
Crow, J. E., 218, *250*
Cummins, W. F., 239, 251, 412, *431*

D

Dagazian, R. Y., 34, *41*
Damm, C. C., 412, 414, *430*
Darwin, C. G., 368, *387*
Daughney, C. C., 422, *431*
Davies, R. C., 204, *209*
Davis, J., 115, *129*
Dawson, J. M., 152, *162*, 167, *209*, 277, *280*, 282, 283, 287, 289, 292, 293, 294, 295, 297, 298, 300, 304, 305, 306, 307, 308, 309, 315, 320, 321, *325*, 328, 362, *365*, 368, *387*, 409, 411, 422, 426, *431*
Dean, S. O., 145
Dei-Cas, R., 166, 197, 204, *208, 209*
de Leon, Moses, *365*
Delmas, M., 166, 197, *209*
De Sacy, S., 204, 208
Dewar, R. L., 254, 255, 257, 258, 273, 275, *279*
Dickman, D., 368, 380, *387*
Di Massa, L. G., 286, *325*
Dimock, D., 166, 190, 193, 197, 200, *208*
Dnestrovskii, Y. N., 132, 134, 138, 144, 145, 147, 153, 154, 155, 157, *162*, 166, *208*
Dobrott, D., 222, 223, *250*
Dory, R. A., 9, *41*, 50, 61, *82*, 132, 133, 139, 143, 163, 154, *163, 164*, 166, 167, 199, 204, *207, 208, 209*, 221, *222, 250*
Douglas, J., 241, *250*
Dozier, C. M., 116, 127, *129*
Drake, J. K., 282, 304, *325*
Druaux, J., 204, *208*
Drummond, W. E., 166, *208*
Düchs, D. F., 8, *40*, 44, *82*, 132, 133, 138, 145, 147, 148, 153, 154, 156, 159, 160, *162, 164*, 166, 192, 193, 199, 202, *208*
Dupree, T. H., 292, *325*
Dunn, D. A., *365*

E

Eckhartt, D., 166, 197, *208*
Ehlers, K. W., 412, *430*
Ellis, R. A., 166, *207*, 422, *431*
Eltgroth, P. G., 126, *129*
England, A. C., 166, *208*
Englehardt, W., 160, *162*
Etzweiler, J. T., 282, 304, *325*
Eubank, H. P., 166, 190, 193, 197, 200, *207, 208*, 422, *431*

F

Fallon, H. P., Jr., 132, *162*
Farmer, O. A., 13, 16, 27, *40*
Farr, W. M., 232, *250*
Feneberg, W., 167, *208*, 223, 224, 225, *250*
Feshback, H., 29, *40*, 220, *251*
Finan, C. A., III, 233, *250*, 421, *431*
Fisher, R. K., 254, 276, *279*
Fisher, S., 128, *128*, 224, 240, 241, *250*
Fix, G., 86, *129*
Foote, J. H., 232, *250*, 412, 414, *430*
Fornberg, B., 11, *40*
Forslund, D. W., 368, 381, *387*
Fowler, R. H., 134, *163*, 204, *207*
Fowler, T. K., 235, *250*, 390, 409, 421, *431*
Freeman, J. R., 8, 16, *40*, 127, *129*
Freis, R. P., 421, 422, *431*
Friedberg, J. P., 167, *208*, 214, *250*, 257, *279*
Friedman, N., 214, *250*
Friedrichs, K. O., 96, *129*
Frieman, E. A., 254, 273, 276, 277, *279*
Fromm, J. E., 22, *40*
Fujisawa, N., 166, *208*
Funahashi, A., 166, *208*
Furth, H. P., 44, 73, *82*, 132, 133, 138, 145, 147, 152, 153, 154, 156, 160, *162*, *164*, 166, 192, 193, 195, 199, 202, *207*, *208*, 233, *249*, 256, *279*, 390, 397, 409, 410, 411, 422, 426, 428, *430*, *431*
Futch, A. H. Jr., 391, 406, 409, 412, 414, 416, *431*
Fuss, D., 233, *250*, 329, 330, *364*, 421, *431*

G

Gajewski, R., 223, *250*
Galeev, A. A., 143, *162*, 172, *208*
Garabedian, P., 51, 55, 56, 59, *82*
Gardner, G. A., 257, *279*
Gardner, J. H., 103, 104, 115, 127, *128*, *129*
Gardner, L. R. T., 257, *279*
Gibson, A., 166, *208*, 213, *250*
Gilleland, J. R., 320, *325*
Ginot, P., 166, 197, *209*
Girard, J. P., 133, 139, 147, 154, *162*, *164*, 166, 197, 199, *208*, *209*
Glasser, A. H., 254, 258, 277, *279*
Goad, W. B., 18, 20, *40*

Godfrey, B. B., 330, 336, 361, 362, *365*, 368, *387*
Goedbloed, J. P., 34, *40*, 167, *208*, 255, 256, *279*
Goldman, S. R., 127, *129*
Goldsmith, G. J., 306, *325*
Goldstein, H., 369, 371, *387*
Goldston, R. J., 422, *431*
Golub, G. H., 379, 380, *387*
Goodman, R. K., 412, 414, *430*
Gorbunov, E. P., 166, *207*
Gordon, F. J., 412, 414, *430*
Gordonova, V. I., 230, *250*
Gourdon, C., 166, 197, *209*
Grad, H., 75, *82*, 137, 144, *162*, 212, 213, 214, 216, 234, 236, 238, 239, 254, 255, 276, *279*
Graf Finck von Finckenstein, F., *164*
Greenspan, E., 154, 156, *162*, 254, 255, 257, 258, 264, 273, 275, 276, 277, *279*, *280*, 320, 321, *325*
Greenwood, J. R., 282, 304, *325*
Grieger, G., 306, *325*
Grimm, R. C., 139, 143, *163*, 166, *208*, 254, 255, 257, 258, 264, 273, 275, 277, *279*, *280*
Grossmann, W., 9, 36, *39*, 167, *207*, 255, *279*, *280*
Grove, D. J., 197, *207*
Gruber, R., 255, 257, 264, *278*
Guest, G. E., 239, *250*
Gurnee, M. N., 306, *325*
Gunn, J., 241, *250*
Guss, W. C., 282, 304, *325*

H

Haber, I., 339, 361, 362, *365*, 368, *387*
Haegi, M., 166, *208*
Hagebeuk, H. J. L., 34, *40*, 167, *208*, 256, *279*
Hain, G., 9, *40*, 44, *82*
Hain, K. H., 9, *40*, 44, *82*, 86, 87, 88, 91, 95, 102, 105, 106, 127, *128*, *129*
Haines, M. G., 77, 78, *83*
Hall, L. S., 212, 217, 233, 235, 237, 238, 239, 240, 244, *250*, 421, 427, *431*
Halstead, A. S., 329, *365*
Hamilton, G. W., 412, 414, *430*
Harding, R. C., 232, *250*
Harlow, F. H., 24, 31, *39*, *40*, 330, *365*

Harries, W. L., 254, *280*, 282, *325*
Hartman, C. W., 233, *249*, *250*
Hasegawa, A., 368, *387*
Hastie, R. J., 232, 238, *251*, 277, *279*
Hazeltine, R. D., 132, *163*, 168, 169, 170, 172, 173, *208*, *209*, 324, *325*
Hendrick, C. L., 239, *250*
Hennion, F., 166, 197, *209*
Hernegger, F., 220, *250*
Hertweck, F., 6, 9, 18, *40*
Hewett, D. W., 384, 385, *387*, *388*
Hicks, H. R., 9, 26, *41*, 167, *209*
Hinnov, E., 132, *162*, 166, 190, 193, 197, 200, *208*
Hinton, F. L., 132, 133, 137, 139, 143, 147, 160, *162*, *163*, *164*, 168, 169, 170, 172, 173, 192, *208*, *209*, 320, 324, *325*
Hirschman, S., 160, *163*
Hirt, C. W., 7, 12, 13, 17, 25, 27, 32, *39*, *40*
Hiskes, J. R., 415, *431*
Hobbs, G. D., 232, *250*
Hockney, R. W., 213, *250*, 282, 286, *325*, 329, 330, 339, *365*
Hofmann, J., 2, *40*
Hogan, J. T., 75, *82*, 133, 134, 139, 143, 144, 145, 152, 153, 154, 155, *162*, *163*, 166, 190, 199, 204, *207*, *208*, *209*, 254, *279*
Holdren, J. P., 329, 363, *365*, 391, 406, 409, 416, *431*
Hooke, W. M., 295, 306, *325*
Hooper, E. B., 412, 414, 422, *430*, *431*
Horton, C. W. Jr., 410, 422, *430*
Howard, J. E., 232, *250*
Hsieh, C. L., 254, 276, *279*
Hsuan, H., 422, *431*
Hu, P. N., 216, *250*
Hughes, M. H., 166, 168, 169, 182, 192, *208*, *209*
Hugill, J., 166, 204, *207*, *208*
Huguet, M., 166, 197, *209*
Hunt, A. L., 412, 414, *430*
Hurwitz, H., 368, *387*

I

Inoue, K., 166, *208*
Irons, F. E., 218, *250*
Isaacson, E., 91, *129*
Israel, M., 86, *129*

Itoh, S., 139, *164*, 166, *208*
Ivanov, D. P., 166, *207*

J

Jackson, J. D., 331, *365*, 368, 370, *387*
Jacobsen, R. A., 166, *207*, 422, *430*, *431*
Jancarik, J., 166, *208*
Janenko, N. N., 147, *163*
Jardin, S. C., 254, 258, *279*
Jassby, D. L., 410, 422, 428, *430*, *431*
Jeffrey, A., 49, *82*
Jensen, T. H., 219, 222, *250*, *251*, 254, 276, *279*
Jernigan, T. C., 204, *209*
Johnson, D. J., 116, *129*
Johnson, J. L., 167, 193, *209*, 254, 255, 257, 258, 264, 273, 275, 276, 277, *279*, *280*
Johnson, L. C., 166, 190, 197, 200, *207*, *208*, 422, *430*
Johnson, T. H., 410, 422, *430*, *431*
Jordan, C., 160, *163*
Judd, D. L., 137, *163*, 389, 392, 393, 394, *431*
Junker, J., 218, *250*

K

Kadomtsev, B. B., 46, 79, 80, 81, *83*, 145, 146, *163*, 172, *208*, 277, *280*, 324, *325*
Kameari, A., 231, *251*
Kamimura, T., 73, *82*, 287, 306, 307, 308, *335*, 329, *364*
Kasai, S., 166, *208*
Kaufman, A. N., 195, 197, *209*, 361, *365*, 368, *387*, 398, *431*
Keeping, P. M., 139, 143, *163*, 166, 168, 169, 192, *208*, *209*
Kelley, G. G., 145, 152, 153, *163*, 166, 204, *208*
Kerner, W., 223, *251*, 256, *279*
Kerst, D. W., 282, 304, *325*
Khelladi, M., 133, 135, 139, 147, 154, 156, *162*, *164*, 199, *208*
Killeen, J., 5, 8, 16, *40*, 44, 73, *82*, 133, 139, 143, 150, 152, *163*, 166, 168, 169, 192, *208*, *209*, 212, 232, 239, 240, 241, 242, *249*, *250*, 251, 363, *364*, *365*, 390, 391, 406, 409, 416, 421, 422, *430*, *431*
Klein, H. H., 361, 262, *365*, 368, *387*
Klib, R. W., 368, *387*
Konstantimov, O. V., 155, *163*

Kooyers, G. P., 329, *366*
Koppendorfer, W., *40*, 44, *82*, 160, *162*
Kostamarov, D. P., 132, 134, 138, 144, 145, 147, 153, 154, 155, 157, *162*, 166, *208*
Kovrizhnick, L. M., 144, *163*
Knorr, G., 170, *208*
Krall, N. A., 126, *129*, 139, 147, *164*, 376, 387
Krause, H., 219, *251*
Kreiss, H. O., *40*
Krommes, J. A., 308, *325*
Kruer, W. L., 283, *325*, 339, *365*
Kruskal, M. D., 50, *82*, 131, 160, *163*, 215, *251*, 277, *279*
Kulsrud, R. M., 50, *82*, 215, *251*, 276, 277, *279*, 410, 422, *430*
Kunieda, S., 166, *208*

L

Lackner, K., 167, *208*, *209*, 223, 224, 225, *250*
Lampe, M., 127, *128*, 348, *365*
Landau, L. D., 5, *40*, 368, *387*
Lane, F. O., 8, 16, *40*
Langdon, A. B., 282, *325*, 330, 336, 337, 339, 348, 350, 361, 362, 363, *365*, 377, 378, *387*
Lapidus, A., 8, 16, *40*
Larkin, F. M., 232, *251*
Lasinski, B. F., 339, 348, *365*
Lauer, E. J., 233, *249*
Launois, D., 166, 197, *209*
Laval, G., 219, *251*, 254, 256, *279*
Lax, P. D., 96, *129*
Lecoustey, P., 166, 197, *209*
Lee, E. P., 422, *431*
Lee, R., 337, 348, 361, 362, *364*, *365*, 368, 378, *387*
Lewis, H. R., 29, *40*, *365*, 369, *387*
Lewis, P. A. W., 341, *365*
Lewy, H., 96, *129*
Liewer, P. C., 126, *129*
Lifshitz, E. M., 5, *40*, 368, *387*
Lin, A. T., 362, *365*
Lindemuth, I. R., 8, 16, *40*, 44, *82*
Lindman, E. L., 335, 336, 338, 344, 345, *365*, 368, 373, 381, 387, *388*
Liu, C. S., 315, *325*, 410, 422, *430*
Low, F. E., 369, *388*
Lui, H. C., 2, 3, 8, 10, *40*
Lüst, R., 267, *279*

M

McAlees, D. G., 204, *207*
McBride, J., *164*
McConnell, A. J., 58, *82*
McCune, J. E., 232, *250*
McDonald, B. E., 129, *129*
MacDonald, W. M., 137, *163*, 389, 392, 393, 394, *431*
McLean, F. A., 116, *129*
McMahon, J. M., 116, *129*
McNally, J. R., 166, 204, *208*, *209*
McNamara, B., 212, 217, 233, 234, 237, 238, 239, 240, *250*, *251*, 421, 422, *431*
Maeno, M., 166, *208*
Mahdavi, M. A., 254, 276, *279*
Manheimer, W. M., 9, *41*, 126, 127, *128*, *129*
Mann, L. W., 256, *279*
Marder, B. M., 225, *251*, 257, 268, *279*
Marty, D., 133, 139, 147, 154, *162*, *164*, 166, 197, 199, 204, *208*, *209*
Martensen, E., 267, *279*
Marx, K. D., 135, 152, *163*, 390, 391, *431*
Mashke, E. K., 219, 220, 223, *250*, *251*
Mastan, C. W., 28, *41*
Masuzaki, M., 231, *251*
Matoba, T., 166, *208*
Matsuda, S., 166, *208*
Matsuda, Y., 319, *325*
Matsui, T., 223, *251*
Mattioli, M., 166, 197, *209*
Max, C. E., 361, *365*
Mazzucato, E., 166, *207*, 422, *430*, *431*
Meade, D., 133, 152, 153, 159, *163*
Medley, S., 156, *163*, 166, *208*
Mercier, C., 132, 133, 138, 144, 145, 147, 154, 155, *163*, *164*, 166, 197, *208*, *209*, 254, *279*
Meservey, E., 166, 190, 193, 197, 200, *208*
Michelis, C. De., 166, 197, *208*
Mikhlin, S. G., 256, *279*, *280*
Mirin, A. A., 363, *364*, *365*, 391, 406, 409, 410, 416, 421, 422, *430*, *431*
Mitchell, R. W., 368, 381, *387*
Moir, R. W., 421, *431*
Molvick, A. W., 239, *251*, 412, *431*
Monticello, D., 9, *41*, 46, 79, 80, 81, *83*, 277, *280*
Moore, T. B., 160, *163*
Morera, J. P., *164*
Moretti, G., 86, *129*

Morgan, O. B., 204, *208*, *209*
Mori, S., 166, *208*
Moriette, P., 139, *164*, 166, 197, *208*
Morozov, V. A., 230, *250*
Morse, P. M., 29, *40*, 220, *251*
Morse, R. L., 330, 331, 336, 347, 348, *365*, 368, 371, 373, 380, 381, *387*, *388*
Morton, K. W., 3, 5, 10, 11, *40*, 177, 178, *209*, 213, *251*, 404, *431*
Mosher, D., 127, *129*
Mostrom, M. A., 361, *365*
Mukhovatov, V. S., 213, *251*
Murakami, M., 166, *208*

N

Nagel, D. J., 116, *129*
Navet, M., 48, *82*
Neidigh, R. V., 166, *208*
Nelson, D. B., 239, *250*
Newcomb, W. A., 167, *209*, 218, *250*, 422, *431*
Nexsen, W. E., Jr., 239, *251*, 412, *431*
Nicholson, D. R., 361, *365*
Nielsen, P., 166, *208*
Nielson, C. W., 330, 331, 335, 336, 337, 338, 347, 348, *365*, 368 371, 373, 378, 379, 380, 381, *387*, *388*
Nilson, E. N., 86, *128*
Ninomiya, H., 231, *251*

O

Oberman, C. R., 276, 277, *279*, 308, *325*
Ohga, T., 166, *208*
Ohkawa, T., 219, 223, *250*, *251*, 254, 276, 279, 282, 304, 305, 320, *325*
Ohlendorf, W., 306, *325*
Ohta, M., 166, *208*, 257, *280*
Okabayashi, M., 219, *251*
Okuda, H., 282, 286, 287, 292, 293, 294, 295, 297, 298, 300, 304, 306, 308, 309, 315, 316, 319, *325*, 362, *365*, 368, *387*
Okuda, T., 223, *251*
Oliphant, T. A., 45, *82*
Orens, J. H., 339, *365*
O'Rourke, P. J., 13, 16, 27, *40*
Orszag, S. A., 86, *129*
Osher, J. E., 412, 414, *430*

Ossakow, S. L., 123, 127, *128*, *129*
Ott, E., 126, *128*, *129*

P

Pacher, H. P., 306, *325*
Panofsky, W. K. H., 331, *366*
Papadopoulos, K., 127, *129*
Parsons, C., 156, *163*
Paul, J. W. M., 166, 204, *207*, *208*
Pavlova, N. L., 134, 138, 144, 145, 147, 153, 154, 155, 157, *162*, 166, *208*
Peaceman, D. W., 8, *40*
Pearlstein, L. D., 232, *250*, 421, *431*
Pellat, R., 219, *251*, 254, 256, *279*
Perel, V. I., 155, *163*
Petrov, M. P., 166, *207*
Pfirsch, D., 143, *163*, 168, 197, *209*, 223, *250*
Phillips, M., 331, *366*
Phillips, P., 166, *208*
Pierre, J. M., 123, *128*
Platz, P., 166, 197, *209*
Plinate, P., 166, 197, *209*
Pogutse, O. P., 145, *163*, 172, *208*, 324, *325*
Porter, G. D., 412, 414, *430*
Post, R. F., 231, *251*, 282, 304, *325*, 398, 409, 421, *431*
Potter, D. E., 3, 5, 6, 8, 29, *41*, 44, 48, 53, 59, 62, 73, 77, 78, *82*, *82*, *83*, 132, *163*, 167, 178, *209*, 255
Pracht, W. E., 10, 18, 19, 26, 27, 31, *39*, *40*
Prater, R., 282, 304, *325*
Pyle, R. V., 412, *430*

R

Rachford, H. H., 8, 40
Raizer, Yu. P., 5, *41*
Rankin, M., 235, *250*, 390, *431*
Rappaz, J., 257, 264, *278*
Rebut, P. H., 166, 197, 204, *208*, *209*
Rees, M., 91, *129*
Rensink, M. E., 363, *365*, 410, 422, *430*, *431*, 422, 422, *431*
Reynolds, P., 166, *208*
Richtmyer, R. D., 3, 5, 8, 10, 11, 12, 13, 16, *40*, 177, 178, *209*, 213, *251*, 404, *431*
Ripin, B. H., 116, *129*
Rivard, W. C., 13, 27, *40*
Riviere, A. C., 204, *209*, 415, *431*

Roache, P., 86, *129*
Roberts, J. E., 390, *430*, *431*
Roberts, K. V., 3, 5, 6, 8, 9, 29, *40*, *41*, 44, 48, *82*, *83*, 132, *163*, 168, 169, 179, 180, 182, *207*, *208*, *209*, 255, *280*, 359, *364*
Roberts, M., 166, 168, *208*
Roberts, P. D., 182, *208*
Roberts, S. J., 9, *40*, 44, *82*
Robinson, D. C., 132, 145, *163*, *251*, 254, *280*
Rome, J. A., 151, 152, *163*, 166, 204, *207*, *208*, *209*
Rose, D. J., 198, *209*
Rosen, B., 254, 258, 264, *279*, *280*, 283, 288, *325*
Rosenbluth, M. N., 9, 34, *41*, 46, 73, 79, 80, 81, *82*, *83*, 132, 137, *162*, *163*, 168, 169, 170, 172, 173, 195, 197, 198, 199, *208*, *209*, 231, 232, *251*, 277, *280*, 315, 320, *325*
Rostler, 368, *387*
Rowlands, G., 234, *251*
Rubin, H., 236, *250*
Rudmin, J. W., 282, 304, *325*
Rudsinski, L. I., 380, *388*
Ruhkadze, R. R., 311, 313, *325*
Rutherford, P. H., 34, *41*, 44, *82*, 132, 133, 145, 147, 153, 154, 156, 160, *162*, *164*, 166, 170, 192, 193, 195, 199, 202, *208*, *209*, 254, 256, 273, *279*, 410, 422, *430*

S

Sagdeev, R. Z., 143, *162*, 172, *208*
Sakharov, A., 154, *163*
Sauthoff, N., 254, *280*
Scannapieco, A. J., 127, *129*
Schlüter, A., 143, *163*, 168, 197, *209*
Schmidt, G. L., 282, 304, *325*
Schneider, W., 6, 9, 10, 18, 36, *39*, *40*, *41*, 167, *207*, 255, *279*
Schulz, W. D., 18, *41*
Schwartz, C., 256, *280*
Selberg, H., 256, *279*
Sen, A., 232, *250*
Shafranov, V. D., 76, *83*, 143, *163*, 167, *209*, 213, 219, 227, *250*, *251*, 254, *279*, *280*, 315, *325*
Shanny, R., 320, 321, *325*
Sharp, W., 167, *209*
Sheffield, G. V., 219, *251*, 254, 258, *279*
Sheffield, J., 166, 204, *207*, *208*
Shimomura, Y., 257, *280*
Sigmar, D. J., 160, *163*
Silin, V. P., 311, 313, *325*
Simonen, T. C., 239, *251*, 412, *431*
Sinclair, R. M., 254, *280*
Sinz, K., 331, 337, *366*
Sledziewski, Z., 166, 197, *209*
Smeulders, P., 166, *197*, *209*
Smith, R. R., 422, *431*
Soubbaramayer, 132, 138, 144, 145, 147, 154, 155, *163*, *164*, 166, 197, *209*
Soule, J. L., *164*, 254, 256, *279*
Speth, E., 166, 204, *207*, *208*
Spitzer, L., Jr., 196, *209*, 398, *431*
Sprangle, P., 123, 126, *128*
Sprott, J. C., 282, 304, *325*
Stallard, B. W., 412, *431*
Stamper, J. A., 116, *129*
Stephanakis, S. J., 127, *129*
Stevens, D. C., 216, 217, *250*, *251*
Stewart, L. D., 204, *208*, *209*
Stirling, W. L., 204, *208*, *209*
Stix, T. H., 204, *209*, 422, *431*
Stodiek, W., 166, 195, 199, *207*, *208*, 254, *280*, 422, 430
Stott, P. E., 166, 204, *207*, *208*
Strang, G., 86, *129*
Strauss, H., 46, 79, 80, 81, *83*, 277, *280*
Strauss, N., *9*, *41*
Strel'kov, V. S., 166, *207*
Sturrock, P. A., 369, *388*
Sugawara, T., 166, *208*
Sulton, A. L., 126, *128*
Suydam, B. G., 219, *251*
Suzuki, N., 166, *208*
Suzuki, V., 223, 224, 225, 227, 230, 231, *251*
Sweetman, D. R., 204, *209*
Sykes, A., 167, *209*, 255, *280*

T

Tachon, J., 166, 197, *209*
Takeda, T., 139, *164*, 166, *208*, 257, *280*
Tamano, T., 222, *250*, 282, 304, 320, *325*
Tanaka, Y., 223, *251*
Tasso, H., 223, *251*, 256, *279*
Tataronis, J., 255, *280*
Taussig, R. T., 44, *82*
Tayana, H., 231, *251*

Taylor, J. B., 197, *209*, 212, 213, 217, 231, 232, 234, 238, *251*
Taylor, J. C., 167, 197, *209*
Temam, R., 147, *163*
Tereshin, V. I., 204, *207*
Tenney, F. H., 152, *162*, 409, 410, 411, 422, 426, *430*, *431*
Thames, F. C., 28, *41*
Thompson, J. F., 28, *41*
Thompson, S. L., 127, *129*
Tikhonov, A. N., 228, *251*
Tidman, D. A., 115, *129*
Toi, K., 166, *208*
Tolnas, E. L., 166, 197, *207*, *208*, 422, *430*
Torossian, A., 166, 197, *209*
Trivelpiece, A. W., 376, *387*
Troyon, F., 257, *278*
Tsang, K., 308, 319, 323, 324, *325*
Tuck, J., 3, *41*
Tull, C. G., 363, *365*, 421, *431*
Tuttle, G. H., 46, 53, 62, 77, 78, 82, *83*, 167, *209*

V

Vaclavik, J., 255, *278*
Van Duzer, J., 328, *366*
Vanek, V., 254, 276, *279*
Van Goeler, S., 254, *280*
von Hagenow, K., 167, *209*

W

Wabig, H., 306, *325*
Wagner, C. E., 9, *41*, 127, *129*, 368, *387*
Walker, H., *250*
Walsh, J. L., 86, *128*
Ware, A. A., 192, *209*
Watkins, M. L., 166, 168, 169, 192, *209*
Weimer, K. E., 254, 258, *279*, *280*
Weitzner, H., 214, 225, *250*, *251*
Welch, J. E., 24, *40*
Welch, P. D., 341, *365*

Welter, H., 351, *364*
Wendroff, B., 96, *129*
Werkoff, F., 391, *431*
Wesson, J. A., 167, *209*, 255, *279*
White, R., 9, *41*, 46, 79, 80, 81, *83*, 277, *280*
Whiteman, K. J., 212, 232, 239, 240, *251*
Whitlock, R. R., 116, *129*
Whitney, K. G., 115, *129*
Widner, M. M., 132, 139, 147, 148, *163*, *164*, 166, *208*
Wilets, L., 276, *279*
Wiley, J. C., 133, 139, 143, 147, *164*, 192, *208*
Wilgus, C. A., 233, 244, *250*, 421, *431*
Wilhelm, R., 219, *251*
Wing, W. R., 166, *208*
Winske, D., 384, 385, *387*, *388*
Winsor, N. K., 115, 116, *129*, 167, *209*, 277, *280*
Wolf, G. H., 306, *325*
Woltjer, J., 217, *251*
Wooten, J., 9, *41*, 167, *209*
Wong, K. L., 282, 304, *325*

Y

Ya'akobi, B., 166, 197, *209*
Yokokura, K., 166, *208*
Yoshikawa, S., 145, *164*, 168, 196, *209*, 254, 257, *280*, 397, *431*
Young, D. M., 29, *41*
Young, F. C., 116, 127, *129*
Young, K. M., 254, *280*
Yu, S. P., 329, *366*
Yushmanov, E. E., 398, *431*

Z

Zakharov, L. E., 227, 228, 229, 230, *251*
Zalesak, S., 103, 104, 115, *128*
Zel'dovich, Ya., B., 5, *41*
Zienkiewicz, O. C., 18, *41*
Zwicker, H., 219, *251*

Subject Index

A

Accumulation point, 272
Action-at-a-distance, 369
Adiabaticy, 69, 232
Alfvén wave, 261, 274
Almost-Lagrangian, 10
Alpha particles, 134, 142, 152, 426
Alternating direction implicit method, 8, 226, 245
Ambipolar potential, 397
Amplification factor, 91, 98, 100, 108
Anisotropic pressure, 231
Anisotropy, 44, 258
Anomalous transport, 282
Antidiffusion, 107
  explicit, 96, 110, 113
  implicit, 99, 103, 108
  phoenical, 100, 103, 108
ASTRON, 224
ATC, 217
Axisymmetric equilibria, 220, 254, 276
Azimuthally symmetric, 390

B

Background gas, 415
Banana regime, 172, 321
Barium clouds, 127
Baseball, 241, 414
Beam trapping, 151
Belt pinch, 219, 225
Bifurcation, 224
Binary encounters, 282
Bounce frequency, 140, 145, 172
Boundary conditions, 25, 30, 338
Breakeven experiment, 426
Bremsstrahlung, 142, 198
Burgers equation, 126

C

Cerenkov radiation, 315
Characteristic methods, 86
Charge continuity, 333
  exchange, 142, 400
  sharing, 329
Classical regime, 172
Clipping, 101, 105, 109, 114
Coherence time, 291
Collision frequency, 140, 175, 355
Collisional radiative model, 159
COMMON blocks, standard, 182
Compression, 86, 96, 104, 115, 396
Conduction, 132, 173
Confinement time, 400
Conservative form of equations, 5, 169, 394
Contact discontinuity, 115
Containment devices, 166
Continuity equation, 5, 85, 105, 119, 127
Continuum, 274
Convection, 3, 86, 96, 104, 115, 132
Convective diffusion, 289
Convergence, 178, 263, 273
Coordinate system, 6, 19, 257, 273, 278
  curvilinear, 102, 116
Coronal equilibrium, 159
Coulomb collisions, 389
Courant stability condition, 333, 338, 363
Crank–Nicholson representation, 147
Current rise, 193
Cyclotron radiation, 198

D

Damping, 91, 99, 109, 300
Darwin model, 364, 368
Density, 86, 102
Diagnostics, 31, 189, 346
Dielectric terms, 298
Diffuse linear pinch, 255, 264
Diffusion, 74, 86, 96, 107, 143, 196, 289, 320
  ambipolar, 170
  nonlinear, 137
  numerical, 91, 101, 106
  physical, 116, 126
Dispersion
  light wave, 332, 361

numerical, 91, 96, 105
plasma wave, 337
Displacement vector, 257
Distribution function, 389
Divertor, 259
Donor-cell algorithm, 91, 102, 109, 114
Doublet II, 219, 221
Douglas–Gunn method, 241
Drift surface, 213, 233, 321
  wave, 145, 311
Dubna Conference, 132
Düchs code, 192

E

Electromagnetic codes, 330, 361
  field momentum, 298
Electron gyro radius, 173
Elliptic cross section, 254, 272
  equations, 369, 379
Energy clamping, 428
  density, 116
  multiplication factor, 410
Equilibration time, 69
Equilibrium, 50, 211, 254, 276, 282
Ergodic magnetic field lines, 294
Error, 89, 100, 107, 360
Eulerian techniques, 5, 88, 103, 115
Euler–Lagrange equation, 256
Expansion function, 256, 263, 272
EXPERT facility, 186

F

Field line curvature, 262
Finite difference methods, 86, 102, 112, 176, 255
  element methods, 86, 257, 271, 276
Firehose instability, 238
Flow energy, 290
  velocity, 87, 92, 115, 125
Fluctuation dissipation theory, 292
Flute instability, 311, 385
Flux, 48, 95, 169
  constraints, 267
  coordinates, 6, 60
  correction, 16, 105
  limiter, 101, 106, 114, 127
  surfaces, 118
  tube equilibria, 237, 246

Fokker–Planck equations, 135, 152, 389
Force-free equilibrium, 218, 223
FORTRAN, 168
Fourier analysis, 89, 106
  transform, 341, 357
Franck–Condon neutrals, 155
Fredholm integral equations, 228
Fusion cross section, 406

G

Galerkin method, 256, 278
Gauge, 369
Gibbs error, 91, 106, 114
Gitterbewegung, 123
Global functions, 257
  iteration, 379
Gradients, 87, 105, 115
Green's function, 259, 268
Grenoble Conference, 133
Group velocity, 316
Guiding-center model, 277, 288

H

Hamiltonian, 371
Harmonics, 87, 97, 106
HBTX, 218
Heat transfer, 282, 307
Helical instabilities, 79
  systems, 276
Hydrodynamics, 115

I

ICARUS, 165
Impact ionization, 159, 415
Implicit differencing, 3, 8, 22, 24, 148, 177, 226, 245
Impurities, 196, 426
Induction matrix, 267
Initial conditions, 168, 349
Injection angle, 420
Instabilities, 34, 79, 122, 195, 238, 254, 381
  numerical, 333, 346, 361
Integral equation, 228
  transform, 223
Inverse aspect ratio, 167
Inverted cascade ionization, 415
Ion waves, 122

SUBJECT INDEX

Isotropy, 390
Iterative methods, 154, 178, 379

J

Jacobian, 259, 268

K

Kármán vortex street, 128
Kelvin–Helmholtz instability, 128
Kinematical equations, 18
Kink modes, 34, 254, 274
Knorr field, 170
Kreiss's theorem, 12
Kruskal–Shafranov current, 219

L

Lagrangian techniques, 5, 9, 45, 86, 103, 122, 256, 261, 369, 374
Landau damping, 308
Lax–Wendroff algorithm, 96, 102
Leapfrog algorithm, 96, 122, 286
Light scattering, 353
Longitudinal invariant, 232
Lorentz ionization, 415
Loss cone angle, 398
  modes, 231
Low-beta approximation, 170
Low phase error (LPE) algorithms, 102, 106, 110, 122
Lower hybrid waves, 300

M

Macros, 246
MACSYMA, 264
Madison Conference, 132
Magnetic axis, 258, 276
  compression, 396
  differential equations, 237, 243
  geometry, 134, 282
  islands, 73
  mirroring, 287
  surface, 44, 50, 254, 272
Magnetosonic mode, 261, 272
Mapping method, 57
MASTER file, 188
Maxwell's equations, 331, 362

MEDUSA, 169
Metric tensor, 261, 276
Minimum-B stability, 234
MIRICLE, 235
Mirror instability, 238
  Levitron, 239
  machine, 231, 287, 389, 417
  ratio, 398
Monte Carlo collision operator, 320
  methods, 156
Moscow Conference, 133
Multipole equilibrium, 223

N

Nearest grid point method, 284
Neoclassical transport, 132, 143, 169, 319
Neutral gas component, 199
  injection, 133, 204
Neutron transport, 156
Noncircular cross section, 134, 254
Nonuniform mesh, 403
  plasma, 310
Normal mode, 261, 391
$n\tau$ figure of merit, 405

O

OLYMPUS programming system, 168
Omnigenity, 233
One-dimensional mode, 134, 139, 167
Orthogonal eigenfunctions, 401
Orthogonalization, 46

P

Paramagnetic equilibria, 237
Particle codes, 367
  momentum, 298
  pushing, 283
  recycling, 199
Pedestal values, 176
Pfirsch–Schlüter diffusion, 143, 197, 320
Phase errors, 91, 108
PIC codes, 328
Plasma magnetization, 237
Plasma–vacuum interface, 29, 259, 267
Plateau regime, 174
Poisson equation, 329, 336
Poloidal magnetic field, 52, 171, 320

Positivity, 87, 92, 96, 101, 105, 114, 122
Postprocessor, 359
Potential energy, 20, 55, 256
Predictor–corrector technique, 149
Pressure profile, 419
  tensor, 234
  waterbag model, 62
Programming techniques, 179
Pseudoclassical transport model, 145, 196

Q

Q figure of merit, 406
Quasi-equilibrium states, 166
Quiet start, 383

R

Radiation, 161, 367
  cooling, 199
Rarefaction wave, 115
Rayleigh–Ritz principle, 256, 263, 272
Rayleigh–Taylor instability, 126
Reflection of charge exchange neutrals, 154
Relativistic particles, 334
Replacement times, 202
Resistive models, 276
  tearing mode, 73, 276
Retarded potential, 370
Reversibility, 103, 108, 373
Rezoning, 26, 104, 118
Rosenbluth potentials, 391
Rotating theta pinch, 31
Rotational transform, 276
  irrational, 305
  rational, 305

S

Safety factor, 197, 259
Scalar potential, 267
Scrambled drift surfaces, 234
Scyllac, 214
Secularity, 87, 114
Sharp boundary equilibria, 223
SHASTA, 96, 102, 109, 116, 126
Shear flow, 289
Sheared magnetic field, 146, 312
Shocks, 115
Singular solutions, 275

Six-regime transport model, 138
Skin current, 193
Sliding zone algorithms, 103, 118
Sound waves, 261
Source term, 399
Spectral methods, 86
Spectrum, MHD, 254, 271, 275
Spitzer conductivity, 173
Splines, 86, 247, 261, 271
Splitting difference scheme, 116, 411
Sputtering, 157
Squarewave test, 89, 96, 105
ST Tokamak, 190
Stability, 11, 87, 91, 98, 101, 105, 178
  conditions, 218, 231, 238, 248, 254, 333, 363
Stellarator, 59, 213
Stuffed cusp, 239
Subtracted dipole scheme, 285
Successive overrelaxation, 25, 221
SUPERLAYER, 363
Surface current, 262

T

Test particles, 288
TFR Tokamak, 197
Thermal instability, 195
Thermalization, 142
Thermonuclear burning, 204
Theta pinch, 29, 126
Thomson scattering, 132
Three-dimensional model, 8, 283
Three-regime transport model, 143
Time-centering, 22
Time-step control, 178
Tokamak, 131, 221, 254, 273, 319, 389
  ordering procedure, 171
Toroidal, 254, 268
  current, 171
  flow, 140
  magnetic field, 52, 168, 319
Trace particles, 31
Transfer function, 91, 98
Transport coefficients, 132, 143, 167, 174, 319
Transverse flow, 290
Trapped particles, 146, 172, 196, 306
Tridiagonal equations, 99, 105, 113, 404
  matrices, 150, 177

Truncation errors, 13, 27
Turbulence, 145, 292
Turbulent heat transport, 307
Two and one-half dimensional model, 283
Two-component torus, 152, 409
Two-dimensional model, 8, 283
Two-plasmon decay, 349
Two-stream instability, 122
2XII, 246, 412

V

Vacuum, 258
   magnetic field, 29, 419
Variational principle, 215
   procedure, 54
Vector coding, 246
   computers, 99, 114
Virtual casing principle, 227

Viscosity, 301
Vlasov equation, 119, 124
   equilibrium, 385

W

Ware pinch effect, 202
Waterbag model, 43, 62
Wave kinetic equation, 315
Weibel instability, 381
Weighting, linear or area, 329
Woltjer's invariant, 217

Z

Z-effective, 427
Z-pinch, 127
Zero residual damping (ZRD), 99, 102, 106
ZETA, 218
ZOHAR, 330

Contents of Previous Volumes

Volume 1: Statistical Physics

The Numerical Theory of Neutron Transport
Bengt G. Carlson

The Calculation of Nonlinear Radiation Transport by a Monte Carlo Method
Joseph A. Fleck, Jr.

Critical-Size Calculations for Neutron Systems by the Monte Carlo Method
Donald H. Davis

A Monte Carlo Calculation of the Response of Gamma-Ray Scintillation Counters
Clayton D. Zerby

Monte Carlo Calculation of the Penetration and Diffusion of Fast Charged Particles
Martin J. Berger

Monte Carlo Methods Applied to Configurations of Flexible Polymer Molecules
Frederick T. Wall, Stanley Windwer, and Paul J. Gans

Monte Carlo Computations on the Ising Lattice
L. D. Fosdick

A Monte Carlo Solution of Percolation in the Cubic Crystal
J. M. Hammersley

AUTHOR INDEX—SUBJECT INDEX

Volume 2: Quantum Mechanics

The Gaussian Function in Calculations of Statistical Mechanics and Quantum Mechanics
Isaiah Shavitt

Atomic Self-Consistent Field Calculations by the Expansion Method
C. C. J. Roothaan and P. S. Bagus

The Evaluation of Molecular Integrals by the Zeta-Function Expansion
M. P. Barnett

Integrals for Diatomic Molecular Calculations
Fernando J. Corbato and Alfred C. Switendick

Nonseparable Theory of Electron-Hydrogen Scattering
A. Temkin and D. E. Hoover

Estimating Convergence Rates of Variational Calculations
Charles Schwartz

AUTHOR INDEX—SUBJECT INDEX

Volume 3: Fundamental Methods in Hydrodynamics

Two-Dimensional Lagrangian Hydrodynamic Difference Equations
William D. Schulz

Mixed Eulerian-Lagrangian Method
R. M. Frank and R. B. Lazarus

The Strip Code and the Jetting of Gas between Plates
John G. Trulio

CEL: A Time-Dependent, Two-Space-Dimensional, Coupled Eulerian-Lagrange Code
W. F. Noh

The Tensor Code
G. Maenchen and S. Sack

Calculation of Elastic-Plastic Flow
Mark. L. Wilkins

Solution by Characteristics of the Equations of One-Dimensional Unsteady Flow
N. E. Hoskin

The Solution of Two-Dimensional Hydrodynamic Equations by the Method of Characteristics
D. J. Richardson

The Particle-in-Cell Computing Method for Fluid Dynamics
Francis H. Harlow

The Time-Dependent Flow of an Incompressible Viscous Fluid
Jacob Fromm

AUTHOR INDEX—SUBJECT INDEX

Volume 4: Applications in Hydrodynamics

Numerical Simulation of the Earth's Atmosphere
Cecil E. Leith

Nonlinear Effects in the Theory of a Wind-Driven Ocean Circulation
Kirk Bryan

Analytic Continuation Using Numerical Methods
Glenn E. Lewis

Numerical Solution of the Complete Krook-Boltzmann Equation for Strong Shock Waves
Moustafa T. Chahine

The Solution of Two Molecular Flow Problems by the Monte Carlo Method
J. K. Haviland

Computer Experiments for Molecular Dynamics Problems
R. A. Gentry, F. H. Harlow, and R. E. Martin

Computation of the Stability of the Laminar Compressible Boundary Layer
Leslie M. Mack

Some Computational Aspects of Propeller Design
William B. Morgan and John W. Wrench, Jr.

Methods of the Automatic Computation of Stellar Evolution
Louis G. Henyey and Richard D. Levée

Computations Pertaining to the Problem of Propagation of a Seismic Pulse in a Layered Solid
F. Abramovici and Z. Alterman

AUTHOR INDEX—SUBJECT INDEX

Volume 5: Nuclear Particle Kinematics

Automatic Retrieval Spark Chambers
J. Bounin, R. H. Miller, and M. J. Neumann

Computer-Based Data Analysis Systems
Robert Clark and W. F. Miller

Programming for the PEPR System
P. L. Bastien, T. L. Watts, R. K. Yamamoto, M. Alston, A. H. Rosenfeld, F. T. Solmitz, and H. D. Taft

A System for the Analysis of Bubble Chamber Film Based upon the Scanning and Measuring Projector (SMP)
Robert I. Hulsizer, John H. Munson, and James N. Snyder

A Software Approach to the Automatic Scanning of Digitized Bubble Chamber Photographs
Robert B. Marr and George Rabinowitz

AUTHOR INDEX—SUBJECT INDEX

Volume 6: Nuclear Physics

Nuclear Optical Model Calculations
Michael A. Melkanoff, Tatsuro Sawada, and Jacques Raynal

Numerical Methods for the Many-Body Theory of Finite Nuclei
Kleber S. Masterson, Jr.

Application of the Matrix Hartree-Fock Method to Problems in Nuclear Structure
R. K. Nesbet

Variational Calculations in Few-Body Problems with Monte Carlo Method
R. C. Herndon and Y. C. Tang

Automated Nuclear Shell-Model Calculations
S. Cohen, R. D. Lawson, M. H. Macfarlane, and M. Soga

Nucleon-Nucleon Phase Shift Analyses by Chi-Squared Minimization
Richard A. Arndt and Malcolm H. MacGregor

AUTHOR INDEX—SUBJECT INDEX

Volume 7: Astrophysics

The Calculation of Model Stellar Atmospheres
Dimitri Mihalas

Computational Methods for Non-LTE Line-Transfer Problems
D. G. Hummer and G. Rybicki

Methods for Calculating Stellar Evolution
R. Kippenhahm, A. Weigert, and Emmi Hofmeister

Computational Methods in Stellar Pulsation
R. F. Christy

Stellar Dynamics and Gravitational Collapse
Michael M. May and Richard H. White

AUTHOR INDEX—SUBJECT INDEX

Volume 8: Energy Bands of Solids

Energy Bands and the Theory of Solids
J. C. Slater

Interpolation Schemes and Model Hamiltonians in Band Theory
J. C. Phillips and R. Sandrock

The Pseudopotential Method and the Single-Particle Electronic Excitation Spectra of Crystals
David Brust

A Procedure for Calculating Electronic Energy Bands Using Symmetrized Augmented Plane Waves
L. F. Mattheiss, J. H. Wood, and A. C. Switendick

Interpolation Scheme for the Band Structure of Transition Metals with Ferromagnetic and Spin-Orbit Interactions
Henry Ehrenreich and Laurent Hodges

Electronic Structure of Tetrahedrally Bonded Semiconductors: Empirically Adjusted OPW Energy Band Calculations
Frank Herman, Richard L. Kortum, Charles D. Kuglin, John P. Van Dyke, and Sherwood Skillman

The Green's Function Method of Korringa, Kohn, and Rostoker for the Calculation of the Electronic Band Structure of Solids
Benjamin Segall and Frank S. Ham

AUTHOR INDEX—SUBJECT INDEX

Volume 9: Plasma Physics

The Electrostatic Sheet Model for a Plasma and Its Modification to Finite-Sized Particles
John M. Dawson

Solution of Vlasov's Equation by Transform Methods
Thomas P. Armstrong, Rollin C. Harding, Georg Knorr, and David Montgomery

The Water-Bag Model
Herbert L. Berk and Keith V. Roberts

The Potential Calculation and Some Applications
R. W. Hockney

Multidimensional Plasma Simulation by the Particle-in-Cell Method
R. L. Morse

Finite-Size Particle Physics Applied to Plasma Simulation
Charles K. Birdsall, A. Bruce Langdon, and H. Okuda

Finite-Difference Methods for Collisionless Plasma Models
Jack A. Byers and John Killeen

Application of Hamilton's Principle to the Numerical Analysis of Vlasov Plasmas
H. Ralph Lewis

Magnetohydrodynamic Calculations
Keith V. Roberts and D. E. Potter

The Solution of the Fokker-Planck Equation for a Mirror-Confined Plasma
John Killeen and Kenneth D. Marx

AUTHOR INDEX—SUBJECT INDEX

Volume 10: Atomic and Molecular Scattering

Numerical Solutions of the Integro-Differential Equations of Electron–Atom Collision Theory
P. G. Burke and M. J. Seaton

Quantum Scattering Using Piecewise Analytic Solutions
Roy G. Gordon

Quantum Calculations in Chemically Reactive Systems
John C. Light

Expansion Methods for Electron–Atom Scattering
Frank E. Harris and H. H. Michels

Calculation of Cross Sections for Rotational Excitation of Diatomic Molecules by Heavy Particle Impact: Solution of the Close-Coupled Equations
William A. Lester, Jr.

Amplitude Densities in Molecular Scattering
Don Secrest

Classical Trajectory Methods
Don L. Bunker

AUTHOR INDEX—SUBJECT INDEX

Volume 11: Seismology: Surface Waves and Earth Oscillations

Finite Difference Methods for Seismic Wave Propagation in Heterogeneous Materials
David M. Boore

Numerical Analysis of Dispersed Seismic Waves
A. M. Dziewonski and A. L. Hales

Fast Surface Wave and Free Mode Computations
F. A. Schwab and L. Knopoff

A Finite Element Method for Seismology
John Lysmer and Lawrence A. Drake

Seismic Surface Waves
H. Takeuchi and M. Saito

AUTHOR INDEX—SUBJECT INDEX

Volume 12: Seismology: Body Waves and Sources

Numerical Methods of Ray Generation in Multilayered Media
F. Hron

Computer Generated Seismograms
Z. Alterman and D. Loewenthal

Diffracted Seismic Signals and Their Numerical Solution
C. H. Chapman and R. A. Phinney

Inversion and Inference for Teleseismic Ray Data
Leonard E. Johnson and Freeman Gilbert

Multipolar Analysis of the Mechanisms of Deep-Focus Earthquakes
M. J. Randall

Computation of Models of Elastic Dislocations in the Earth
Ari Ben-Menahem and Sarva Jit Singh

AUTHOR INDEX—SUBJECT INDEX

Volume 13: Geophysics

Signal Processing and Frequency-Wavenumber Spectrum Analysis for a Large Aperture Seismic Array
Jack Capon

Models of the Sources of the Earth's Magnetic Field
Charles O. Stearns and Leroy R. Alldredge

Computations with Spherical Harmonics and Fourier Series in Geomagnetism
D. E. Winch and R. W. James

Inverse Methods in the Interpretation of Magnetic and Gravity Anomalies
M. H. P. Bott

Analysis of Geoelectromagnetic Data
S. H. Ward, W. J. Peeples, and J. Ryu

Nonlinear Spherical Harmonic Analysis of Paleomagnetic Data
J. M. Wells

Harmonic Analysis of Earth Tides
Paul Melchior

Computer Usage in the Computation of Gravity Anomalies
Manik Talwani

Analysis of Irregularities in the Earth's Rotation
D. E. Smylie, G. K. C. Clarke, and T. J. Ulrych

Convection in the Earth's Mantle
D. L. Turcotte, K. E. Torrance, and A. T. Hsui

AUTHOR INDEX—SUBJECT INDEX

Volume 14: Radio Astronomy

Radioheliography
N. R. Labrum, D. J. McLean, and J. P. Wild

Pulsar Signal Processing
Timothy H. Hankins and Barney J. Rickett

Aperture Synthesis
W. N. Brouw

Computations in Radio-Frequency Spectroscopy
John A. Ball

AUTHOR INDEX—SUBJECT INDEX

Volume 15: Vibrational Properties of Solids

The Calculation of Phonon Frequencies
G. Dolling

The Use of Computers in Scattering Experiments with Slow Neutrons
R. Pynn

Group Theory of Lattice Dynamics by Computer
John L. Warren and Thomas G. Worlton

Lattice Dynamics and Related Properties of Point Defects
R. F. Wood

Lattice Dynamics of Surfaces of Solids
F. W. de Wette and G. P. Alldredge

Vibrational Properties of Amorphous Solids
R. J. Bell

Lattice Dynamics of Quantum Crystals
T. R. Koehler

Methods of Brillouin Zone Integration
G. Gilat

Computer Studies of Transport Properties in Simple Models of Solids
William M. Visscher

AUTHOR INDEX—SUBJECT INDEX